# Artificial Intelligence for Multimedia Signal Processing

# Artificial Intelligence for Multimedia Signal Processing

Editors

**Byung-Gyu Kim**
**Dongsan Jun**

MDPI • Basel • Beijing • Wuhan • Barcelona • Belgrade • Manchester • Tokyo • Cluj • Tianjin

*Editors*
Byung-Gyu Kim
Department of IT
Engineering, Sookmyung
Women's University,
Seoul 04310, Korea

Dongsan Jun
Department of Computer Engineering,
Dong-A University,
Busan 49315, Korea

*Editorial Office*
MDPI
St. Alban-Anlage 66
4052 Basel, Switzerland

This is a reprint of articles from the Special Issue published online in the open access journal *Applied Sciences* (ISSN 2076-3417) (available at: https://www.mdpi.com/journal/applsci/special_issues/AI_Multimedia_Signal).

For citation purposes, cite each article independently as indicated on the article page online and as indicated below:

LastName, A.A.; LastName, B.B.; LastName, C.C. Article Title. *Journal Name* **Year**, *Volume Number*, Page Range.

**ISBN 978-3-0365-4965-1 (Hbk)**
**ISBN 978-3-0365-4966-8 (PDF)**

© 2022 by the authors. Articles in this book are Open Access and distributed under the Creative Commons Attribution (CC BY) license, which allows users to download, copy and build upon published articles, as long as the author and publisher are properly credited, which ensures maximum dissemination and a wider impact of our publications.

The book as a whole is distributed by MDPI under the terms and conditions of the Creative Commons license CC BY-NC-ND.

# Contents

**About the Editors** . . . . . . . . . . . . . . . . . . . . . . . . . . . . . . . . . . . . . . . . . . . . . . . . . . . . . . . . vii

**Preface to "Artificial Intelligence for Multimedia Signal Processing"** . . . . . . . . . . . . . . ix

**Byung-Gyu Kim and Dong-San Jun**
Artificial Intelligence for Multimedia Signal Processing
Reprinted from: *Appl. Sci.* **2022**, *12*, 7358, doi:10.3390/app12157358 . . . . . . . . . . . . . . . . . 1

**Yooho Lee, Sang-hyo Park, Eunjun Rhee, Byung-Gyu Kim and Dongsan Jun**
Reduction of Compression Artifacts Using a Densely Cascading Image Restoration Network
Reprinted from: *Appl. Sci.* **2021**, *11*, 7803, doi:10.3390/app11177803 . . . . . . . . . . . . . . . . . 5

**Prasad Hettiarachchi, Rashmika Nawaratne, Damminda Alahakoon, Daswin De Silva and Naveen Chilamkurti**
Rain Streak Removal for Single Images Using Conditional Generative Adversarial Networks
Reprinted from: *Appl. Sci.* **2021**, *11*, 2214, doi:10.3390/app11052214 . . . . . . . . . . . . . . . . . 19

**Lifang Wu, Heng Zhang, Sinuo Deng, Ge Shi and Xu Liu**
Discovering Sentimental Interaction via Graph Convolutional Network for Visual Sentiment Prediction
Reprinted from: *Appl. Sci.* **2021**, *11*, 1404, doi:10.3390/app11041404 . . . . . . . . . . . . . . . . . 31

**Seonjae Kim, Dongsan Jun, Byung-Gyu Kim, Hunjoo Lee and Eunjun Rhee**
Single Image Super-Resolution Method Using CNN-Based Lightweight Neural Networks
Reprinted from: *Appl. Sci.* **2021**, *11*, 1092, doi:10.3390/app11031092 . . . . . . . . . . . . . . . . . 45

**Woon-Ha Yeo, Young-Jin Heo, Young-Ju Choi, and Byung-Gyu Kim**
Place Classification Algorithm Based on Semantic Segmented Objects
Reprinted from: *Appl. Sci.* **2020**, *10*, 9069, doi:10.3390/app10249069 . . . . . . . . . . . . . . . . . 57

**Dong-seok Lee, Jong-soo Kim, Seok Chan Jeong and Soon-kak Kwon**
Human Height Estimation by Color Deep Learning and Depth 3D Conversion
Reprinted from: *Appl. Sci.* **2020**, *10*, 5531, doi:10.3390/app10165531 . . . . . . . . . . . . . . . . . 69

**Yu-Kai Lin, Mu-Chun Su and Yi-Zeng Hsieh**
The Application and Improvement of Deep Neural Networks in Environmental Sound Recognition
Reprinted from: *Appl. Sci.* **2020**, *10*, 5965, doi:10.3390/app10175965 . . . . . . . . . . . . . . . . . 87

**Hyung Yong Kim, Ji Won Yoon, Sung Jun Cheon, Woo Hyun Kang, and Nam Soo Kim**
A Multi-Resolution Approach to GAN-Based Speech Enhancement
Reprinted from: *Appl. Sci.* **2021**, *11*, 721, doi:10.3390/app11020721 . . . . . . . . . . . . . . . . . . 103

**Yun Kyung Lee and Jeon Gue Park**
Multimodal Unsupervised Speech Translation for Recognizing and Evaluating Second Language Speech
Reprinted from: *Appl. Sci.* **2021**, *11*, 2642, doi:10.3390/app11062642 . . . . . . . . . . . . . . . . . 119

**Debashis Das Chakladar, Pradeep Kumar, Shubham Mandal, Partha Pratim Roy, Masakazu Iwamura and Byung-Gyu Kim**
3D Avatar Approach for Continuous Sign Movement Using Speech/Text
Reprinted from: *Appl. Sci.* **2021**, *11*, 3439, doi:10.3390/app11083439 . . . . . . . . . . . . . . . . . 135

**Krishna Kumar Thirukokaranam Chandrasekar and Steven Verstockt**
Context-Based Structure Mining Methodology for Static Object Re-Identification in Broadcast Content
Reprinted from: *Appl. Sci.* **2021**, *11*, 7266, doi:10.3390/app11167266 . . . . . . . . . . . . . . . . . **149**

**Haiying Liu, Sinuo Deng, Lifang Wu, Meng Jian, Bowen Yang and Dai Zhang**
Recommendations for Different Tasks Based on the Uniform Multimodal Joint Representation
Reprinted from: *Appl. Sci.* **2020**, *10*, 6170, doi:10.3390/app10186170 . . . . . . . . . . . . . . . . . **167**

**Ian-Christopher Tanoh, Paolo Napoletano**
A Novel 1-D CCANet for ECG Classification
Reprinted from: *Appl. Sci.* **2021**, *11*, 2758, doi:10.3390/app11062758 . . . . . . . . . . . . . . . . . **187**

# About the Editors

**Byung-Gyu Kim**

Byung-Gyu Kim has received his BS degree from Pusan National University, Korea, in 1996 and an MS degree from Korea Advanced Institute of Science and Technology (KAIST) in 1998. In 2004, he received a PhD degree in the Department of Electrical Engineering and Computer Science from Korea Advanced Institute of Science and Technology (KAIST). In March 2004, he joined in the real-time multimedia research team at the Electronics and Telecommunications Research Institute (ETRI), Korea where he was a senior researcher. In ETRI, he developed so many real-time video signal processing algorithms and patents and received the Best Paper Award in 2007. From February 2009 to February 2016, he was associate professor in the Division of Computer Science and Engineering at SunMoon University, Korea. In March 2016, he joined the Department of Information Technology (IT) Engineering at Sookmyung Women's University, Korea where he is currently a professor.

In 2007, he is serving as an associate editor of Circuits, Systems and Signal Processing (Springer), The Journal of Supercomputing (Springer), The Journal of Real-Time Image Processing (Springer), and International Journal of Image Processing and Visual Communication (IJIPVC). From March 2018, he is serving as the Editor-in-Chief of The Journal of Multimedia Information System and associate editor of IEEE Access Journal. From 2019, he was appointed as the associate editor and topic editor of Heliyon-Computer Science (Elsevier), Journal of Image Science and Technology (IS & T), Electronics (MDPI), Applied Sciences (MDPI), and Sensors (MDPI), respectively. He also served or serves as Organizing Committee of CSIP 2011, a Co-organizer of CICCAT2016/2017, and Program Committee Members of many international conferences. He has received the Special Merit Award for Outstanding Paper from the IEEE Consumer Electronics Society, at IEEE ICCE 2012, Certification Appreciation Award from the SPIE Optical Engineering in 2013, and the Best Academic Award from the CIS in 2014. He has been honored as an IEEE Senior member in 2015. He is serving as a professional reviewer in many academic journals including IEEE, ACM, Elsevier, Springer, Oxford, SPIE, IET, MDPI, and so on.

He has published over 250 international journal and conference papers, patents in his field. His research interests include image and video signal processing for the content-based image coding, video coding techniques, 3D video signal processing, deep/reinforcement learning algorithm, embedded multimedia system, and intelligent information system for image signal processing. He is a senior member of IEEE and a professional member of ACM, and IEICE.

**Dongsan Jun**

Dongsan Jun received his BS degree in electrical engineering and computer science from Pusan National University, South Korea in 2002, and his MS and PhD degree in electrical engineering from the Korea Advanced Institute of Science and Technology (KAIST), Daejeon, South Korea in 2004 and 2011, respectively. In 2004, he joined in the realistic AV research group at the Electronics and Telecommunications Research Institute (ETRI), Daejeon, South Korea, where he was a principal researcher. From March 2018 to August 2021, he was an assistant professor with the Department of Information and Communication Engineering at Kyungnam University, Changwon, South Korea. In September 2021, he joined the Department of Computer Engineering at DONG-A University, Busan, South Korea, where he is currently an assistant professor. His research interests include image/video computing systems, pattern recognition, video compression, real-time processing, deep learning, and realistic broadcasting systems.

# Preface to "Artificial Intelligence for Multimedia Signal Processing"

Recent developments in image/video-based deep learning technology have enabled new services in the field of multimedia and recognition technology. Artificial intelligence technologies are also actively applied to broadcasting and multimedia processing technologies. A substantial amount of research has been conducted in a wide variety of fields, such as content creation, transmission, and security, and these attempts have been made in the past two to three years to improve image, video, speech, and other data compression efficiency in areas related to MPEG media processing technology. Additionally, technologies such as media creation, processing, editing, and creating scenarios are very important areas of research in multimedia processing and engineering. To accommodate these needs, many researchers are studying various signal and image processing technologies to provide a variety of new or future multimedia processing and services. In this issue, we have gathered several well-written and researched papers for advanced signal/image, video data processing, and text/content information mining including deep learning approaches. This book comprises thirteen peer-reviewed articles covering a review of the development of deep learning-based approaches, the original research articles for the learning mechanism and multimedia signal processing. This book also covers topics including computer vision field, speech/sound/text processing, and content analysis/information mining. This volume will be of good use to designers and engineers in both academia and industry who would like to develop an understanding of emerging multimedia signal processing, as well as to students.

**Byung-Gyu Kim and Dongsan Jun**
*Editors*

*Editorial*

# Artificial Intelligence for Multimedia Signal Processing

**Byung-Gyu Kim [1],\* and Dong-San Jun [2]**

1. Department of IT Engineering, Sookmyung Women's University, Seoul 04310, Korea
2. Department of Computer Engineering, Dong-A University, Busan 49315, Korea; dsjun@dau.ac.kr
\* Correspondence: bg.kim@sookmyung.ac.kr

**Citation:** Kim, B.-G.; Jun, D.-S. Artificial Intelligence for Multimedia Signal Processing. *Appl. Sci.* **2022**, *12*, 7358. https://doi.org/10.3390/app12157358

Received: 15 July 2022
Accepted: 21 July 2022
Published: 22 July 2022

**Publisher's Note:** MDPI stays neutral with regard to jurisdictional claims in published maps and institutional affiliations.

**Copyright:** © 2022 by the authors. Licensee MDPI, Basel, Switzerland. This article is an open access article distributed under the terms and conditions of the Creative Commons Attribution (CC BY) license (https://creativecommons.org/licenses/by/4.0/).

## 1. Introduction

At the ImageNet Large Scale Visual Re-Conversion Challenge (ILSVRC), a 2012 global image recognition contest, the University of Toronto Supervision team led by Prof. Geoffrey Hinton took first and second place by a landslide, sparking an explosion of interest in deep learning. Since then, global experts and companies such as Google, Microsoft, nVidia, and Intel have been competing to lead artificial intelligence technologies, such as deep learning. Now, they are developing deep-learning-based technologies that can applied to all industries to solve many classification and recognition problems.

These artificial intelligence technologies are also actively applied to broadcasting and multimedia processing technologies based on recognition and classification [1–3]. A vast amount of research has been conducted in a wide variety of fields, such as content creation, transmission, and security, and attempts have been made in the past two to three years to improve image, video, speech, and other data compression efficiency in areas related to MPEG media processing technology [4–6]. Additionally, technologies such as media creation, processing, editing, and creating scenarios are very important areas of research in multimedia processing and engineering. In this issue, we present excellent papers related to advanced computational intelligence algorithms and technologies for emerging multimedia processing.

## 2. Emerging Multimedia Signal Processing

Thirteen papers related to artificial intelligence for multimedia signal processing have been published in this Special Issue. They deal with a broad range of topics concerning advanced computational intelligence algorithms and technologies for emerging multimedia signal processing.

We present the following works in relation to the computer vision field. Lee et al. propose a densely cascading image restoration network (DCRN) consisting of an input layer, a densely cascading feature extractor, a channel attention block, and an output layer [7]. The densely cascading feature extractor has three densely cascading (DC) blocks, and each DC block contains two convolutional layers. From this design, they achieved better quality measures for the compressed joint photographic experts group (JPEG) images compared with the existing methods. In [8], an image de-raining approach is developed using the generative capabilities of recently introduced conditional generative adversarial networks (cGANs). This method could be very useful to recover visual quality when degraded due to diverse weather conditions, recording conditions, or motion blur.

Additionally, Wu et al. suggest a framework to leverage the sentimental interaction characteristic based on a graph convolutional network (GCN) [9]. They first utilize an off-the-shelf tool to recognize the objects and build a graph over them. Visual features are represented as nodes, and the emotional distances between the objects act as edges. Then, they employ GCNs to obtain the interaction features among the objects, which are fused with the CNN output of the whole image to predict the result. This approach is very useful to analyze human sentiment analysis. In [10], two lightweight neural networks with a hybrid residual and dense connection structure are suggested by Kim et al. to improve super-resolution performance. They show that the proposed methods could significantly reduce both the inference speed and the memory required to store parameters and intermediate feature maps, while maintaining similar image quality compared to the previous methods.

Kim et al. propose an efficient scene classification algorithm for three different classes by detecting objects in the scene [11]. The authors utilize a pre-trained semantic segmentation model to extract objects from an image. After that, they construct a weighting matrix to better determine the scene class. Finally, this classifies an image into one of three scene classes (i.e., indoor, nature, city) using the designed weighting matrix. This technique can be utilized for semantic searches in multimedia databases.

Lastly, an estimation method for human height is proposed by Lee et al. using color and depth information [12]. They use color images for deep learning by mask R-CNN to detect a human body and a human head separately. If color images are not available for extracting the human body region due to a low light environment, then the human body region is extracted by comparison with the current frame in the depth video.

For speech, sound, and text processing, Lin et al. improve the raw-signal-input network from other research using deeper network architectures [13]. They also propose a network architecture that can combine different kinds of network feeds with different features. In the experiment, the proposed scheme achieves an accuracy of 73.55% in the open audio dataset, "Dataset for Environmental Sound Classification 50" (ESC50). A multi-scale discriminator that discriminates between real and generated speech at various sampling rates is devised by Kim et al. to stabilize GAN training [14]. In this paper, the proposed structure is compared with conventional GAN-based speech enhancement algorithms using the VoiceBank-DEMAND dataset. They show that the proposed approach can make the training faster and more stable.

To translate the speech, a multimodal unsupervised scheme is proposed by Lee and Park [15]. They make a variational autoencoder (VAE)-based speech conversion network by decomposing the spectral features of the speech into a speaker-invariant content factor and a speaker-specific style factor to estimate diverse and robust speech styles. This approach can help second language (L2) speech education. To develop a 3D avatar-based sign language learning system, Chakladar et al. suggest a system that converts the input speech/text into corresponding sign movements for Indian Sign Language (ISL) [16]. The translation module achieves a 10.50 SER (sign error rate) score in the actual test.

Two papers concern content analysis and information mining. The first one, by Krishna Kumar Thirukokaranam Chandrasekar and Steven Verstockt, regards a context-based structure mining pipeline [17]. The proposed scheme not only attempts to enrich the content, but also simultaneously splits it into shots and logical story units (LSU). They demonstrate quantitatively that the pipeline outperforms existing state-of-the-art methods for shot boundary detection, scene detection, and re-identification tasks. The other paper outlines a framework which can learn the multimodal joint representation of pins, including text representation, image representation, and multimodal fusion [18]. In this work, the authors combine image representations and text representations in a multimodal form. It is shown that the proposed multimodal joint representation outperforms unimodal representation in different recommendation tasks.

For ECG signal processing, Tanoh and Napoletano propose a 1D convolutional neural network (CNN) that exploits a novel analysis of the correlation between the two leads of the noisy electrocardiogram (ECG) to classify heartbeats [19]. This approach is one-dimensional, enabling complex structures while maintaining reasonable computational complexity.

I hope that the technical papers published in this Special Issue can help researchers and readers to understand the emerging theories and technologies in the field of multimedia signal processing.

**Funding:** This research received no external funding.

**Acknowledgments:** We thank all authors who submitted excellent research work to this Special Issue. We are grateful to all reviewers who contributed evaluations of scientific merits and quality of the manuscripts and provided countless valuable suggestions to improve their quality and the overall value for the scientific community. Our special thanks go to the editorial board of MDPI Applied Sciences journal for the opportunity to guest edit this Special Issue, and to the Applied Sciences Editorial Office staff for the hard and precise work required to keep to a rigorous peer-review schedule and complete timely publication.

**Conflicts of Interest:** The authors declare no conflict of interest.

## References

1. Kim, J.-H.; Hong, G.-S.; Kim, B.-G.; Dogra, D.P. deepGesture: Deep Learning-based Gesture Recognition Scheme using Motion Sensors. *Displays* **2018**, *55*, 38–45. [CrossRef]
2. Kim, J.-H.; Kim, B.-G.; Roy, P.P.; Jeong, D.-M. Efficient Facial Expression Recognition Algorithm Based on Hierarchical Deep Neural Network Structure. *IEEE Access* **2019**, *7*, 2907327. [CrossRef]
3. Jeong, D.; Kim, B.-G.; Dong, S.-G. Deep Joint Spatio-Temporal Network (DJSTN) for Efficient Facial Expression Recognition. *Sensors* **2020**, *20*, 1936. [CrossRef] [PubMed]
4. Lee, Y.; Jun, D.; Kim, B.-G.; Lee, H. Enhanced Single Image Super Resolution Method using a Lightweight Multi-scale Channel Dense Network for Small Object Detection. *Sensors* **2021**, *21*, 3351. [CrossRef] [PubMed]
5. Park, S.-J.; Kim, B.-G.; Chilamkurti, N. A Robust Facial Expression Recognition Algorithm Based on Multi-Rate Feature Fusion Scheme. *Sensors* **2021**, *21*, 6954. [CrossRef] [PubMed]
6. Choi, Y.-J.; Lee, Y.-W.; Kim, B.-G. Residual-based Graph Convolutional Network (RGCN) for Emotion Recognition in Conversation (ERC) for Smart IoT. *Big Data* **2021**, *9*, 279–288. [CrossRef] [PubMed]
7. Lee, Y.; Park, S.-H.; Rhee, E.; Kim, B.-G.; Jun, D. Reduction of Compression Artifacts Using a Densely Cascading Image Restoration Network. *Appl. Sci.* **2021**, *11*, 7803. [CrossRef]
8. Hettiarachchi, P.; Nawaratne, R.; Alahakoon, D.; De Silva, D.; Chilamkurti, N. Rain Streak Removal for Single Images Using Conditional Generative Adversarial Networks. *Appl. Sci.* **2021**, *11*, 2214. [CrossRef]
9. Wu, L.; Zhang, H.; Deng, S.; Shi, G.; Liu, X. Discovering Sentimental Interaction via Graph Convolutional Network for Visual Sentiment Prediction. *Appl. Sci.* **2021**, *11*, 1404. [CrossRef]
10. Kim, S.; Jun, D.; Kim, B.-G.; Lee, H.; Rhee, E. Single Image Super-Resolution Method Using CNN-Based Lightweight Neural Networks. *Appl. Sci.* **2021**, *11*, 1092. [CrossRef]
11. Yeo, W.-H.; Heo, Y.-J.; Choi, Y.-J.; Kim, B.-G. Place Classification Algorithm Based on Semantic Segmented Objects. *Appl. Sci.* **2020**, *10*, 9069. [CrossRef]
12. Lee, D.-S.; Kim, J.-S.; Jeong, S.C.; Kwon, S.-K. Human Height Estimation by Color Deep Learning and Depth 3D Conversion. *Appl. Sci.* **2020**, *10*, 5531. [CrossRef]
13. Lin, Y.-K.; Su, M.-C.; Hsieh, Y.-Z. The Application and Improvement of Deep Neural Networks in Environmental Sound Recognition. *Appl. Sci.* **2020**, *10*, 5965. [CrossRef]
14. Kim, H.Y.; Yoon, J.W.; Cheon, S.J.; Kang, W.H.; Kim, N.S. A Multi-Resolution Approach to GAN-Based Speech Enhancement. *Appl. Sci.* **2021**, *11*, 721. [CrossRef]
15. Lee, Y.K.; Park, J.G. Multimodal Unsupervised Speech Translation for Recognizing and Evaluating Second Language Speech. *Appl. Sci.* **2021**, *11*, 2642. [CrossRef]
16. Das Chakladar, D.; Kumar, P.; Mandal, S.; Roy, P.P.; Iwamura, M.; Kim, B.-G. 3D Avatar Approach for Continuous Sign Movement Using Speech/Text. *Appl. Sci.* **2021**, *11*, 3439. [CrossRef]
17. Thirukokaranam Chandrasekar, K.K.; Verstockt, S. Context-Based Structure Mining Methodology for Static Object Re-Identification in Broadcast Content. *Appl. Sci.* **2021**, *11*, 7266. [CrossRef]
18. Liu, H.; Deng, S.; Wu, L.; Jian, M.; Yang, B.; Zhang, D. Recommendations for Different Tasks Based on the Uniform Multimodal Joint Representation. *Appl. Sci.* **2020**, *10*, 6170. [CrossRef]
19. Tanoh, I.-C.; Napoletano, P. A Novel 1-D CCANet for ECG Classification. *App. Sci.* **2021**, *11*, 2758. [CrossRef]

*Article*

# Reduction of Compression Artifacts Using a Densely Cascading Image Restoration Network

Yooho Lee [1], Sang-hyo Park [2], Eunjun Rhee [3], Byung-Gyu Kim [4,*] and Dongsan Jun [1,*]

1. Department of Convergence IT Engineering, Kyungnam University, Changwon 51767, Korea; yhlee@kyungnam-ispl.kr
2. Department of Computer Science and Engineering, Kyungpook National University, Daegu 41566, Korea; s.park@knu.ac.kr
3. Intelligent Convergence Research Laboratory, Electronics and Telecommunications Research Institute (ETRI), Daejeon 34129, Korea; ejrhee@etri.re.kr
4. Department of IT Engineering, Sookmyung Women's University, Seoul 04310, Korea
* Correspondence: bg.kim@sookmyung.ac.kr (B.-G.K.); dsjun9643@kyungnam.ac.kr (D.J.)

**Abstract:** Since high quality realistic media are widely used in various computer vision applications, image compression is one of the essential technologies to enable real-time applications. Image compression generally causes undesired compression artifacts, such as blocking artifacts and ringing effects. In this study, we propose a densely cascading image restoration network (DCRN), which consists of an input layer, a densely cascading feature extractor, a channel attention block, and an output layer. The densely cascading feature extractor has three densely cascading (DC) blocks, and each DC block contains two convolutional layers, five dense layers, and a bottleneck layer. To optimize the proposed network architectures, we investigated the trade-off between quality enhancement and network complexity. Experimental results revealed that the proposed DCRN can achieve a better peak signal-to-noise ratio and structural similarity index measure for compressed joint photographic experts group (JPEG) images compared to the previous methods.

**Keywords:** computer vision; deep learning; convolutional neural network; image processing; image restoration; single image artifacts reduction; dense networks; residual networks; channel attention networks

## 1. Introduction

As realistic media are widespread in various image processing areas, image compression is one of the key technologies to enable real-time applications with limited network bandwidth. While image compression techniques, such as joint photographic experts group (JPEG) [1], web picture [2], and high-efficiency video coding main still picture [3], can achieve significant compression performances for efficient image transmission and storage [4], they lead to undesired compression artifacts due to lossy coding because of quantization. These artifacts generally affect the performance of image restoration methods in terms of super-resolution [5–10], contrast enhancement [11–14], and edge detection [15–17].

Reduction methods for compression artifacts were initially studied by developing a specific filter inside the compression process [18]. Although these approaches can efficiently remove ringing artifacts [19], the improvement in image regions is limited at high frequencies. Examples of such approaches include deblocking-oriented approaches [20,21], wavelet transforms [22,23], and shape-adaptive discrete cosine transforms [24]. Recently, artifacts reduction (AR) networks using deep learning have been developed with various deep neural networks (DNNs), such as convolutional neural networks (CNNs), recurrent neural networks (RNNs), long short-term memory (LSTM), and generative adversarial networks (GANs). Because CNN [25] can efficiently extract feature maps with deep and cascading structures, CNN-based artifact reduction (AR) methods can achieve visual enhancement in terms of peak signal-to-noise ratio (PSNR) [26], PSNR including blocking effects (PSNR-B) [27,28], and structural similarity index measures (SSIM) [29].

Despite the developments of AR, most CNN-based approaches tend to design the heavy network architecture by increasing the number of network parameters and operations. Because it is difficult to deploy such heavy models on hand-held devices operated on low complexity environments, it is necessary to design the lightweight AR networks. In this paper, we propose a lightweight CNN-based artifacts reduction model to reduce the memory capacity as well as network parameters. The main works of this study are summarized as follows:

- To reduce the coding artifacts of the compressed images, we propose a CNN based densely cascading image restoration network (DCRN) with two essential parts, densely cascading feature extractor and channel attention block.
- Through a various ablation study, the proposed network is designed to guarantee the optimal trade-off between the PSNR and the network complexity.
- Compared to the previous method, the proposed network is designed to obtain comparable AR performance while utilizing the small number of network parameters and memory size. In addition, it can provide the fastest inference speed, except for initial AR network [30].
- Compared to the latest methods to show the highest AR performances (PSNR, SSIM, and PSNR-B), the proposed method can reduce the number of parameters and total memory size maximum by 2% and 5%, respectively.

The remainder of this paper is organized as follows: in Section 2, we review previous studies related to CNN-based artifact reduction methods. In Section 3, we describe the proposed method. Finally, in Sections 4 and 5, we present the experimental results and conclusions, respectively.

## 2. Related Works

Due to the advancements in deep learning technologies, research of low-level computer vision, such as super-resolution (SR) and image denoising, has been combined with a variety of CNN architectures to provide higher image restoration than that of conventional image processing. Dong et al. proposed an artifact reduction convolutional neural network (ARCNN) [30], which consists of four convolutional layers and trains an end-to-end mapping from a compressed image to a reconstructed image. After the advent of ARCNN, Mao et al. [31] proposed a residual encoder–decoder network, which conducts encoding and decoding processes with symmetric skip connections in stacking convolutional and deconvolutional layers. Chen et al. [32] proposed a trainable nonlinear reaction diffusion, which is simultaneously learned from training data through a loss-based approach with all parameters, including filters and influence functions. Zhang et al. [33] proposed a denoising convolutional neural network (DnCNN), which is composed of a combination of 17 convolutional layers with a rectified linear unit (ReLU) [34] activation function and batch normalization for removing white Gaussian noise. Cavigelli et al. [35] proposed a deep CNN for image compression artifact suppression, which consists of 12 convolutional layers with hierarchical skip connections and a multi-scale loss function.

Guo et al. [36] proposed a one-to-many network, which is composed of many stacked residual units, with each branch containing five residual units and the aggregation sub-network comprising 10 residual units. Each residual unit uses batch normalization, ReLU activation function, and convolutional layer twice. The architecture of residual units is found to improve the recovery quality. Tau et al. [37] proposed a very deep persistent memory network with a densely recursive residual architecture-based memory block that adaptively learns the different weights of various memories. Dai et al. [38] proposed a variable-filter-size residual-learning CNN, which contains six convolutional layers and concatenates variable-filter-size convolutional layers. Zhang et al. [39] proposed a dual-domain multi-scale CNN with an auto-encoder, dilated convolution, and discrete cosine transform (DCT) unit. Liu et al. [40] designed a multi-level wavelet CNN that builds a u-net architecture with a four-layer fully convolutional network (FCN) without pooling and takes all sub-images as inputs. Each layer of a CNN block is composed of 3 × 3 kernel filters, batch normalization, and ReLU. A dual-pixel-wavelet domain deep CNN-based soft decoding network for JPEG-compressed images [41] is composed of two parallel branches, each serving as the pixel domain soft decoding branch and wavelet domain soft decoding branch. Fu et al. [42] proposed a deep convolutional sparse coding (DCSC) network that has dilated convolutions to extract multi-scale features with the same filter for three different scales. The implicit dual-domain convolutional network (IDCN) for robust color image compression AR [43] consists of a feature encoder, correction baseline and feature decoder. Zhang et al. [44] proposed a residual dense network (RDN), which consists of 16 residual dense blocks, and each dense block contains eight dense layers with local residual learning.

Although most of the aforementioned methods demonstrate better AR performance, they tend to possess more complicated network structures on account of the large number of network parameters needed and heavy memory consumption. Table 1 lists the properties of the various AR networks and compares their advantages and disadvantages.

**Table 1.** Properties among the artifact reduction networks.

| Method | AR Performance | Complexity |
| --- | --- | --- |
| ARCNN [30] | Low PSNR | Low network complexity |
| DnCNN [33] | Medium PSNR | Medium network complexity |
| DCSC [42] | Medium PSNR (High PSNR-B) | Medium network complexity |
| IDCN [43] | High PSNR and PSNR-B | High network complexity |
| RDN [44] | High PSNR and PSNR-B | High network complexity |

For the network component, a residual network [45] was designed for shortcut connections to simplify identity mapping, and outputs were added to the outputs of the stacked layers. A densely connected convolutional network [46] directly connects all layers with one another based on equivalent feature map sizes. The squeeze-and-excitation (SE) network [47] is composed of global average pooling and a 1 × 1 convolutional layer. These networks use the weights of previous feature maps, and such weights are applied to previous feature maps to generate the output of the SE block, which can be provided to subsequent layers of the network. In this study, we propose an AR network to combine with those networks [45–47] for better image restoration performance than the previous methods.

## 3. Proposed Method

### 3.1. Overall Architecture of DCRN

Figure 1 shows the overall architecture of the proposed DCRN to remove compression artifacts caused by JPEG compression. The DCRN consists of the input layer, a densely cascading feature extractor, a channel attention block, and the output layer. In particular, the densely cascading feature extractor contains three densely cascading blocks to exploit the intermediate feature maps within sequential dense networks. In Figure 1, $W \times H$ and $C$ are the spatial two-dimensional filter size and the number of channels, respectively. The convolution operation of the $i$-th layer is denoted as $H_i$ and calculates the output feature maps ($F_i$) from the previous feature maps ($F_{i-1}$), as shown in Equation (1):

$$F_i = H_i(F_{i-1}) = \delta(W_i * F_{i-1} + B_i), \tag{1}$$

where $\delta$, $W_i$, $B_i$, and $*$ represent the parametric ReLU function as an activation function, filter weights, biases, and the notation of convolution operation, respectively. After extracting the feature maps of the input layer, densely cascading feature extractor generates $F_5$, as expressed in Equation (2). As shown in Figure 2, a densely cascading (DC) block has two convolutional layers, five dense layers, and a bottleneck layer. To train the network effectively and reduce overfitting, we designed dense layers that consist of a variable number of channels. Dense layers 1 to 4 consist of 16 channels and the final dense layer consists of 64 channels. The DC block operation $H_i^{DC}$ is presented in Equation (2):

$$F_3 = H_3^{DC}(F_2) = H_3^{DC}(H_2^{DC}(H_1^{DC}(F_0))). \tag{2}$$

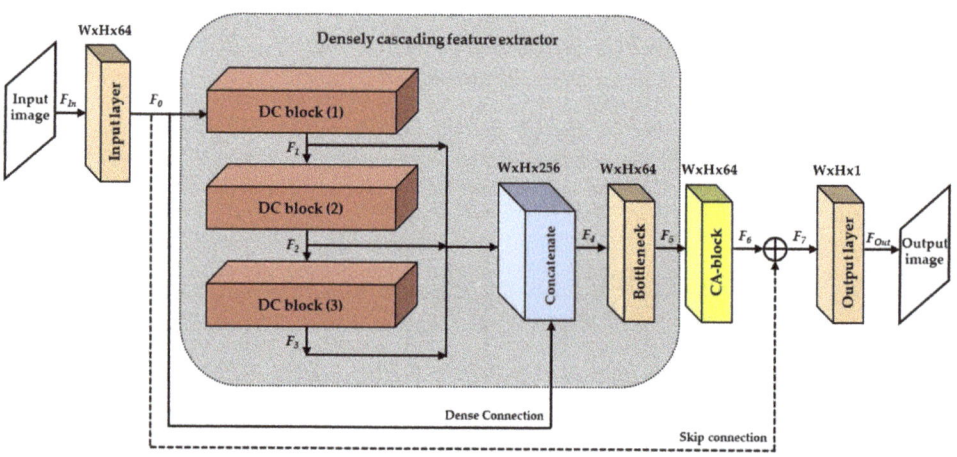

**Figure 1.** Overall architecture of the proposed DCRN. Symbol '+' indicates the element-wise sum.

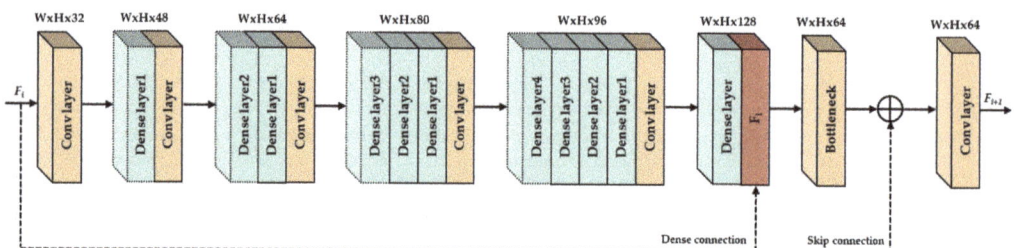

**Figure 2.** The architecture of a DC block.

Then, each DC block output is concatenated with the output of the input layer feature map operations. After concatenating both the output feature maps from all DC blocks and the input layer, the bottleneck layer calculates $F_5$ to reduce the number of channels of $F_4$, as in Equation (3):

$$F_5 = H_5(F_4) = H_5([F_3, F_2, F_1, F_0]). \tag{3}$$

As shown in Figure 3, a channel attention (CA) block performs the global average pooling (GAP) followed by two convolutional layers and the sigmoid function after the output from the densely cascading feature extractor is passed to it. The CA block can discriminate the more important feature maps, and it assigns different weights to each feature map in order to adapt feature responses. After generating $F_6$ through the CA block, an output image is generated from the element-wise sum between the skip connection ($F_0$) and the feature maps ($F_6$).

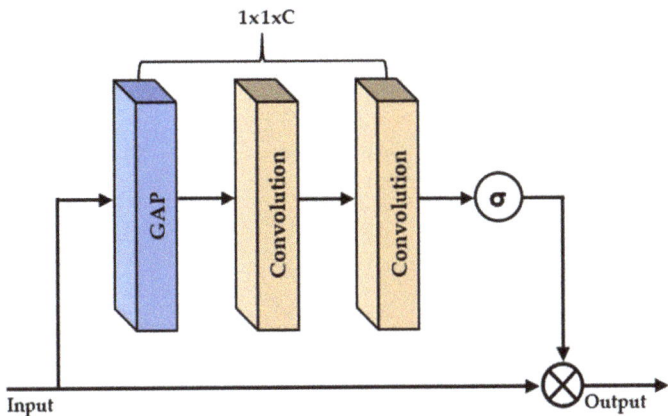

**Figure 3.** The architecture of a CA-block. '$\sigma$' and '$\otimes$' indicate sigmoid function and channel-wise product, respectively.

*3.2. Network Training*

In the proposed DCRN, we set the filter size as $3 \times 3$ except for the CA block, whose kernel size is $1 \times 1$. Table 2 shows the selected hyper parameters in the DCRN. We used zero padding to allow all feature maps to have the same spatial resolution between the different convolutional layers. We defined L1 loss [48] as the loss function using Adam optimizer [49] with a batch size of 128. The learning rate was decreased from $10^{-3}$ to $10^{-5}$ for 50 epochs.

**Table 2.** Hyper parameters of the proposed DCRN.

| Hyper Parameters | Options |
| --- | --- |
| Loss function | L1 loss |
| Optimizer | Adam |
| Batch size | 128 |
| Num. of epochs | 50 |
| Learning rate | $10^{-3}$ to $10^{-5}$ |
| Initial weight | Xavier |
| Activation function | Parametric ReLU |
| Padding mode | Zero padding |

To design a lightweight architecture, we first studied the relationship between network complexity and performance according to the number of dense layer feature maps within the DC block. Second, we checked the performance of various activation functions. Third, we studied the performance of loss functions. Fourth, we investigated the relationship between network complexity and performance based on the number in each dense layers of DC block and the number of DC blocks. Finally, we studied the performance of the tool-off test (skip connection, channel attention block).

Table 3 lists the PSNR obtained according to the number of concatenated feature maps within the DC block. We set the optimal number of concatenated feature maps to 16 channels. Moreover, we conducted verification tests to determine the most suitable activation function for the proposed network, the results of which are shown in Figure 4. After measuring the PSNR and SSIM obtained via various activation functions, such as ReLU [34], leaky ReLU [50], and parametric ReLU [51], parametric ReLU was chosen for the proposed DCRN. Table 4 summarizes the results of the verification tests concerning loss functions, in terms of the L1 and mean square error (MSE) losses. As shown in Table 4, the L1 loss exhibits marginally improved PSNR, SSIM, and PSNR-B compared to those exhibited by the MSE loss. In addition, we verified the effectiveness the of skip connection and channel attention block mechanisms. Through the results of tool-off tests on the proposed DCRN, which are summarized in Figure 5, we confirmed that both skip connection and channel attention block affect the AR performance of the proposed method.

**Table 3.** Verification test on the number of concatenated feature maps within the DC block.

| Category | PSNR (dB) | Num of Parameter | Total Memory Size (MB) |
| --- | --- | --- | --- |
| 4 channel | 29.58 | 316 K | 33.56 |
| 8 channel | 29.61 | 366 K | 36.39 |
| 16 channel | 29.64 | 479 K | 42.10 |
| 32 channel | 29.68 | 770 K | 53.75 |
| 64 channel | 29.69 | 1600 K | 78.01 |

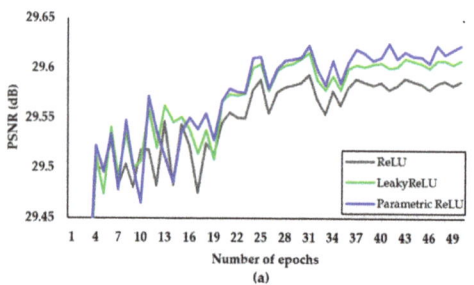 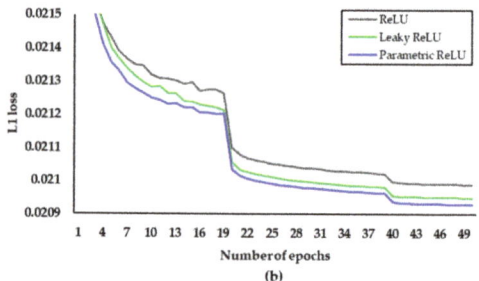

**Figure 4.** Verification of activation functions. (a) PSNR per epoch. (b) L1 loss per epoch.

**Table 4.** Verification tests for loss functions.

| Category | PSNR (dB) | SSIM | PSNR-B (dB) |
| --- | --- | --- | --- |
| L1 loss | 29.64 | 0.825 | 29.35 |
| MSE loss | 29.62 | 0.824 | 29.33 |

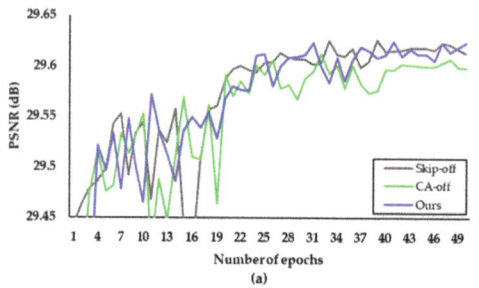 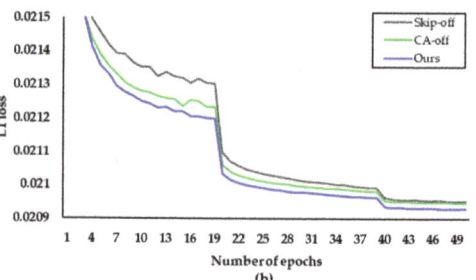

**Figure 5.** Verification of the skip connection off (skip-off), channel attention blocks off (CA-off) and proposed method in terms of AR performance. (a) PSNR per epoch. (b) L1 loss per epoch.

Note that the higher the number of DC blocks and dense layers, the more the memory required to store the network parameters. Finally, we performed a variety of verification tests on the validation dataset to optimize the proposed method. In this paper, we denote the number of DC blocks and the number of dense layers per DC block as DC and L, respectively. The performance comparison between the proposed and existing methods in terms of the AR performance (i.e., PSNR), model size (i.e., number of parameters), and total memory size is displayed in Figures 6 and 7. We set the value of DC and L to three and five, respectively.

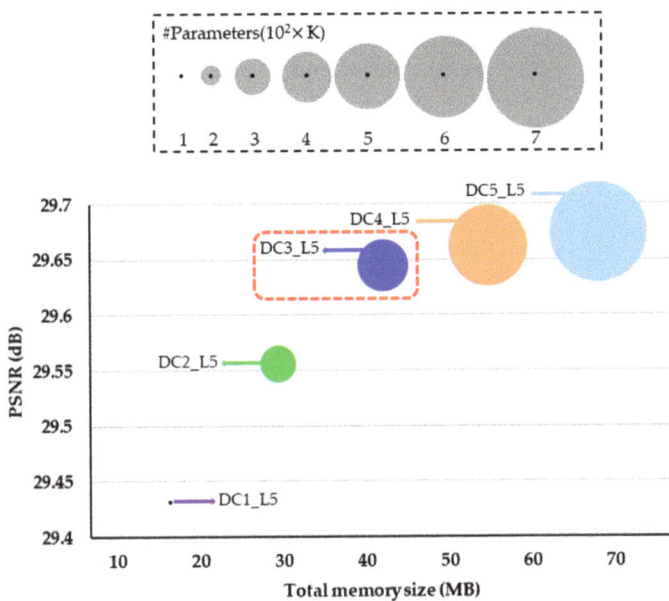

**Figure 6.** Verification of the number of DC blocks (DC) in terms of AR performance and complexity by using the Classic5 dataset. The circle size represents the number of parameters. The x and y-axis denote the total memory size and PSNR, respectively.

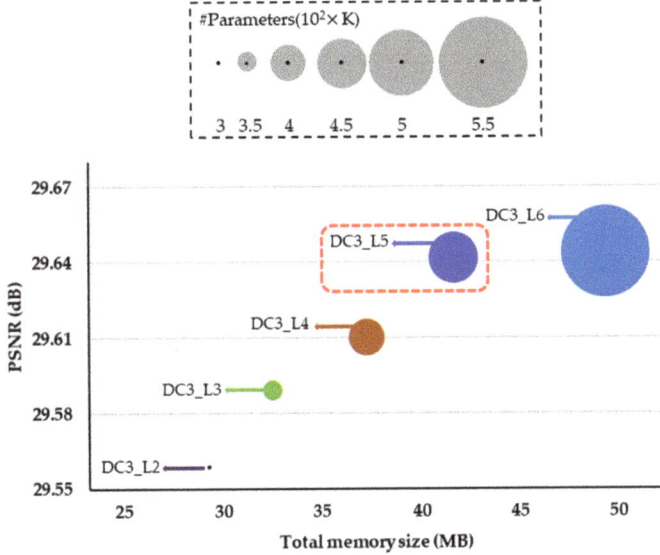

**Figure 7.** Verification of the number of dense layers (L) per DC block (DC) in terms of AR performance and complexity by using the Classic5 dataset. The circle size represents the number of parameters. The x and y-axis denote the total memory size and PSNR, respectively.

## 4. Experimental Results

We used 800 images from DIV2K [52] as the training images. After they were converted into YUV color format, only Y components were encoded and decoded by the JPEG codec under three image quality factors (10, 20, and 30). Through this process, we collected 1,364,992 patches of a 40 × 40 size from the original and reconstructed images. To evaluate the proposed method, we used Classic5 [24] (five images) and LIVE1 [53] (29 images) as the test datasets and Classic5 as the validation dataset.

All experiments were performed on an Intel Xeon Gold 5120 (14 cores @ 2.20 GHz) with 177 GB RAM and two NVIDIA Tesla V100 GPUs under the experimental environment described in Table 5.

**Table 5.** Experimental environments.

| Experimental Environments | Options |
|---|---|
| Input size ($F_{In}$) | 40 × 40 × 1 |
| Label size ($F_{Out}$) | 40 × 40 × 1 |
| CUDA version | 10.1 |
| Linux version | Ubuntu 16.04 |
| Deep learning frameworks | Pytorch 1.4.0 |

In terms of the performance of image restoration, we compared the proposed DCRN with JPEG, ARCNN [30], DnCNN [33], DCSC [42], IDCN [43] and RDN [44]. In terms of the AR performance (i.e., PSNR and SSIM), the number of parameters and total memory size, the performance comparisons between the proposed and existing methods are depicted in Figure 8.

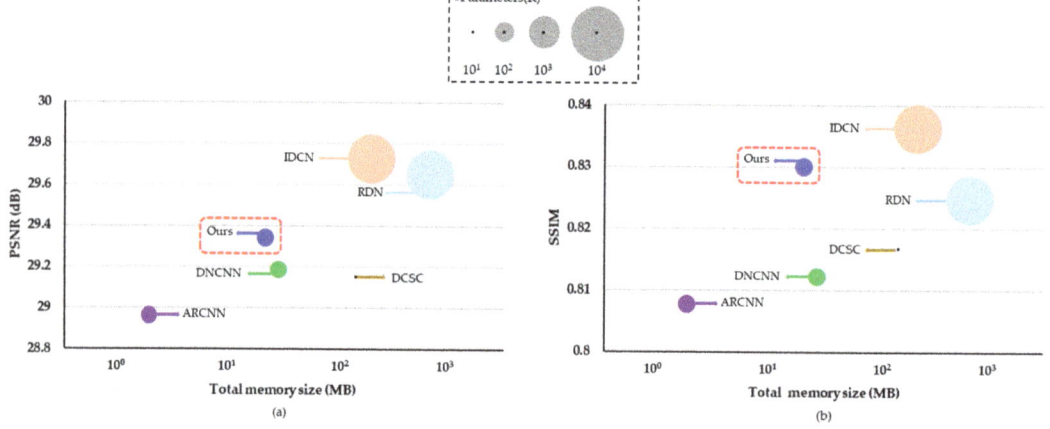

**Figure 8.** Comparisons of the network performance and complexity between the proposed DCRN and existing methods for the LIVE1 dataset. The circle size represents the number of parameters. (**a**) The x and y-axis denote the total memory size and PSNR, respectively. (**b**) The x and y-axis denote the total memory size and SSIM, respectively.

Tables 6–8 enumerate the results of PSNR, SSIM, and PSNR-B, respectively, for each of the methods studied. As per the results in Table 7, it is evident that the proposed method is superior to the others in terms of SSIM. However, RDN [44] demonstrate higher PSNR values. While DCRN shows a better PSNR-B compared to that of DnCNN, it has comparable performance with DCSC in terms of PSNR-B using the Classic5 dataset. While the RDN was likely to improve AR performance by increasing the number of network parameters, the proposed method was focused to design the lightweight network with the small number of network parameters.

Table 6. PSNR (dB) comparisons on the test datasets. The best results of dataset are shown in bold.

| Dataset | Quality Factor | JPEG | ARCNN [30] | DnCNN [33] | DCSC [42] | RDN [44] | Ours |
|---|---|---|---|---|---|---|---|
| Classic5 | 10 | 27.82 | 29.03 | 29.40 | 29.25 | **30.00** | 29.64 |
|  | 20 | 30.12 | 31.15 | 31.63 | 31.43 | **32.15** | 31.87 |
|  | 30 | 31.48 | 32.51 | 32.91 | 32.68 | **33.43** | 33.15 |
| LIVE1 | 10 | 27.77 | 28.96 | 29.19 | 29.17 | **29.67** | 29.34 |
|  | 20 | 30.07 | 31.29 | 31.59 | 31.48 | **32.07** | 31.74 |
|  | 30 | 31.41 | 32.67 | 32.98 | 32.83 | **33.51** | 33.16 |

Table 7. SSIM comparisons on the test datasets. The best results of dataset are shown in bold.

| Dataset | Quality Factor | JPEG | ARCNN [30] | DnCNN [33] | DCSC [42] | RDN [44] | Ours |
|---|---|---|---|---|---|---|---|
| Classic5 | 10 | 0.780 | 0.793 | 0.803 | 0.803 | 0.819 | **0.825** |
|  | 20 | 0.854 | 0.852 | 0.861 | 0.860 | 0.867 | **0.880** |
|  | 30 | 0.884 | 0.881 | 0.886 | 0.885 | 0.893 | **0.903** |
| LIVE1 | 10 | 0.791 | 0.808 | 0.812 | 0.815 | 0.825 | **0.830** |
|  | 20 | 0.869 | 0.873 | 0.880 | 0.880 | 0.888 | **0.895** |
|  | 30 | 0.900 | 0.904 | 0.909 | 0.909 | 0.915 | **0.922** |

Table 8. PSNR-B (dB) comparisons on the test datasets. The best results of dataset are shown in bold.

| Dataset | Quality Factor | JPEG | ARCNN [30] | DnCNN [33] | DCSC [42] | Ours |
|---|---|---|---|---|---|---|
| Classic5 | 10 | 25.20 | 28.78 | 29.10 | 29.24 | **29.35** |
|  | 20 | 27.50 | 30.60 | 31.19 | **31.41** | 31.40 |
|  | 30 | 28.93 | 32.00 | 32.36 | **32.66** | 32.52 |
| LIVE1 | 10 | 25.33 | 28.77 | 28.91 | **29.17** | 29.03 |
|  | 20 | 27.56 | 30.79 | 31.08 | **31.47** | 31.21 |
|  | 30 | 28.92 | 32.22 | 32.35 | **32.81** | 32.43 |

Table 9 classifies the network complexity in terms of the number of network parameters and total memory size (MB). The proposed DCRN reduced the number of parameters to as low as 72%, 5% and 2% of those needed in DnCNN, IDCN and RDN, respectively. In addition, the total memory size was as low as 91%, 41%, 17% and 5% of that required for DnCNN, DCSC, IDCN and RDN, respectively. Since the same network parameters were repeated 40 times in DCSC, the total memory size was large even though the number of network parameters was smaller than that of the other methods. As shown in Figure 9, the inference speed of the proposed method is greater than that of all networks, except for ARCNN. Although the proposed method is slower than ARCNN, it is clearly better than ARCNN in terms of PSNR, SSIM, and PSNR-B, as per the results in Tables 6–8. Figure 10 shows examples of the visual results of DCRN and the other methods on the test datasets. Based on the results, we were able to confirm that DCRN can recover more accurate textures than other methods.

Table 9. Comparisons of the network complexity between the proposed DCRN and the previous methods.

| Category | Number of Parameters | Total Memory Size (MB) |
|---|---|---|
| ARCNN [30] | 106 K | 3.16 |
| DnCNN [33] | 667 K | 46.31 |
| DCSC [42] | 93 K | 102.34 |
| IDCN [43] | 11 M | 254.13 |
| RDN [44] | 22 M | 861.97 |
| Ours | 479 K | 42.10 |

250.00
200.00
150.00
100.00
50.00
0.00
Inference speed (sec)
2.69
14.41
23.79
93.37
204.87
13.85
ARCNN [30]
DnCNN [33]
DCSC [42]
IDCN [43]
RDN [44]
Ours

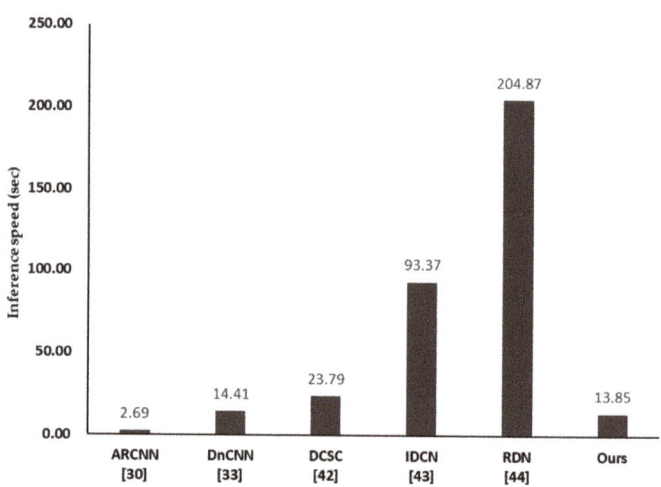

**Figure 9.** Inference speed on Classic5.

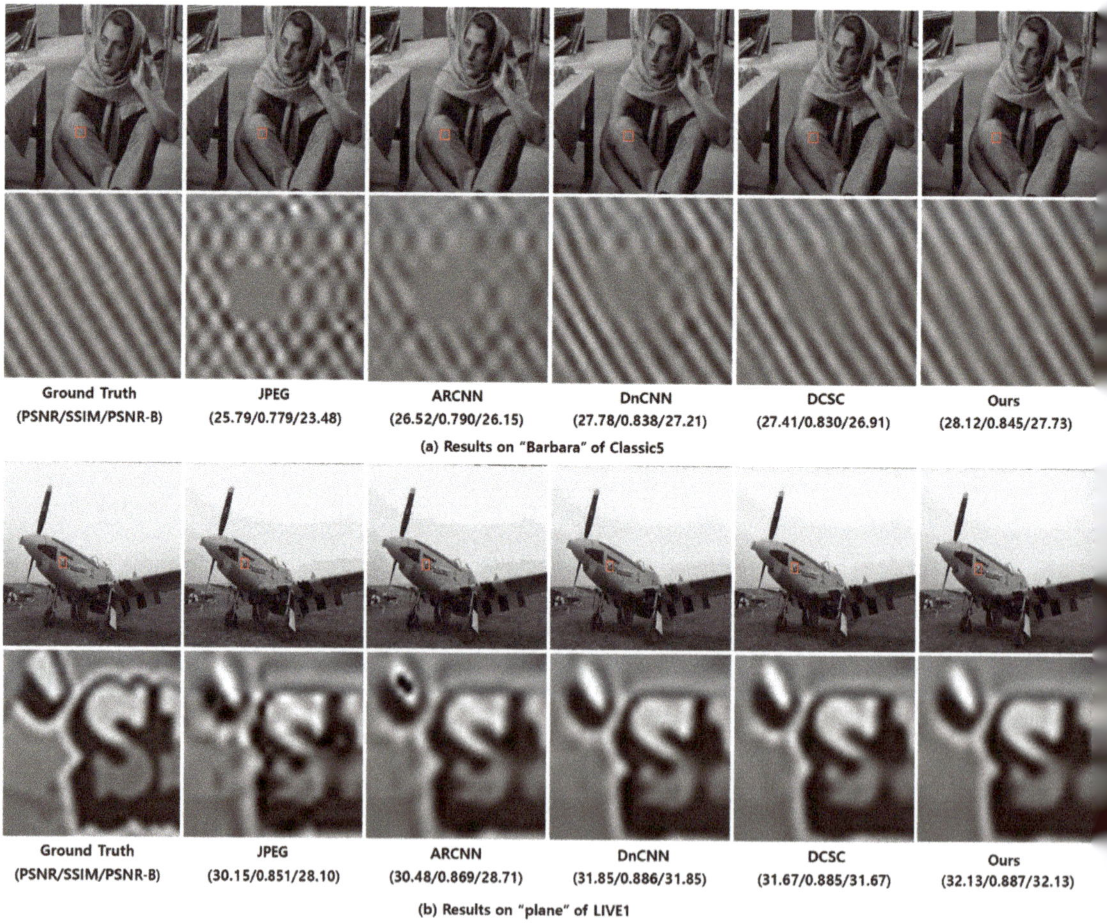

Ground Truth (PSNR/SSIM/PSNR-B)
JPEG (25.79/0.779/23.48)
ARCNN (26.52/0.790/26.15)
DnCNN (27.78/0.838/27.21)
DCSC (27.41/0.830/26.91)
Ours (28.12/0.845/27.73)
(a) Results on "Barbara" of Classic5
Ground Truth (PSNR/SSIM/PSNR-B)
JPEG (30.15/0.851/28.10)
ARCNN (30.48/0.869/28.71)
DnCNN (31.85/0.886/31.85)
DCSC (31.67/0.885/31.67)
Ours (32.13/0.887/32.13)
(b) Results on "plane" of LIVE1

**Figure 10.** *Cont.*

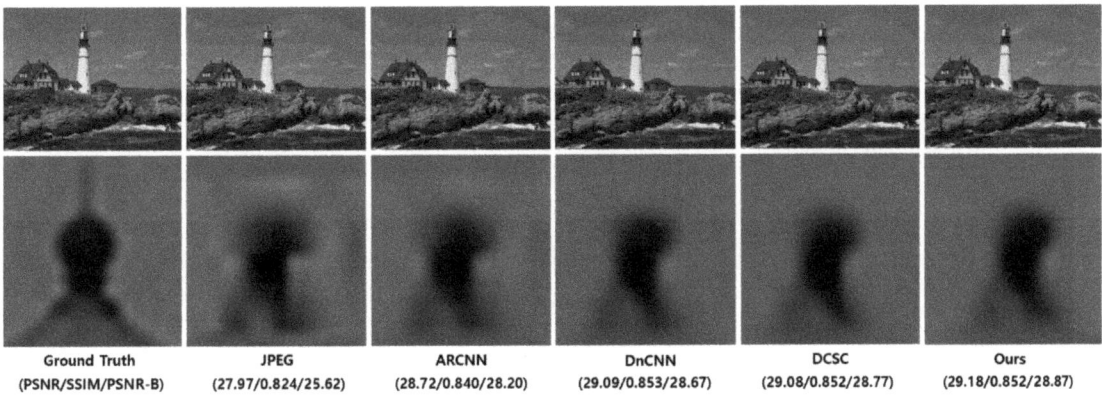

**Figure 10.** Visual comparisons on a JPEG compressed images where the figures of the second row represent the zoom-in for the area represented by the red box.

## 5. Conclusions

Image compression leads to undesired compression artifacts due to the lossy coding that occurs through quantization. These artifacts generally degrade the performance of image restoration techniques, such as super-resolution and object detection. In this study, we propose a DCRN, which consists of the input layer, a densely cascading feature extractor, a channel attention block, and the output layer. The DCRN aims to recover compression artifacts. To optimize the proposed network architecture, we extracted 800 training images from the DIV2K dataset and investigated the trade-off between the network complexity and quality enhancement achieved. Experimental results showed that the proposed DCRN can lead to the best SSIM for compressed JPEG images compared to that of other existing methods, except for IDCN. In terms of network complexity, the proposed DCRN reduced the number of parameters by as low as 72%, 5% and 2% compared to DnCNN, IDCN and RDN, respectively. In addition, the total memory size was as low as 91%, 41%, 17% and 5% of that required for DnCNN, DCSC, IDCN and RDN, respectively. Even though the proposed method was slower than ARCNN, it's PSNR, SSIM, and PSNR-B are clearly better than those of ARCNN.

**Author Contributions:** Conceptualization, Y.L., B.-G.K. and D.J.; methodology, Y.L., B.-G.K. and D.J.; software, Y.L.; validation, S.-h.P., E.R., B.-G.K. and D.J.; formal analysis, Y.L., B.-G.K. and D.J.; investigation, Y.L., B.-G.K. and D.J.; resources, B.-G.K. and D.J.; data curation, Y.L., S.-h.P. and E.R.; writing—original draft preparation, Y.L.; writing—review and editing, B.-G.K. and D.J.; visualization, Y.L.; supervision, B.-G.K. and D.J.; project administration, B.-G.K. and D.J.; funding acquisition, E.R. All authors have read and agreed to the published version of the manuscript.

**Funding:** This work is supported by the Korea Agency for Infrastructure Technology Advancement (KAIA) grant funded by the Ministry of Science and ICT (Grant 21PQWO-B153349-03).

**Institutional Review Board Statement:** Not applicable.

**Informed Consent Statement:** Not applicable.

**Data Availability Statement:** Not applicable.

**Conflicts of Interest:** The authors declare no conflict of interest.

## References

1. Wallace, G. The JPEG still picture compression standard. *IEEE Trans. Consum. Electron.* **1992**, *38*, 18–34. [CrossRef]
2. Google. Webp—A New Image Format for the Web. Google Developers Website. Available online: https://developers.google.com/speed/webp/ (accessed on 16 August 2021).
3. Sullivan, G.; Ohm, J.; Han, W.; Wiegand, T. Overview of the High Efficiency Video Coding (HEVC) Standard. *IEEE Trans. Circuits Syst. Video Technol.* **2012**, *12*, 1649–1668. [CrossRef]
4. Aziz, S.; Pham, D. Energy Efficient Image Transmission in Wireless Multimedia Sensor Networks. *IEEE Commun. Lett.* **2013**, *17*, 1084–1087. [CrossRef]
5. Lee, Y.; Jun, D.; Kim, B.; Lee, H. Enhanced Single Image Super Resolution Method Using Lightweight Multi-Scale Channel Dense Network. *Sensors* **2021**, *21*, 3351. [CrossRef]
6. Kim, S.; Jun, D.; Kim, B.; Lee, H.; Rhee, E. Single Image Super-Resolution Method Using CNN-Based Lightweight Neural Networks. *Appl. Sci.* **2021**, *11*, 1092. [CrossRef]
7. Peled, S.; Yeshurun, Y. Super resolution in MRI: Application to Human White Matter Fiber Visualization by Diffusion Tensor Imaging. *Magn. Reason. Med.* **2001**, *45*, 29–35. [CrossRef]
8. Shi, W.; Caballero, J.; Ledig, C.; Zhang, X.; Bai, W.; Bhatia, K.; Marvao, A.; Dawes, T.; Regan, D.; Rueckert, D. Cardiac Image Super-Resolution with Global Correspondence Using Multi-Atlas PatchMatch. *Med. Image Comput. Comput. Assist. Interv.* **2013**, *8151*, 9–16.
9. Thornton, M.; Atkinson, P.; Holland, D. Sub-pixel mapping of rural land cover objects from fine spatial resolution satellite sensor imagery using super-resolution pixel-swapping. *Int. J. Remote Sens.* **2006**, *27*, 473–491. [CrossRef]
10. Zhang, L.; Zhang, H.; Shen, H.; Li, P. A super-resolution reconstruction algorithm for surveillance images. *Signal Process.* **2010**, *90*, 848–859. [CrossRef]
11. Li, Y.; Guo, F.; Tan, R.; Brown, M. A Contrast Enhancement Framework with JPEG Artifacts Suppression. In Proceedings of the European Conference on Computer Vision, Zurich, Switzerland, 6–12 September 2014; pp. 174–188.
12. Tung, T.; Fuh, C. ICEBIM: Image Contrast Enhancement Based on Induced Norm and Local Patch Approaches. *IEEE Access* **2021**, *9*, 23737–23750. [CrossRef]
13. Srinivas, K.; Bhandari, A.; Singh, A. Exposure-Based Energy Curve Equalization for Enhancement of Contrast Distorted Images. *IEEE Trans. Circuits Syst. Video Technol.* **2020**, *12*, 4663–4675. [CrossRef]
14. Wang, J.; Hu, Y. An Improved Enhancement Algorithm Based on CNN Applicable for Weak Contrast Images. *IEEE Access* **2020**, *8*, 8459–8476. [CrossRef]
15. Liu, Y.; Xie, Z.; Liu, H. An Adaptive and Robust Edge Detection Method Based on Edge Proportion Statistics. *IEEE Trans. Image Process.* **2020**, *29*, 5206–5215. [CrossRef]
16. Ofir, N.; Galun, M.; Alpert, S.; Brandt, S.; Nadler, B.; Basri, R. On Detection of Faint Edges in Noisy Images. *IEEE Trans. Pattern Anal. Mach. Intell.* **2020**, *42*, 894–908. [CrossRef]
17. He, J.; Zhang, S.; Yang, M.; Shan, Y.; Huang, T. BDCN: Bi-Directional Cascade Network for Perceptual Edge Detection. *IEEE Trans. Pattern Anal. Mach. Intell.* **2020**, *10*, 1–14. [CrossRef] [PubMed]
18. Shen, M.; Kuo, J. Review of Postprocessing Techniques for Compression Artifact Removal. *J. Vis. Commun. Image Represent.* **1998**, *9*, 2–14. [CrossRef]
19. Gonzalez, R.; Woods, R. *Digital Image Processing*; Pearson Education: London, UK, 2002.
20. List, P.; Joch, A.; Lainema, J.; Bjøntegaard, G.; Karczewicz, M. Adaptive Deblocking Filter. *IEEE Trans. Circuits Syst. Video Technol.* **2003**, *13*, 614–619. [CrossRef]
21. Reeve, H.; Lim, J. Reduction of Blocking Effect in Image Coding. In Proceedings of the IEEE International Conference on Acoustics, Speech, and Signal Processing, Boston, MA, USA, 14–16 April 1983; pp. 1212–1215.
22. Liew, A.; Yan, H. Blocking Artifacts Suppression in Block-Coded Images Using Overcomplete Wavelet Representation. *IEEE Trans. Circuits Syst. Video Technol.* **2004**, *14*, 450–461. [CrossRef]
23. Foi, A.; Katkovnik, V.; Egiazarian, K. Pointwise Shape-Adaptive DCT for High-Quality Denoising and Deblocking of Grayscale and Coloe Images. *IEEE Trans. Image Process.* **2007**, *16*, 1–17. [CrossRef]
24. Chen, K.H.; Guo, J.I.; Wang, J.S.; Yeh, C.W.; Chen, J.W. An Energy-Aware IP Core Design for the Variable-Length DCT/IDCT Targeting at MPEG4 Shape-Adaptive Transforms. *IEEE Trans. Circuits Syst. Video Technol.* **2005**, *15*, 704–715. [CrossRef]
25. Lecun, Y.; Boser, B.; Denker, J.; Henderson, D.; Howard, R.; Hubbard, W.; Jackel, L. Backpropagation Applied to handwritten Zip code Recognition. *Neural Comput.* **1989**, *1*, 541–551. [CrossRef]
26. Hore, A.; Ziou, D. Image Quality Metrics: PSNR vs. SSIM. In Proceedings of the International Conference on Pattern Recognition, Istanbul, Turkey, 23–26 August 2010; pp. 2366–2369.
27. Silpa, K.; Mastani, A. New Approach of Estimation PSNR-B for De-blocked Images. *arXiv* **2013**, arXiv:1306.5293.
28. Yim, C.; Bovik, A. Quality Assessment of Deblocked Images. *IEEE Trans. Image Process.* **2011**, *20*, 88–98.
29. Wang, Z.; Bovik, A.; Sheikh, H.; Simoncelli, E. Image Quality Assessment: From Error Visibility to Structural Similarity. *IEEE Trans. Image Process.* **2004**, *13*, 600–612. [CrossRef]
30. Dong, C.; Deng, Y.; Loy, C.; Tang, X. Compression Artifacts Reduction by a Deep Convolutional Network. In Proceedings of the International Conference on Computer Vision, Las Condes Araucano Park, Chile, 11–18 December 2015; pp. 576–584.

31. Mao, X.; Shen, C.; Yang, Y. Image Restoration Using Very Deep Convolutional Encoder-Decoder Networks with Symmetric Skip Connections. *arXiv* **2016**, arXiv:1603.09056.
32. Chen, Y.; Pock, T. Trainable Nonlinear Reaction Diffusion: A Flexible Framework for Fast and Effective Image Restoration. *IEEE Trans. Pattern Anal. Mach. Intell.* **2017**, *39*, 1256–1272. [CrossRef]
33. Zhang, K.; Zuo, W.; Chen, Y.; Meng, D.; Zhang, L. Beyond a Gaussian Denoiser: Residual Learning of Deep CNN for Image Denoising. *IEEE Trans. Image Process.* **2017**, *26*, 3142–3155. [CrossRef]
34. Glorot, X.; Bordes, A.; Bengio, Y. Deep Sparse Rectifier Neural Networks. In Proceedings of the Fourteenth International Conference on Artificial Intelligence and Statistics, Fort Lauderdale, FL, USA, 11–13 April 2011; pp. 315–323.
35. Cavigelli, L.; Harger, P.; Benini, L. CAS-CNN: A Deep Convolutional Neural Network for Image Compression Artifact Suppression. In Proceedings of the International Joint Conference on Neural Networks, Anchorage, AK, USA, 14–19 May 2017; pp. 752–759.
36. Guo, J.; Chao, H. One-to-Many Network for Visually Pleasing Compression Artifacts Reduction. In Proceedings of the Conference on Computer Vision and Pattern Recognition Workshops, Honolulu, HI, USA, 21–26 July 2017; pp. 3038–3047.
37. Tai, Y.; Yang, J.; Liu, X.; Xu, C. MemNet: A Persistent Memory Network for Image Restoration. In Proceedings of the IEEE International Conference on Computer Vision, Venice, Italy, 22–29 October 2017; pp. 4549–4557.
38. Dai, Y.; Liu, D.; Wu, F. A Convolutional Neural Network Approach for Post-Processing in HEVC Intra Coding. In Proceedings of the International Conference on Multimedia Modeling, Reykjavik, Iceland, 4–6 January 2017; pp. 28–39.
39. Zhang, X.; Yang, W.; Hu, Y.; Liu, J. DMCNN: Dual-Domain Multi-Scale Convolutional Neural Network for Compression Artifact Removal. In Proceedings of the IEEE International Conference on Image Processing, Athens, Greece, 7–10 October 2018; pp. 390–394.
40. Liu, P.; Zhang, H.; Zhang, K.; Lin, L.; Zuo, W. Multi-Level Wavelet-CNN for Image Restoration. In Proceedings of the Conference on Computer Vision and Pattern recognition Workshops, Salt Lake City, UT, USA, 18–22 June 2018; pp. 886–895.
41. Chen, H.; He, X.; Qing, L.; Xiong, S.; Nguyen, T. DPW-SDNet: Dual Pixel-Wavelet Domain Deep CNNs for Soft Decoding of JPEG-Compressed Images. In Proceedings of the Conference on Computer Vision and Pattern recognition Workshops, Salt Lake City, UT, USA, 18–22 June 2018; pp. 824–833.
42. Fu, X.; Zha, Z.; Wu, F.; Ding, X.; Paisley, J. JPEG Artifacts Reduction via Deep Convolutional Sparse Coding. In Proceedings of the IEEE International Conference on Computer Vision, Seoul, Korea, 27 October–2 November 2019; pp. 2501–2510.
43. Zheng, B.; Chen, Y.; Tian, X.; Zhou, F.; Liu, X. Implicit Dual-domain Convolutional Network for Robust Color Image Compression Artifact Reduction. *IEEE Trans. Circuits Syst. Video Technol.* **2020**, *30*, 3982–3994. [CrossRef]
44. Zhang, Y.; Tian, Y.; Kong, Y.; Zhong, B.; Fu, Y. Residual Dense Network for Image Super-Resolution. *IEEE Trans. Pattern Anal. Mach. Intell.* **2020**, *43*, 2480–2495. [CrossRef]
45. He, K.; Zhang, X.; Ren, S.; Sun, J. Deep Residual Learning for Image Recognition. In Proceedings of the Conference on Computer Vision and Pattern Recognition, Las Vegas, NY, USA, 27–30 June 2016; pp. 770–778.
46. Huang, G.; Liu, Z.; Van Der Maaten, L.; Weinberger, K. Densely Connected Convolutional Networks. In Proceedings of the Conference on Computer Vision and Pattern Recognition, Honolulu, HI, USA, 21–26 July 2017; pp. 4700–4708.
47. Hu, J.; Shen, L.; Albanie, S.; Sun, G.; Wu, E. Squeeze-and-Excitation Networks. In Proceedings of the Conference on Computer Vision and Pattern Recognition, Salt Lake City, UT, USA, 18–22 June 2018; pp. 1–13.
48. Zhao, H.; Gallo, O.; Frosio, I.; Kautz, J. Loss Functions for Image Restoration with Neural Networks. *IEEE Trans. Comput. Imaging* **2017**, *3*, 47–57. [CrossRef]
49. Kingma, D.; Ba, J. Adam: A method for stochastic optimization. *arXiv* **2014**, arXiv:1412.6980.
50. Redford, A.; Metz, L.; Chintala, S. Unsupervised representation learning with deep convolutional generative adversarial networks. *arXiv* **2015**, arXiv:1511.06434.
51. He, K.; Zhang, X.; Ren, S.; Sun, J. Delving Deep into Rectifiers: Surpassing Human-Level Performance on Imagenet classification. In Proceedings of the IEEE International Conference on Computer Vision, Santiago, Chile, 13–16 December 2015; pp. 1026–1034.
52. Agustsson, E.; Timofte, R. NTIRE 2017 Challenge on Single Image Super-Resolution: Dataset and Study. In Proceedings of the Conference on Computer Vision and Pattern Recognition Workshops, Honolulu, HI, USA, 21–26 July 2017.
53. Yang, J.; Wright, J.; Huang, T.; Ma, Y. Image Super-Resolution Via Sparse Representation. *IEEE Trans. Image Process.* **2010**, *19*, 2861–2873. [CrossRef]

*Article*

# Rain Streak Removal for Single Images Using Conditional Generative Adversarial Networks

Prasad Hettiarachchi [1,2], Rashmika Nawaratne [1,*], Damminda Alahakoon [1], Daswin De Silva [1] and Naveen Chilamkurti [3]

1. Research Centre for Data Analytics and Cognition, La Trobe University, Melbourne, VIC 3086, Australia; prasad.12@cse.mrt.ac.lk (P.H.); D.Alahakoon@latrobe.edu.au (D.A.); D.DeSilva@latrobe.edu.au (D.D.S.)
2. Department of Computer Science and Engineering, University of Moratuwa, Moratuwa 10400, Sri Lanka
3. Department of Computer Science and Information Technology, La Trobe University, Melbourne, VIC 3086, Australia; N.Chilamkurti@latrobe.edu.au
* Correspondence: B.Nawaratne@latrobe.edu.au

**Abstract:** Rapid developments in urbanization and smart city environments have accelerated the need to deliver safe, sustainable, and effective resource utilization and service provision and have thereby enhanced the need for intelligent, real-time video surveillance. Recent advances in machine learning and deep learning have the capability to detect and localize salient objects in surveillance video streams; however, several practical issues remain unaddressed, such as diverse weather conditions, recording conditions, and motion blur. In this context, image de-raining is an important issue that has been investigated extensively in recent years to provide accurate and quality surveillance in the smart city domain. Existing deep convolutional neural networks have obtained great success in image translation and other computer vision tasks; however, image de-raining is ill posed and has not been addressed in real-time, intelligent video surveillance systems. In this work, we propose to utilize the generative capabilities of recently introduced conditional generative adversarial networks (cGANs) as an image de-raining approach. We utilize the adversarial loss in GANs that provides an additional component to the loss function, which in turn regulates the final output and helps to yield better results. Experiments on both real and synthetic data show that the proposed method outperforms most of the existing state-of-the-art models in terms of quantitative evaluations and visual appearance.

**Keywords:** deep learning; generative adversarial networks; traffic surveillance image processing; image de-raining

**Citation:** Hettiarachchi, P.; Nawaratne, R.; Alahakoon, D.; De Silva, D.; Chilamkurti, N. Rain Streak Removal for Single Images Using Conditional Generative Adversarial Networks. *Appl. Sci.* **2021**, *11*, 2214. https://doi.org/10.3390/app11052214

Academic Editor: Rubén Usamentiaga

Received: 4 February 2021
Accepted: 25 February 2021
Published: 3 March 2021

**Publisher's Note:** MDPI stays neutral with regard to jurisdictional claims in published maps and institutional affiliations.

**Copyright:** © 2021 by the authors. Licensee MDPI, Basel, Switzerland. This article is an open access article distributed under the terms and conditions of the Creative Commons Attribution (CC BY) license (https://creativecommons.org/licenses/by/4.0/).

## 1. Introduction

Rain is a common weather condition that negatively impacts computer vision systems. Raindrops appear as bright streaks in images due to their high velocity and light scattering. Since image recognition and detection algorithms are designed for clean inputs, it is essential to develop an effective mechanism for rain streak removal.

A number of research efforts have been reported in the literature focusing on restoring rain images, and different approaches have been taken. Some have attempted to remove rain streaks using video [1–3], while other researchers have focused on rain image recovery from a single image by considering the image as a signal separation task [4–6].

Since rain streaks overlap with background texture patterns, it is quite a challenging task to remove the rain streaks while maintaining the original texture in the background. Most of the times, this results in over-smoothed regions that are visible in the background after the de-raining process. De-raining algorithms [7,8] tend to over de-rain or under de-rain the original image. A key limitation in the traditional, handcrafted methods is that the feature learning is manual and designed to deal only with certain types of rain streaks, and they do not perform well with varying scales, shapes, orientations, and densities of raindrops [9,10]. In contrast, by using convolutional neural networks (CNNs), the feature learning process becomes an integral part of the algorithm and is able to unveil many hidden features. Convolutional neural network-based methods [11–13] have gained huge improvements in image de-raining during the last few years. These methods try to figure out a nonlinear mapping between the input rainy image and the expected ground truth image.

Still, there is potential for improvements and optimizations within CNN-based image de-raining algorithms, which could lead to more visually appealing and accurate results. Instead of being just constrained to characterizing rain streaks, visual quality should also be considered when defining the optimization functions, which will result in improving the visual appeal of test results. When defining the objective function, it should consider the fact that the performance of vision algorithms, such as classification/detection, should not be affected by the presence of rain streaks. The addition of this discriminative information ensures that the output is indistinguishable from its original counterpart.

Generative modeling is an unsupervised learning task in machine learning that involves automatically discovering and learning the patterns in input data in such a way that the model can be used to generate new examples that are indistinguishable from reality. The concept of generative adversarial networks (GANs) was originally presented in [14] and has gained a high level of interest, with several successful applications and directions reported within a short period in the machine learning community. Existing CNN-based mechanisms only consider either $L_1$(Least Absolute Deviations) or $L_2$ (Least Square Errors) errors, whereas in conditional GANs, they have additional adversarial loss components, which result in very good, qualitative, visually appealing image outputs.

In our approach, we propose a conditional generative adversarial network-based framework for rain streak removal. Our model consists of a densely connected generator (G) network and a CNN-based discriminator (D) network. The generator network converts rainy images to de-rained images in such a way that it fools the discriminator network. In certain scenarios, traditional GANs tend to make output images more artificial and visually displeasing. To mitigate this issue, we have introduced a conditional CNN with skip connections for the generator. Skip connections guarantee better convergence by efficiently leveraging features from different layers of the network. The proposed model is based on the Pix2Pix framework by Isola et al. [15] and the conditional generative adversarial networks originally proposed by Fu et al. [16]. We have also used the source codes provided by authors of LPNet [17] and GMM [18] for quantitative and qualitative comparisons of the proposed model.

This paper makes the following contributions:

1. Propose a conditional, GAN-based deep learning architecture to remove rain streaks from images by adapting U-Net architecture-based CNN for single image de-raining.
2. Develop a classifier to identify whether the generated image is real or fake based on intra-convolutional "PatchGAN" architecture.
3. Due to the lack of access to the ground truth of rainy images, we present a new dataset synthesizing rainy images using real-world clean images, which are used as the ground truth counterpart in this research.

The paper is organized as follows: In Section 2, we provide an overview of related methods for image de-raining and the basic concepts behind cGANs. Section 3 describes the proposed model (CGANet—Conditional Generative Adversarial Network model) in detail with its architecture. Section 4 describes the experimental details with evaluation results. Section 5 provides the conclusion. Implementation details and the dataset used for the experiments are publicly available at GitHub (https://github.com/prasadmaduranga/CGANet (accessed on 11 December 2020)).

## 2. Related Work

In the past, numerous methods and research approaches have been proposed for image de-raining. These methods can be categorized as single image-based methods and video-based methods. With the evolution of neural networks, deep learning-based methods have become more dominant and efficient compared to past state-of-the-art methods.

## 2.1. Single Image-based Methods

Single image-based methods have limited access to information compared to video-based methods, which makes it more challenging to remove the rain streaks. Single image-based methods include low-rank approximations [3,19], dictionary learning [4,5,20], and kernel-based methods [21]. In [4], the authors decomposed the image into high- and low-frequency components and recognized the rain streaks by processing the high-frequency components. Other mechanisms have used gradients [22] and mixture models [18] to model and remove rain streaks. In [18], the authors introduced a patch-based prior for both clean and rainy layers using Gaussian mixture models (GMM). The GMM prior for rainy layers was learned from rainy images, while for the clean images, it was learned from natural images. Nonlocal mean filtering and kernel regression were used to identify rain streaks in [21].

## 2.2. Video-based Methods

With the availability of inter-frame information, video-based image de-raining is relatively more effective and easier compared to single image de-raining. Most research studies [1,23,24] have focused on detecting potential rain streaks using their physical characteristics and removing them using image restoration algorithms. In [25], the authors divided rain streaks into dense and sparse groups and removed the streaks using a matrix decomposition algorithm. Other methods have focused on de-raining in the Fourier domain [1] using Gaussian mixture models [23], matrix completions [24], and low-rank approximations [3].

## 2.3. Deep Learning based Methods

Deep learning-based methods have gained much popularity and success in a variety of high-level computer vision tasks in the recent past [26–28] as well as in image processing problems [29–31]. Deep learning was introduced for de-raining in [11] where a three-layer CNN was used for removing rain streaks and dirt spots in an image that had been taken through glass. In [12], a CNN was proposed for video-based de-raining, while a recurrent neural network was adopted by Liu in [32]. The authors in [33] proposed a residual-guide feature fusion network for single image de-raining. A pyramid of networks was proposed in [17], which used the domain-specific knowledge to reinforce the learning process.

CNNs learn to minimize a loss function, and the loss value itself decides the quality of output results. Significant design efforts and domain expertise are required to define an effective loss function. In other words, it is necessary to provide the CNN with what the user requires to minimize. Instead, if it is possible to set a high-level, general goal such as "make the output image indistinguishable from the target images", then the CNN can automatically learn a loss function to satisfy the goal. This is the basic underlying concept behind generative adversarial networks (GANs).

## 2.4. Generative Adversarial Networks

Generative adversarial networks [14] are unsupervised generative models that contain two deep neural networks. The two neural networks are named as the generator ($G$) and discriminator ($D$) and are trained parallelly during the training process. GAN training can be considered to be a two-player min-max game where the generator and discriminator compete with each other to achieve each other's targeted goal. The generator is trained to learn a mapping from a random noise vector ($z$) in latent space to an image ($x$) in a target domain: $G(z) \rightarrow x$. The discriminator ($D$) learns to classify a given image as a real (output close to 1) image or a fake (output close to 0) image from the generator ($G$): $D(x) \rightarrow [0.1]$. Both the generator and decimator can be considered as two separate neural networks trained from backpropagation, and they have separate loss functions. Figure 1 shows the high-level architecture of the proposed conditional GAN model. The generator will try to generate synthetic images that resemble real images to fool the discriminator. The discriminator learns how to identify the real images from the generated synthetic images from the generator.

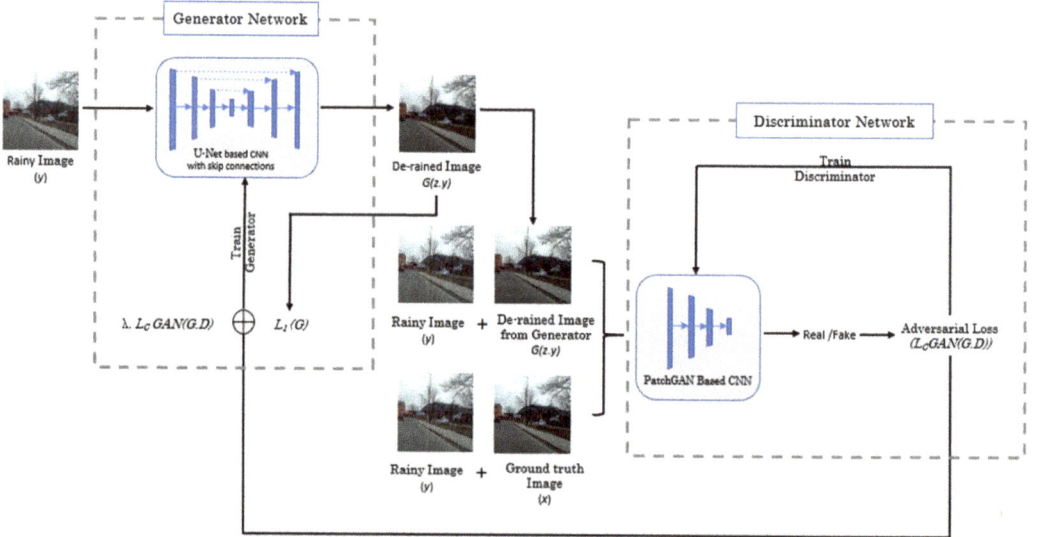

**Figure 1.** High-level architecture of the proposed model (CGANet).

The widest adaptation of GANs is for data augmentation, or that is to say, to learn from existing real-world samples and generate new samples consistent with the distribution. Generative modeling has been used in a wide range of application domains including computer vision, natural language processing, computer security, medicine, etc.

Xu et al. [34] used GANs for synthesizing image data to train and validate perception systems for autonomous vehicles. In addition to that, [35,36] used GANs for data fusion for developing image classification models while mitigating the issue of having smaller datasets. Furthermore, GANs were used for augmenting datasets for adversarial training in [37]. To increase the resolution of images, a super-resolution GAN was proposed by Ledig et al. [38], which took a low-resolution image as the input and generated a high-resolution image with 4× upscaling. To convert the image content from one domain to another, an image-to-image translation approach was proposed by Isola et al. [15] using CGANs. Roy et al. [39] proposed a TriGAN, which could solve the problem of image translation by adapting multiple source domains. Experiments showed that the SeqGAN proposed in [40] outperformed the traditional methods used for music and speech generation. In the computer security domain, Hu and Tan [41] proposed a GAN-based model to generate malware. For private product customization, Hwang et al. [42] proposed GANs to manufacture medical products.

### 3. Proposed Model

The proposed approach uses image-to-image translation for the image de-raining task. In a GAN, the generator produces the output based on the latent variable or the noise variable (z). However, in the proposed approach, it is necessary for a correlation to exist between the source image and the generator output image. We have applied the conditional GAN [16], which is a variant of the traditional GAN that takes additional information, y, as the input. In this case, we provide a source image with rain streaks as additional information for both the generator and discriminator. $x$ represents the target image.

The objective of a conditional GAN is as follows:

$$L_{cGAN}(G.D) = E_{x \sim pdata(X)}[log(D(x.y))] + E_{z \sim pz(Z)}[log(1 - D(G(z.y).y))] \quad (1)$$

where $pdata(X)$ denotes the real data probability distribution defined in the data space $X$, and $pz(Z)$ denotes the probability distribution of the latent variable $z$ defined on the latent space $Z$. $E_{x \sim pdata(X)}$ and $E_{z \sim pz(Z)}$ represent the expectations over the data spaces $X$ and $Z$ respectively. $G(.)$ and $D(.)$ represent the non-linear mappings of the generator and discriminator networks respectively.

In an image de-raining task, the higher-order color and texture information has to be preserved during the image translation. This has a significant impact on the visual performance of the output. Adversarial loss alone is not sufficient for this task. The loss function should be optimized so that it penalizes the perceptual differences between the output image and the target image.

Our implementation architecture is based on the work of Isola's [15] Pix2Pix framework. It learns a mapping from an input image to an output image along with the objective function to train the model. In Pix2Pix, it suggests $L_1$ (mean absolute error) loss instead of $L_2$ (mean squared error) loss for the GAN objective function, since it encourages less blurring in the generator output. $L_1$ loss averages the pixel level absolute difference between the target image and the generated image $G(z.y)$ over the image space $x.y.z$.

$$L_1(G) = Ex.y.z[\| x - G(z.y) \|] \quad (2)$$

Finally, the loss function for this work is as follows:

$$L(G.D) = L_{cGAN}(G.D) + \lambda L_1(G) \quad (3)$$

Lambda ($\lambda$) is a hyperparameter that controls the weights of the terms. In this case, we kept lambda = 100 [15]. When training the model, lambda was increased to train a discriminator and minimized to train a generator. The final objective was to identify the generator $G^*$ by solving the following optimization problem:

$$G^* = \arg \min_G \max_D (L_{cGAN}(G,D) + \lambda L_1(G)) \quad (4)$$

*Model Overview*

- Generator Network

In image-to-image translations, it is necessary to map a high-resolution input grid to a high-resolution output grid. Though the input and output images differ in appearance, they share the same underlying structure, and as such, it is necessary to consider this factor when designing the generator architecture. Most previous work used the encoder-decoder network [43] for such scenarios. In encoder-decoder CNN, the input is progressively downsampled until the bottleneck layer, where the process gets reversed and starts to upsample the input data. Convolutional layers use 4 × 4 filters and strides with size 2 for downsampling. The same size of kernel is used for transpose convolution operation during upsampling. Each convolution/deconvolution operation is followed by batch normalization and Rectified Linear Unit (ReLU) activation. Weights of the generator are updated depending on the adversarial loss of the discriminator and the $L_1$ loss of the generator. Architecture details are shown in Table 1.

Table 1. Generator architecture of the CGANet model.

| Generator Architecture |
|---|
| Input(256 × 256), Num_c = 3 |
| Downsampling: 4 × 4 Convolution + BN + ReLu, Output: 128 × 128, Num_c: 64 |
| Downsampling 4 × 4 Convolution + BN + ReLu, Output: 64 × 64, Num_c: 128 |
| Downsampling 4 × 4 Convolution + BN + ReLu, Output: 64 × 64, Num_c: 128 |
| Downsampling: 4 × 4 Convolution + BN + ReLu, Output: 32 × 32, Num_c: 256 |
| Downsampling: 4 × 4 Convolution + BN + ReLu, Output: 16 × 16, Num_c: 512 |
| Downsampling 4 × 4 Convolution + BN + ReLu, Output: 128 × 128, Num_c: 512 |
| Downsampling 4 × 4 Convolution + BN + ReLu, Output: 8×8, Num_c: 512 |
| Downsampling 4 × 4 Convolution + BN + ReLu, Output: 4 × 4, Num_c: 512 |
| Downsampling 4 × 4 Convolution + BN + ReLu, Output: 2 × 2, Num_c: 512 |
| Downsampling 4 × 4 Convolution + BN + ReLu, Output: 1 × 1, Num_c: 512 |
| Upsampling 4 × 4 Convolution + BN + ReLu, Output: 2 × 2, Num_c: 512 |
| Concatenation: Input (2×2×512), (2 × 2 × 512), Output (2 × 2 × 1024) |
| Upsampling 4 × 4 Convolution + BN + ReLu, Output: 4 × 4, Num_c: 512 |
| Concatenation: Input (4 × 4 × 512), (4 × 4 × 512), Output (4 × 4 × 1024) |
| Upsampling 4 × 4 Convolution + BN + ReLu, Output: 8 × 8, Num_c: 512 |
| Concatenation: Input (8×8×512), (8×8×512), Output (8 × 8 × 1024) |
| Upsampling: 4 × 4 Transpose Convolution + BN + ReLu, Output: 16 × 16, Num_c: 512 |
| Concatenation: Input (16 × 16 × 512), (16 × 16 × 512), Output (16 × 16 × 1024) |
| Upsampling: 4 × 4 Transpose Convolution + BN + ReLu, Output: 32 × 32, Num_c: 256 |
| Concatenation: Input (32 × 32 × 256), (32 × 32 × 256), Output (32 × 32 × 512) |
| Upsampling: 4 × 4 Transpose Convolution + BN + ReLu, Output: 64 × 64, Num_c: 128 |
| Concatenation: Input (64 × 64 × 128), (64 × 64 × 128), Output (64 × 64 × 256) |
| Upsampling: 4 × 4 Transpose Convolution + BN + ReLu, Output: 128 × 128, Num_c: 64 |
| Concatenation: Input (128 × 128 × 64), (128 × 128 × 64), Output (128 × 128 × 128) |
| Upsampling: 4 × 4 Transpose Convolution, Output: 256×256, Num_c: 3 |

These networks require all the input information to pass through each of the middle layers. In most of the image-to-image translation problems, it is desirable to share the feature maps across the network since both input and output images represent the same underlying structure. For this purpose, we added a skip connection while following the general configuration of a "U-Net" [44]. Skip connections simply concatenate the channels at the ith layer with the channels at the (n–i)th layer.

- Discriminator Network

We adapted PatchGAN architecture [45] for the discriminator, which penalized the structure at the scale of patches. It tried to classify each N × N patch as either real or fake. Final output of the discriminator (D) was calculated by averaging the received responses by running the discriminator convolutionally across the image. In this case, the patch was 30 × 30 in size, and each convolutional layer was followed by a ReLU activation and batch normalization. Zero-padding layers were used to preserve the edge details of the input feature maps during the convolution. Discriminator architecture is described in Table 2.

**Table 2.** Discriminator architecture of the CGANet model.

| Discriminator Architecture |
| --- |
| Input Image (256 × 256 × 3) + Target Image (256 × 256 × 3) |
| Concatenation: Input (256 × 256 × 3), (256 × 256 × 3), Output (2 × 2 × 1024) |
| Downsample 4 × 4 Convolution + BN + ReLu, Output: 128 × 128, Num_c: 64 |
| Downsample 4 × 4 Convolution + BN + ReLu, Output: 64 × 64, Num_c: 128 |
| Downsample 4 × 4 Convolution + BN + ReLu, Output: 32 × 32, Num_c: 256 |
| Zero Padding 2D: Output: 34 × 34, Num_c: 256 |
| Downsample 4 × 4 Convolution + BN + ReLu, Output: 31 × 31, Num_c: 512 |
| Zero Padding 2D: Output: 33×33, Num_c: 512 |
| Downsample 4 × 4 Convolution + BN + ReLu, Output: 30 × 30, Num_c: 1 |

## 4. Experimental Details

This section discusses the experimental details of our proposed CGANet model and the quality matrices used to evaluate the performance of the proposed model. CGANet performance is compared with two other state-of-the-art methods: the Gaussian mixture model [18] and lightweight pyramid networks [17]. The algorithm implementation was conducted using Python and TensorFlow 2.0 [46]. CGANet was trained on a computer with a 2.2 GHz, 6-core Intel core i7 processor, 16 GB memory, and an AMD Radeon Pro 555X GPU.

### 4.1. Dataset

The training set consisting of 1500 images was chosen from a global road damage detection challenge dataset [47]. Rain streaks of different angles and intensities have been added to those images using Photoshop to create a synthesized rainy image set. Corresponding clean images become the target ground truth image set for the synthesized rainy image set. The test set consists of both synthesized and real-world rainy images. Three hundred synthesized images were chosen from the global road damage detection challenge dataset and pre-processed similarly when preparing the training set. Test dataset outputs are shown in Figure 2 as a comparison between the proposed CGANet model and the state-of-the-art de-raining methods. Real-world rainy images were taken from the internet, and they were considered only for demonstrating the effectiveness of the CGANet model. Since ground truth images were not available for the real-world rainy images, they were not taken into the account when training the model. Test results of real-world images are shown in Figure 3.

### 4.2. Evaluation Matrix and Results

The peak signal-to-noise ratio (PSNR) and structural similarity index (SSIM) [48] were used to evaluate and compare the performance of the model. PSNR measures how far the de-rained image is distorted from its real ground truth image by using the mean squared error at the pixel level. As shown in Table 1, the proposed CGANet model obtained the best PSNR value compared to the other two methods. The structural similarity index (SSIM) is a perception-based index that evaluates image degradation as the perceived difference in structural information while also incorporating both luminance masking and contrast masking terms. Table 3 shows the SSIM value comparison between the proposed CGANet model and the other two state-of-the art methods. By referring to this comparison, we could verify that the proposed method performed well compared to other de-raining mechanisms, and this is also visually verifiable in Figures 2 and 3.

Input      GMM [18]      Pyramid [17]      CGANet      Ground Truth

**Figure 2.** Qualitative comparison between GMM, pyramid networks, and proposed CGANet methods.

**Figure 3.** CGANet on real-world dataset ((**Left**) Input image; (**Right**) de-rained output).

Table 3. Quantitative comparison between different de-raining methods (mean ± STD).

| Index | Pyramid | GMM | CGANet |
|---|---|---|---|
| PSNR | 23.48 ± 2.09 | 24.37 ± 2.15 | 25.85 ± 1.57 |
| SSIM | 0.731 ± 0.06 | 0.762 ± 0.06 | 0.768 ± 0.04 |

*4.3. Parameter Settings*

To optimize the proposed model, we followed the findings provided in the original GAN paper [14]. Instead of training the generator to minimize *log(1 − D(x; G(x; z))*, we trained it to maximize *log D(x; G(x; z))*. Since the discriminator could be trained much faster compared to the generator, we divided the discriminator loss by 2 while optimizing the discriminator. As such, the discriminator training speed slowed down compared to the generator. Both the discriminator and generator models were trained with an Adam optimizer [49] with a learning rate of 0.0002 and a momentum parameter $\beta 1$ of 0.5 [15]. The model was trained using 150 epochs and updated after each image, and as such, the batch size was 1.

## 5. Conclusions

In this paper, we have proposed a single image de-raining model based on conditional generative adversarial networks and a Pix2Pix framework. The model consists of two neural networks: a generator network to map rainy images to de-rained images, and a discriminator network to classify real and generated de-rained images. Different performance matrices were used to evaluate the performance of the new model using both synthesized and real-world image data. The evaluations proved that the proposed CGANet model outperformed the state-of-the-art methods for image de-raining. The new CGANet model is presented as a high-potential approach for successful de-raining of images.

This paper is focused on image de-raining; however, the proposed model applies equally well to any other image translation problem in a different domain. In future developments, further analysis can be carried out to optimize the loss function by incorporating more comprehensive components with local and global perceptual information.

**Author Contributions:** Conceptualization, P.H., D.A., and R.N.; methodology, P.H. and R.N.; investigation, P.H.; data curation, P.H.; writing—original draft preparation, P.H.; writing—review and editing, R.N., D.A., D.D.S., and N.C.; supervision, D.A., D.D.S., and N.C. All authors have read and agreed to the published version of the manuscript.

**Funding:** This research received no external funding.

**Institutional Review Board Statement:** Not applicable.

**Informed Consent Statement:** Not applicable.

**Data Availability Statement:** Data and source code for the experiments are publicly available at https://github.com/prasadmaduranga/CGANet (accessed on 11 December 2020).

**Acknowledgments:** This work was supported by a La Trobe University Postgraduate Research Scholarship.

**Conflicts of Interest:** The authors declare no conflict of interest.

## References

1. Barnum, P.C.; Narasimhan, S.; Kanade, T. Analysis of Rain and Snow in Frequency Space. *Int. J. Comput. Vis.* **2009**, *86*, 256–274. [CrossRef]
2. Brewer, N.; Liu, N. Using the Shape Characteristics of Rain to Identify and Remove Rain from Video. In *Constructive Side-Channel Analysis and Secure Design*; Springer: Berlin/Heidelberg, Germany, 2008; pp. 451–458.
3. Chen, Y.-L.; Hsu, C.-T. A Generalized Low-Rank Appearance Model for Spatio-temporally Correlated Rain Streaks. In Proceedings of the 2013 IEEE International Conference on Computer Vision, Sydney, Australia, 1–8 December 2013; pp. 1968–1975.
4. Kang, L.-W.; Lin, C.-W.; Fu, Y.-H. Automatic Single-Image-Based Rain Streaks Removal via Image Decomposition. *IEEE Trans. Image Process.* **2012**, *21*, 1742–1755. [CrossRef]

5. Huang, D.-A.; Kang, L.-W.; Wang, Y.-C.F.; Lin, C.-W. Self-Learning Based Image Decomposition with Applications to Single Image Denoising. *IEEE Trans. Multimedia* **2013**, *16*, 83–93. [CrossRef]
6. Sun, S.-H.; Fan, S.-P.; Wang, Y.-C.F. Exploiting image structural similarity for single image rain removal. In Proceedings of the 2014 IEEE International Conference on Image Processing (ICIP), Paris, France, 28 October 2014; pp. 4482–4486.
7. Yang, W.; Tan, R.T.; Feng, J.; Liu, J.; Guo, Z.; Yan, S. Deep Joint Rain Detection and Removal from a Single Image. In Proceedings of the 2017 IEEE Conference on Computer Vision and Pattern Recognition (CVPR), Honolulu, HI, USA, 21–26 July 2017; pp. 1685–1694.
8. Fu, X.; Huang, J.; Zeng, D.; Huang, Y.; Ding, X.; Paisley, J. Removing Rain from Single Images via a Deep Detail Network. In Proceedings of the 2017 IEEE Conference on Computer Vision and Pattern Recognition (CVPR), Honolulu, HI, USA, 21–26 July 2017; pp. 1715–1723.
9. Zhang, X.; Li, H.; Qi, Y.; Leow, W.K.; Ng, T.K. Rain Removal in Video by Combining Temporal and Chromatic Properties. In Proceedings of the 2006 IEEE International Conference on Multimedia and Expo, Toronto, Canada, 9–12 July 2006; pp. 461–464.
10. Liu, P.; Xu, J.; Liu, J.; Tang, X. Pixel Based Temporal Analysis Using Chromatic Property for Removing Rain from Videos. *Comput. Inf. Sci.* **2009**, *2*, 53. [CrossRef]
11. Eigen, D.; Krishnan, D.; Fergus, R. Restoring an Image Taken through a Window Covered with Dirt or Rain. In Proceedings of the 2013 IEEE International Conference on Computer Vision, Sydney, Australia, 1–8 December 2013; pp. 633–640.
12. Chen, J.; Tan, C.-H.; Hou, J.; Chau, L.-P.; Li, H. Robust Video Content Alignment and Compensation for Rain Removal in a CNN Framework. In Proceedings of the 2018 IEEE/CVF Conference on Computer Vision and Pattern Recognition, Salt Lake City, UT, USA, 18–23 June 2018; pp. 6286–6295.
13. Fu, X.; Huang, J.; Ding, X.; Liao, Y.; Paisley, J. Clearing the Skies: A Deep Network Architecture for Single-Image Rain Removal. *IEEE Trans. Image Process.* **2017**, *26*, 2944–2956. [CrossRef]
14. Goodfellow, I.; Pouget-Abadie, J.; Mirza, M.; Xu, B.; Warde-Farley, D.; Ozair, S.; Courville, A.; Bengio, Y. Generative adversarial nets. In Proceedings of the Advances in Neural Information Processing Systems, Montreal, QC, Canada, 8–13 December 2014; pp. 2672–2680.
15. Isola, P.; Zhu, J.-Y.; Zhou, T.; Efros, A.A. Image-to-Image Translation with Conditional Adversarial Networks. In Proceedings of the IEEE Conference on Computer Vision and Pattern Recognition, Honolulu, HI, USA, 21–26 July 2017; pp. 5967–5976.
16. Mirza, M.; Osindero, S. Conditional generative adversarial nets. *arXiv* **2014**, arXiv:1411.1784.
17. Fu, X.; Liang, B.; Huang, Y.; Ding, X.; Paisley, J. Lightweight pyramid networks for image deraining. *arXiv* **2019**, arXiv:1805.06173. [CrossRef] [PubMed]
18. Li, Y.; Tan, R.T.; Guo, X.; Lu, J.; Brown, M.S. Rain Streak Removal Using Layer Priors. In Proceedings of the 2016 IEEE Conference on Computer Vision and Pattern Recognition (CVPR), Las Vegas, NV, USA, 27–30 June 2016; pp. 2736–2744.
19. Chang, Y.; Yan, L.; Zhong, S. Transformed Low-Rank Model for Line Pattern Noise Removal. In Proceedings of the 2017 IEEE International Conference on Computer Vision (ICCV), Venice, Italy, 22–29 October 2017; pp. 1735–1743.
20. Luo, Y.; Xu, Y.; Ji, H. Removing Rain from a Single Image via Discriminative Sparse Coding. In Proceedings of the 2015 IEEE International Conference on Computer Vision (ICCV), Santiago, Chile, 7–13 December 2015; pp. 3397–3405.
21. Kim, J.H.; Lee, C.; Sim, J.Y.; Kim, C.S. Single-image de-raining using an adaptive nonlocal means filter. In Proceedings of the 2013 IEEE International Conference on Image Processing, Melbourne, Australia, 15–18 September 2013; pp. 914–917.
22. Zhu, L.; Fu, C.-W.; Lischinski, D.; Heng, P.-A. Joint Bi-layer Optimization for Single-Image Rain Streak Removal. In Proceedings of the 2017 IEEE International Conference on Computer Vision (ICCV), Venice, Italy, 22–29 October 2017; pp. 2545–2553.
23. Bossu, J.; Hautière, N.; Tarel, J.-P. Rain or Snow Detection in Image Sequences through Use of a Histogram of Orientation of Streaks. *Int. J. Comput. Vis.* **2011**, *93*, 348–367. [CrossRef]
24. Kim, J.H.; Sim, J.Y.; Kim, C.S. Video de-raining and desnowing using temporal correlation and low-rank matrix completion. *IEEE Trans. Image Process.* **2015**, *24*, 2658–2670. [CrossRef]
25. Ren, W.; Tian, J.; Han, Z.; Chan, A.; Tang, Y. Video desnowing and de-raining based on matrix decomposition. In Proceedings of the IEEE Conference on Computer Vision and Pattern Recognition, Honolulu, HI, USA, 21–26 July 2017; pp. 4210–4219.
26. Krizhevsky, A.; Sutskever, I.; Hinton, G.E. Imagenet classification with deep convolutional neural networks. In Proceedings of the Advances in Neural Information Processing Systems, Lake Tahoe, NV, USA, 3–6 December 2012; pp. 1097–1105.
27. Gong, M.; Zhao, J.; Liu, J.; Miao, Q.; Jiao, L. Change Detection in Synthetic Aperture Radar Images Based on Deep Neural Networks. *IEEE Trans. Neural Netw. Learn. Syst.* **2016**, *27*, 125–138. [CrossRef]
28. He, K.; Zhang, X.; Ren, S.; Sun, J. Deep residual learning for image recognition. *arXiv* **2015**, arXiv:1512.03385.
29. Hou, W.; Gao, X.; Tao, D.; Li, X. Blind Image Quality Assessment via Deep Learning. *IEEE Trans. Neural Netw. Learn. Syst.* **2014**, *26*, 1275–1286. [CrossRef]
30. Nawaratne, R.; Kahawala, S.; Nguyen, S.; De Silva, D. A Generative Latent Space Approach for Real-time Road Surveillance in Smart Cities. *IEEE Trans. Ind. Inform.* **2020**, *1*, 1. [CrossRef]
31. Nawaratne, R.; Alahakoon, D.; De Silva, D.; Yu, X. Spatiotemporal Anomaly Detection Using Deep Learning for Real-Time Video Surveillance. *IEEE Trans. Ind. Inform.* **2020**, *16*, 393–402. [CrossRef]
32. Liu, J.; Yang, W.; Yang, S.; Guo, Z. Erase or Fill? Deep Joint Recurrent Rain Removal and Reconstruction in Videos. In Proceedings of the 2018 IEEE/CVF Conference on Computer Vision and Pattern Recognition, Salt Lake City, UT, USA, 18–23 June 2018; pp. 3233–3242.

33. Fan, Z.; Wu, H.; Fu, X.; Hunag, Y.; Ding, X. Residual-guide feature fusion network for single image de-raining. *arXiv* **2018**, arXiv:1804.07493.
34. Xu, W.; Souly, N.; Brahma, P.P. Reliability of GAN Generated Data to Train and Validate Perception Systems for Autonomous Vehicles. In Proceedings of the IEEE/CVF Winter Conference on Applications of Computer Vision, Online Conference. 9 January 2021; pp. 171–180.
35. Saha, S.; Sheikh, N. Ultrasound Image Classification using ACGAN with Small Training Dataset. *arXiv* **2021**, arXiv:2102.01539.
36. Roy, S.; Sangineto, E.; Sebe, N.; Demir, B. Semantic-fusion gans for semi-supervised satellite image clas-sification. In Proceedings of the 2018 25th IEEE International Conference on Image Processing (ICIP), Athens, Greece, 7–10 October 2018; pp. 684–688.
37. Cohen, J.; Rosenfeld, E.; Kolter, Z. Certified adversarial robustness via randomized smoothing. In Proceedings of the International Conference on Machine Learning, Long Beach, CA, USA, 10–15 June 2019; pp. 1310–1320.
38. Ledig, C.; Theis, L.; Huszar, F.; Caballero, J.; Cunningham, A.; Acosta, A.; Aitken, A.; Tejani, A.; Totz, J.; Wang, Z.; et al. Photo-Realistic Single Image Super-Resolution Using a Generative Adversarial Network. In Proceedings of the 2017 IEEE Conference on Computer Vision and Pattern Recognition (CVPR), Honolulu, HI, USA, 21–26 July 2017; pp. 5892–5900.
39. Roy, S.; Siarohin, A.; Sangineto, E.; Sebe, N.; Ricci, E. TriGAN: Image-to-image translation for multi-source domain adaptation. *Mach. Vis. Appl.* **2021**, *32*, 1–12. [CrossRef]
40. Yu, L.; Zhang, W.; Wang, J.; Yu, Y. SeqGAN: Sequence generative adversarial nets with policy gradient. In Proceedings of the AAAI Conference on Artificial Intelligence, San Francisco, CA, USA, 4–10 February 2017; pp. 2852–2858.
41. Hu, W.; Tan, Y. Generating Adversarial Malware Examples for Black-Box Attacks Based on GAN. Available online: https://arxiv.org/abs/1702.05983 (accessed on 11 December 2020).
42. Hwang, J.-J.; Azernikov, S.; Efros, A.A.; Yu, S.X. Learning beyond human expertise with generative models for dental restorations. *arXiv* **2018**, arXiv:1804.00064.
43. Hinton, E.G.; Salakhutdinov, R.R. Reducing the dimensionality of data with neural networks. *Science* **2006**, *313*, 504–507. [CrossRef]
44. Ronneberger, O.; Fischer, P.; Brox, T. U-Net: Convolutional Networks for Biomedical Image Segmentation. In Proceedings of the International Conference on Medical Image Computing and Computer-Assisted Intervention, Munich, Germany, 5–9 October 2015; pp. 234–241.
45. Li, C.; Wand, M. Precomputed Real-Time Texture Synthesis with Markovian Generative Adversarial Networks. In Proceedings of the 14th European Conference on Computer Vision (ECCV) Amsterdam, The Netherlands, 11–14 October 2016; pp. 702–716.
46. Abadi, M.; Barham, P.; Chen, J.; Chen, Z.; Davis, A.; Dean, J.; Zheng, X. Tensorflow: A system for large-scale machine learning. In Proceedings of the 12th {USENIX} Symposium on Operating Systems Design and Implementation, Savannah, GA, USA, 2–4 November 2016; pp. 265–283.
47. Arya, D.; Maeda, H.; Ghosh, S.K.; Toshniwal, D.; Mraz, A.; Kashiyama, T.; Sekimoto, Y. Transfer learning-based road damage detection for multiple countries. *arXiv* **2020**, arXiv:2008.13101.
48. Wang, Z.; Bovik, A.C.; Sheikh, H.R.; Simoncelli, E.P. Image quality assessment: From error visibility to structural similarity. *IEEE Trans. Image Process.* **2004**, *13*, 600–612. [CrossRef] [PubMed]
49. Da, K. A method for stochastic optimization. *arXiv* **2014**, arXiv:1412.6980.

Article

# Discovering Sentimental Interaction via Graph Convolutional Network for Visual Sentiment Prediction

**Lifang Wu, Heng Zhang, Sinuo Deng, Ge Shi * and Xu Liu**

Faculty of Information Technology, Beijing University of Technology, Beijing 100124, China; lfwu@bjut.edu.cn (L.W.); zhangheng2018@emails.bjut.edu.cn (H.Z.); dsn0w@emails.bjut.edu.cn (S.D.); liuxu91@bjut.edu.cn (X.L.)
* Correspondence: tinkersxy@gmail.com

**Abstract:** With the popularity of online opinion expressing, automatic sentiment analysis of images has gained considerable attention. Most methods focus on effectively extracting the sentimental features of images, such as enhancing local features through saliency detection or instance segmentation tools. However, as a high-level abstraction, the sentiment is difficult to accurately capture with the visual element because of the "affective gap". Previous works have overlooked the contribution of the interaction among objects to the image sentiment. We aim to utilize interactive characteristics of objects in the sentimental space, inspired by human sentimental principles that each object contributes to the sentiment. To achieve this goal, we propose a framework to leverage the sentimental interaction characteristic based on a Graph Convolutional Network (GCN). We first utilize an off-the-shelf tool to recognize objects and build a graph over them. Visual features represent nodes, and the emotional distances between objects act as edges. Then, we employ GCNs to obtain the interaction features among objects, which are fused with the CNN output of the whole image to predict the final results. Experimental results show that our method exceeds the state-of-the-art algorithm. Demonstrating that the rational use of interaction features can improve performance for sentiment analysis.

**Keywords:** visual sentiment analysis; sentiment classification; convolutional neural networks; graph convolutional networks

Citation: Wu, L.; Zhang, H.; Deng, S.; Shi, G.; Liu, X. Discovering Sentimental Interaction via Graph Convolutional Network for Visual Sentiment Prediction. *Appl. Sci.* **2021**, *11*, 1404. https://doi.org/10.3390/app11041404

Academic Editor: Byung-Gyu Kim

Received: 30 December 2020
Accepted: 29 January 2021
Published: 4 February 2021

**Publisher's Note:** MDPI stays neutral with regard to jurisdictional claims in published maps and institutional affiliations.

**Copyright:** © 2021 by the authors. Licensee MDPI, Basel, Switzerland. This article is an open access article distributed under the terms and conditions of the Creative Commons Attribution (CC BY) license (https://creativecommons.org/licenses/by/4.0/).

## 1. Introduction

With the vast popularity of social networks, people tend to express their emotions and share their experiences online through posting images [1], which promotes the study of the principles of human emotion and the analysis and estimation of human behavior. Recently, with the wide application of convolution neural networks (CNNs) in emotion prediction, numerous studies [2–4] have proved the excellent ability of CNN to recognize the emotional features of images.

Based on the theory that the emotional cognition of a stimulus attracts more human attention [5], some researchers enriched emotional prediction with saliency detection or instance segmentation to extract more concrete emotional features [6–8]. Yang et al. [9] put forward the "Affective Regions" which are objects that convey significant sentiments, and proposed three fusion strategies for image features from the original image and "Affective Regions". Alternatively, Wu et al. [8] utilized saliency detection to enhance the local features, improving the classification performance to a large margin.

"Affective Regions" or Local features in images play a crucial role in image emotion, and the above methods can effectively improve classification accuracy. However, although these methods have achieved great success, there are still some drawbacks. They focused on improving visual representations and ignored emotional effectiveness of objects, which leads to a non-tendential feature enhancement. For example, in an image expressing a positive sentiment, positivity is generated by interaction among objects. Separating objects and directly merging the features will lose much of the critical information of image.

Besides, they also introduce a certain degree of noise, which leads to the limited performance improvement obtained through visual feature enhancement. For example, in human common sense, "cat" tends to be a positive categorical word. As shown in Figure 1a,b, when "cat" forms the image with other neutral or positive objects, the image tends to be more positive, consistent with the conclusion that local features can improve accuracy. In the real world, however, there are complex images, as shown in Figure 1c,d, "cat" can be combined with other objects to express opposite emotional polarity, reflecting the effect of objects on emotional interactions. Specifically, in Figure 1d, the negative sentiment is not directly generated by the "cat" and "injector", but the result of the interaction between the two in the emotional space. Indiscriminate feature fusion of such images will affect the performance of the classifier.

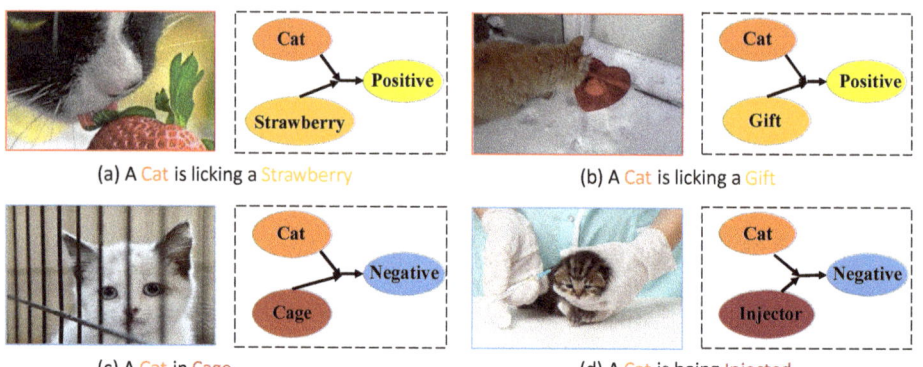

(a) A Cat is licking a Strawberry  (b) A Cat is licking a Gift

(c) A Cat in Cage  (d) A Cat is being Injected

**Figure 1.** Examples from EmotionROI dataset and social media: We use a graph model to describe the sentimental interaction between objects and the double arrow means the interaction in the process of human emotion reflection.

To address the abovementioned problems, we design a framework with two branches, one of which uses a deep network to extract visual emotional features in images. The other branch uses GCN to extract emotional interaction features of objects. Specially, we utilize Detectron2 to obtain the object category, location, and additional information in images. And then, SentiWordNet [10] is selected as an emotional dictionary to mark each category word with emotional intensity value. Based on the above information, we use the sentimental value of objects and visual characteristics in each image to build the corresponding graph model. Finally, we employ GCN to update and transmit node features, generate features after object interaction, which, together with visual components, serve as the basis for sentiment classification.

The contributions of this paper can be summarized as follows:

1. We propose an end-to-end image sentiment analysis framework that employs GCN to extract sentimental interaction characteristics among objects. The proposed model makes extensive use of the interaction between objects in the emotional space rather than directly integrating the visual features.

2. We design a method to construct graphs over images by utilizing Detectron2 and SentiWordNet. Based on the public datasets analysis, we leverage brightness and texture as the features of nodes and the distances in emotional space as edges, which can effectively describe the appearance characteristics of objects.

3. We evaluate our method on five affective datasets, and our method outperforms previous high-performing approaches.

We make all programs of our model publicly available for research purposes https://github.com/Vander111/Sentimental-Interaction-Network.

## 2. Related Work

### 2.1. Visual Sentiment Prediction

Existing methods can be classified into two groups: dimensional spaces and categorical states. Dimensional spaces methods employ valence-arousal space [11] or activity-weight-heat space [12] to represent emotions. On the contrary, categorical states methods classify emotions into corresponding categories [13,14], which is easier for people to understand, and our work falls into categorical states group.

Feature extraction is of vital importance to emotion analysis, various kinds of features may contribute to the emotion of images [15]. Some researchers have been devoting themselves to exploring emotional features and bridging the "affective gap", which can be defined as the lack of coincidence between image features and user emotional response to the image [16]. Inspired by art and psychology, Machajdik and Hanbury [14] designed low-level features such as color, texture, and composition. Zhao et al. [17] proposed extensive use of visual image information, social context related to the corresponding users, the temporal evolution of emotion, and the location information of images to predict personalized emotions of a specified social media user.

With the availability of large-scale image datasets such as ImageNet and the wide application of deep learning, the ability of convolutional neural networks to learn discriminative features has been recognized. You et al. [3] fine-tuned the pre-trained AlexNet on ImageNet to classify emotions into eight categories. Yang et al. [18] integrated deep metric learning with sentiment classification and proposed a multi-task framework for affective image classification and retrieval.

Sun et al. [19] discovered affective regions based on an object proposal algorithm and extracted corresponding in-depth features for classification. Later, You et al. [20] adopted an attention algorithm to utilize localized visual features and got better emotional classification performance than using global visual features. To mine emotional features in images more accurately, Zheng et al. [6] combined the saliency detection method with image sentiment analysis. They concluded that images containing prominent artificial objects or faces, or indoor and low depth of field images, often express emotions through their saliency regions. To enhance the work theme, photographers blurred the background to emphasize the main body of the picture [14], which led to the birth of close-up or low-depth photographs. Therefore, the focus area in low-depth images fully expresses the information that the photographer and forwarder want to tell, especially emotional information.

On the other hand, when natural objects are more prominent than artificial objects or do not contain faces, or open-field images, emotional information is usually not transmitted only through their saliency areas. Based on these studies, Fan et al. [7] established an image dataset labeled with statistical data of eye-trackers on human attention to exploring the relationship between human attention mechanisms and emotional characteristics. Yang et al. [9] synthetically considered image objects and emotional factors and obtained better sentiment analysis results by combining the two pieces of information.

Such methods make efforts in extracting emotional features accurately to improve classification accuracy. However, as an integral part of an image, objects may carry emotional information. Ignoring the interaction between objects is unreliable and insufficient. This paper selects the graph model and graph convolution network to generate sentimental interaction information and realize the sentiment analysis task.

### 2.2. Graph Convolutional Network(GCN)

The notion of graph neural networks was first outlined in Gori et al. [21] and further expound in Scarselli et al. [22]. However, these initial methods required costly neural "message-passing" algorithms to convergence, which was prohibitively expensive on massive data. More recently, there have been many methods based on the notion of GCN, which originated from the graph convolutions based on the spectral graph theory of Bruna et al. [23]. Based on this work, a significant number of jobs were published and attracted the attention of researchers.

Compared with the deep learning model introduced above, the graph model virtually constructs relational models. Chen et al. [24] combined GCN with multi-label image recognition to learn inter-dependent object information from labels. A novel re-weighted strategy was designed to construct the correlation matrix for GCN, and they got a higher accuracy compared with many previous works. However, this method is based on the labeled objects information from the dataset, which needs many human resources.

In this paper, we employ the graph structure to capture and explore the object sentimental correlation dependency. Specifically, based on the graph, we utilize GCN to propagate sentimental information between objects and generate corresponding interaction features, which is further applied to the global image representation for the final image sentiment prediction. Simultaneously, we also designed a method to build graph models from images based on existing image emotion datasets and describe the relationship features of objects in the emotional space, which can save a lot of workforce annotation.

## 3. Method

### 3.1. Framework

This section aims to develop an algorithm to extract interaction feature without manual annotation and combine it with holistic representation for image sentiment analysis. As shown in Figure 2, given an image with sentiment label, we employ a panoptic segmentation model, i.e., Detectron2, to obtain category information of objects and based on which we build a graph to represent the relationships among objects. Then, we utilize the GCN to leverage the interaction feature of objects in the emotional space. Finally, the interactive features of objects are concatenated with the holistic representation (CNN branch) to generate the final predictions. In the application scenario, given an image, we first use the panoramic segmentation model for data preprocessing to obtain the object categories and location information and establish the graph model. The graph model and the image are input into the corresponding branch to get the final sentiment prediction result.

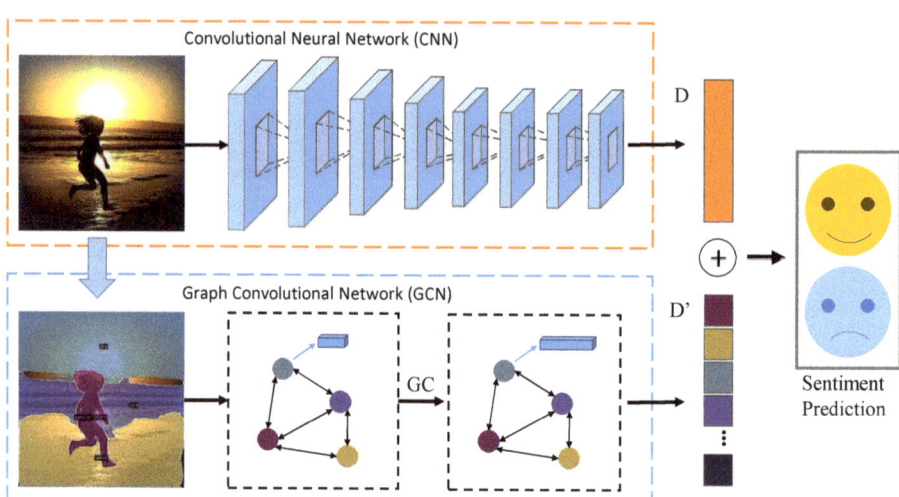

**Figure 2.** Pipline of proposed approach framework.

### 3.2. Graph Construction

#### 3.2.1. Objects Recognition

Sentiment is a complex logical response, to which the relations among objects in the image have a vital contribution. To deeply comprehend the interaction, we build a graph structure (relations among objects) to realize interaction features. And we take the categories of objects as the node and the hand-crafted feature as the representation of the object. However, existing image sentiment datasets, such as Flickr and Instagram (FI) [3], EmotionROI [25], etc., do not contain the object annotations. Inspired by the previous work [9], we employ the panoptic segmentation algorithm to detect objects.

We choose the R101-FPN model of Detectron2, containing 131 common object categories, such as "person", "cat","bird", "tree" etc., to realize recognition automatically. As shown in Figure 3, through the panoptic segmentation model, we process the original image Figure 3a to obtain the image Figure 3b containing the object category and location information.

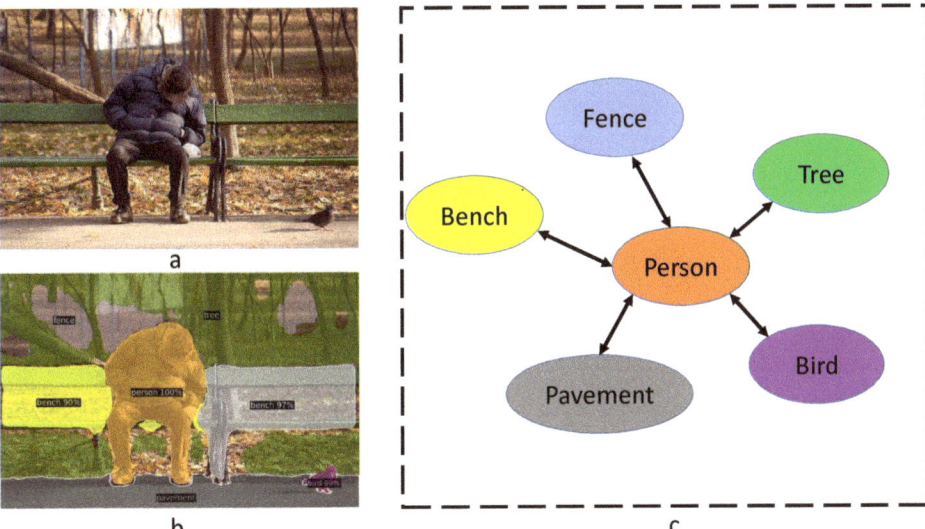

**Figure 3.** Example of building graph model. Given the input image (**a**), Detectron2 can detect the region and categories of objects and (**b**) is the segmentation result. Based on the detection information, we build a graph (**c**) over the corresponding image.

#### 3.2.2. Graph Representation

As a critical part of the graph structure, edges determine the weights of node information propagation and aggregation. In other fields, some researchers regard semantic relationship or co-occurrence frequency of objects as edges [1,26]. However, as a basic feature, there is still a gap between object semantics and sentiment, making it hard to accurately describe the sentimental relationship. Further, it is challenging to label abstract sentiments non-artificially due to the "affective gap" between low-level visual features and high-level sentiment. To solve this problem, we use the semantic relationship of objects in emotional space as the edges of the graph structure. Given the object category, we employ SentiWordNet as a sentiment annotation to label each category with sentimental information. SentiWordNet is a lexical resource for opinion mining that annotates the positive and negative values in the range [0,1] to words.

As shown in Equations (1) and (2), we retrieve words related to the object category in SentiWordNet, and judge the sentimental strength of the current word $W$ with the average value of related words $W'$, where $W_p$ is the positive emotional strength, $W_n$ is the negative emotion strength.

$$W_n = \frac{\sum_{i=1}^{n} W'_{in}}{n} \quad (1)$$

$$W_p = \frac{\sum_{i=1}^{n} W'_{ip}}{n} \quad (2)$$

In particular, we stipulate that sentimental polarity of a word is determined by positive and negative strength. As shown in Equation (3), sentiment value S is the difference between the two sentimental intensity of words. In this way, positive words have a positive sentiment value, and negative words are the opposite. And S is in [−1, 1] because of the intensity of sentiments is between 0–1 in SentiWordNet.

$$S = W_p - W_n \tag{3}$$

Based on this, we design the method described in Equation (4). We can use a sentimental tendency of objects to measure the sentimental distance $L_{ij}$ between words $W_i$ and $W_j$. When two words have the same sentimental tendency, we define the difference between the two sentiment values $S_i$ and $S_j$ as the distance in the sentimental space. On the contrary, we specify that two words with opposite emotional tendencies are added by one to enhance the sentimental difference. Further, we build the graph over the sentimental values and the object information. In Figure 3c, we show the relationship among node "person" and adjacent nodes, and the length of the edge reflects the distance between nodes.

$$L_{ij} = \begin{cases} ||S_i| - |S_j|| + 1, & if \ S_i * S_j > 0 \\ 0.5, & if \ S_i = 0, S_j = 0 \\ ||S_i| - |S_j||, & otherwise \end{cases} \tag{4}$$

#### 3.2.3. Feature Representation

The graph structure describes the relationship between objects. And the nodes of the graph aim to describe the features of each object, where we select hand-crafted feature, intensity distribution, and texture feature as the representation of objects. Inspired by Machajdik [14], we calculate and analyze the image intensity characteristics on image datasets EmotionROI and FI. In detail, we quantify the intensity of each pixel to 0–10 and make histograms of intensity distribution. As shown in Figure 4, we find that the intensity of positive emotions (joy, surprise, etc.) is higher than that of negative emotions (anger, sadness, etc.) when the brightness is 4–6, while the intensity of negative emotions is higher on 1–2.

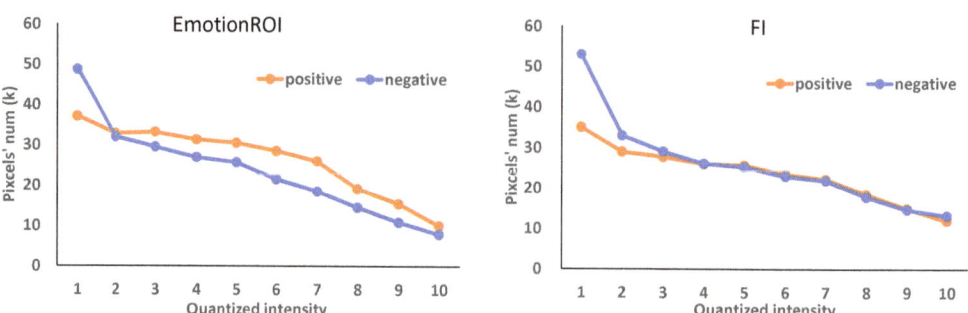

**Figure 4.** The distribution curve of the number of brightness pixels of different emotion categories in the EmotionROI and Flickr and Instagram (FI) dataset.

The result shows that the intensity distribution can distinguish the sentimental polarity of the images to some extent. At the same time, we use the Gray Level Co-occurrence Matrix(GLCM) to describe the texture feature of each object in the image as a supplement to the image detail feature. Specifically, we quantified the luminance values as 0–255 and calculated a 256-dimensional eigenvector with 45 degrees as the parameter of GLCM. The node feature in the final graph model is a 512-dimensional eigenvector.

## 3.3. Interaction Graph Inference

Sentiment contains implicit relationships among the objects. Graph structure expresses low-level visual features and the relationship among objects, which is the source of interaction features, and inference is the process of generating interaction features. To simulate the interaction process, we employ GCN to propagate and aggregate the low-level features of objects under the supervision of sentimental distances. We select the stacked GCNs, in which the input of each layer is the output $H^l$ from the previous layer, and generate the new node feature $H^{l+1}$.

The feature update process of the layer l is shown in Equation (5), $\tilde{A}$ is obtained by adding the edges of the graph model, namely the adjacency matrix and the identity matrix. $H^l$ is the output feature of the previous layer, $H^{l+1}$ is the output feature of the current layer, $W^l$ is the weight matrix of the current layer, and $\sigma$ is the nonlinear activation function. $\tilde{D}$ is the degree matrix of $\tilde{A}$, which is obtained by Equation (6). The first layer's input is the initial node feature $H^0$ of 512 dimensions generated from the brightness histogram and GLCM introduced above. Also, the final output of the model is the feature vector of 2048 dimensions.

$$H^{l+1} = \sigma(\tilde{D}^{-\frac{1}{2}} \tilde{A} \tilde{D}^{-\frac{1}{2}} H^l W^l) \tag{5}$$

$$\tilde{D}_{ii} = \sum_j \tilde{A}_{ij} \tag{6}$$

## 3.4. Visual Feature Representation

As a branch of machine learning, deep learning has been widely used in many fields, including sentiment image classification. Previous studies have proved that CNN network can effectively extract visual features in images, such as appearance and position, and map them to emotional space. In this work, we utilize CNN to realize the expression of visual image features. To make a fair comparison with previous works, we select the popularly used model VGGNet [27] as the backbone to verify the effectiveness of our method. For VGGNet, we adopt a fine-tuning strategy based on a pre-trained model on ImageNet and change the output number of the last fully connected layer from 4096 to 2048.

## 3.5. Gcn Based Classifier Learning

In the training process, we adopt the widely used concatenation method for feature fusion. In the visual feature branch, we change the last fully connected layer output of the VGG model to 2048 to describe the visual features extracted by the deep learning model. For the other branch, we process the graph model features in an average operation. In detail, the Equation (7) is used to calculate interaction feature $F_g$, where n is the number of nodes in a graph model, $F'$ is the feature of each node after graph convolution.

$$F_g = \frac{\sum_{i=1}^{n} F'}{n} \tag{7}$$

After the above processing, we employ the fusion method described in Equation (8) to calculate the fusion feature of visual and relationship, which is fed into the fully connected layer and realize the mapping between features and sentimental polarity. And the traditional cross entropy function is taken as the loss function, as shown in Equation (9), N is the number of training images, $y_i$ is the labels of images, and $P_i$ is the probability of prediction that 1 represents a positive sentiment and 0 means negative.

$$F = [F_d; F_g] \tag{8}$$

$$L = -\frac{1}{N} \sum_{i=1}^{N} (y_i * \log P_i + (1 - y_i) * \log(1 - P_i)) \tag{9}$$

Specifically, $P_i$ is defined as Equation (10), where $c$ is the number of classes. In this work, $c$ is defined as 2, and $f_j$ is the output of the last fully connected layer.

$$P_i = \frac{e^{f_i}}{\sum_{j=1}^{c} e^{f_j}} \tag{10}$$

## 4. Experiment Results
### 4.1. Datasets

We evaluate our framework on five public datasets: FI, Flickr [28], EmotionROI [25], Twitter I [29], Twitter II [28]. Figure 5 shows examples of these datasets. FI dataset is collected by querying with eight emotion categories (i.e., amusement, anger, awe, contentment, disgust, excitement, fear, sadness) as keywords from Flickr and Instagram, and ultimately gets 90,000 noisy images. The original dataset is further labeled by 225 Amazon Mechanical Turk (AMT) workers and resulted in 23,308 images receiving at least three agreements. The number of images in each emotion category is larger than 1000. Flickr contains 484,258 images in total, and the corresponding ANP automatically labeled each image. EmotionROI consists of 1980 images with six sentiment categories assembled from Flickr and annotated with 15 regions that evoke sentiments. Twitter I and Twitter II datasets are collected from social websites and labeled with two categories (i.e., positive and negative) by AMT workers, consisting of 1296 and 603 images. Specifically, we conducted training and testing on the three subsets of Twitter I: "Five agree", "At least four agree" and "At least three agree", which are filtered according to the annotation. For example, "Five agree" indicates that all the Five AMT workers rotate the same sentiment label for a given image. As shown in Table 1.

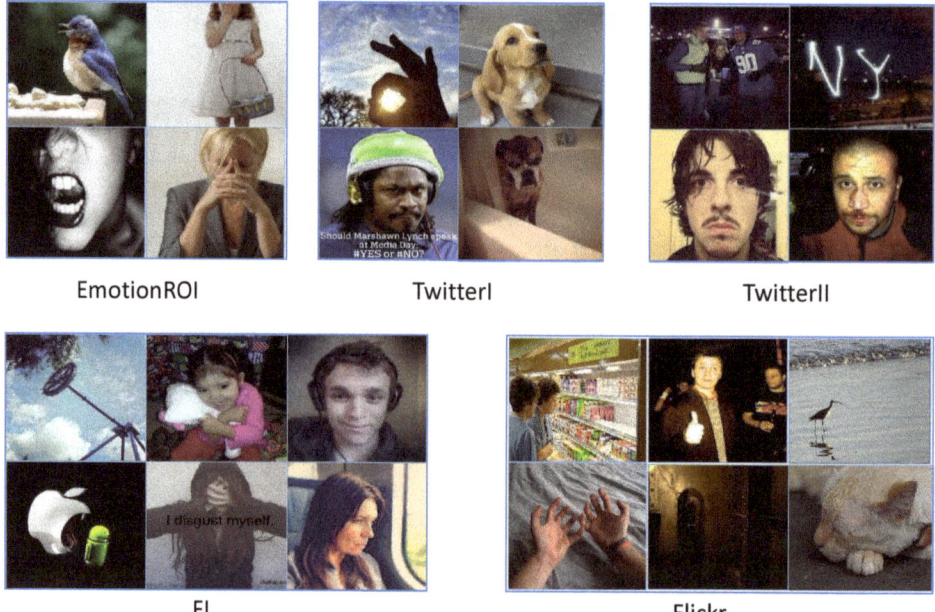

**Figure 5.** Some examples in the five datasets.

**Table 1.** Released and freely available datasets, where #Annotators respectively represent the number of annotators.

| Dataset | Images Number | Source | #Annotators | Emotion Model |
|---|---|---|---|---|
| FI | 23,308 | social media | 225 | Mikels |
| Flickr | 484,258 | social media | - | Ekman |
| TwitterI | 1269 | social media | 5 | Sentiment |
| TwitterII | 603 | social media | 9 | Sentiment |
| EmotionROI | 1980 | social media | 432 | Ekman |

According to the affective model, the multi-label datasets EmotionROI and FI are divided into two parts: positive and negative, to achieve the sentimental polarity classification. EmotionROI has six emotion categories: anger, disgust, fear, joy, sadness, and surprise. Images with labels of anger, disgust, fear, sadness are relabeled as negative, and those with joy and surprise are labeled as positive. In the FI dataset, we divided Mikel's eight emotion categories into binary labels based on [30], suggesting that amusement, contentment, excitement, and awe are mapped to the positive category, and sadness, anger, fear, and disgust are labeled as negative.

### 4.2. Implementation Details

Following previous works [9], we select VGGNet with 16 layers [25] as the backbone of the visual feature extraction and initialize it with the weights pre-trained on ImageNet. At the same time, we remove the last fully connected layer of the VGGNet. We randomly crop and resize the input images into 224 × 224 with random horizontal flip for data enhancement during the training. On FI, we select SGD as the optimizer and set Momentum to 0.9. The initial learning rate is 0.01, which drops by a factor of 10 per 20 epoch. And Table 2 shows the specific training strategy on the five datasets. In the relational feature branch, we use two GCN layers whose output dimensions are 1024 and 2048. 512-dimension vector characterizes each input node feature in the graph model. We adopted the same split and test method for the data set without specific division as Yang et al. [9]. For small-scale data sets, we refer to the strategy of Yang et al. [9], take the model parameters trained on the FI as initial weights, and fine-tune the model on the training set.

**Table 2.** Setting of training parameters on the dataset of FI, Flickr, EmotionROI, Twitter I, Twitter II.

| Dataset | Learning Rate | Drop Factor | Croped Size | Momentum | Optimizer |
|---|---|---|---|---|---|
| FI | 0.01 | 20 | 224 × 224 | 0.9 | SGD |
| Flickr | 0.01 | 5 | 224 × 224 | 0.9 | SGD |
| TwitterI | 0.02 | 30 | 224 × 224 | 0.9 | SGD |
| TwitterII | 0.03 | 20 | 224 × 224 | 0.9 | SGD |
| EmotionROI | 0.03 | 30 | 224 × 224 | 0.9 | SGD |

### 4.3. Evaluation Settings

To demonstrate the validity of our proposed framework for sentiment analysis, we evaluate the framework against several baseline methods, including methods using traditional features, CNN-based methods, and CNN-based methods combined with instance segmentation.

- The global color histograms (GCH) consists of 64-bin RGB histogram, and the local color histogram features (LCH) divide the image into 16 blocks and generate a 64-bin RGB histogram for each block [31].
- Borth et al. [28] propose SentiBank to describe the sentiment concept by 1200 adjectives noun pairs (ANPs), witch performs better for images with rich semantics.

- DeepSentibank [32] utilizes CNNs to discover ANPs and realizes visual sentiment concept classification. We apply the pre-trained DeepSentiBank to extract the 2089-dimension features from the last fully connected layer and employ LIBSVM for classification.
- You et al. [29] propose to select a potentially cleaner training dataset and design the PCNN, which is a progressive model based on CNNs.
- Yang et al. [9] employ object detection technique to produce the "Affective Regions" and propose three fusion strategy to generate the final predictions.
- Wu et al. [8] utilize saliency detection to enhance the local features, improving the classification performance to a large margin. And they adopt an ensemble strategy, which may contribute to performance improvement.

*4.4. Classification Performance*

We evaluate the classification performance on five affective datasets. Table 3 shows that the result of depth feature is higher than that of the hand-crafted feature and CNNs outperform the traditional methods. The VGGNet achieves significant performance improvements over the traditional methods such as DeepSentibank and PCNN on FI datasets of good quality and size. Simultaneously, due to the weak in annotation reliability, VGGNet does not make such significant progress on the Flickr dataset, indicating the dependence of the depth model on high-quality data annotation. Furthermore, our proposed method performs well compared with single model methods. For example, we achieve about 1.7% improvement on FI and 2.4% on EmotionROI dataset, which means that the sentimental interaction features extracted by us can effectively complete the image sentiment classification task. Besides, we adopt a simple ensemble strategy and achieve a better performance than state-of-the-art method.

**Table 3.** Sentiment classification accuracy on FI, Flickr, Twitter I, Twitter II, EmotionROI. Results with bold indicate the best results compared with other algorithms.

| Method | FI | Flickr | Twitter I | | | Twitter II | EmotionROI |
| --- | --- | --- | --- | --- | --- | --- | --- |
| | | | Twitter I-5 | Twitter I-4 | Twitter I-3 | | |
| GCH | - | - | 67.91 | 97.20 | 65.41 | 77.68 | 66.53 |
| LCH | - | - | 70.18 | 68.54 | 65.93 | 75.98 | 64.29 |
| SentiBank | - | - | 71.32 | 68.28 | 66.63 | 65.93 | 66.18 |
| DeepSentiBank | 61.54 | 57.83 | 76.35 | 70.15 | 71.25 | 70.23 | 70.11 |
| VGGNet [27] | 70.64 | 61.28 | 83.44 | 78.67 | 75.49 | 71.79 | 72.25 |
| PCNN | 75.34 | 70.48 | 82.54 | 76.50 | 76.36 | 77.68 | 73.58 |
| Yang [9] | 86.35 | 71.13 | 88.65 | 85.10 | 81.06 | 80.48 | 81.26 |
| Ours-single | 88.12 | 72.31 | 89.24 | 85.19 | 81.25 | 80.59 | 83.62 |
| Wu [8] | 88.84 | 72.39 | 89.50 | **86.97** | 81.65 | 80.97 | 83.04 |
| Ours-ensemble | 88.71 | 73.11 | 89.65 | 84.48 | 81.72 | 82.68 | 84.29 |

*4.5. the Role of Gcn Branch*

As shown in Table 4, compared with the fine-tuned VGGNet, our method has an average performance improvement of 4.2%, which suggests the effectiveness of sentimental interaction characteristics in image emotion classification task.

**Table 4.** The model performance comparison across image datasets. Results with bold indicate the best results compared with other algorithms.

| Method | FI | Flickr | Twitter I | | | Twitter II | EmotionROI |
| --- | --- | --- | --- | --- | --- | --- | --- |
| | | | Twitter I-5 | Twitter I-4 | Twitter I-3 | | |
| Fine-tuned VGGNet | 83.05 | 70.12 | 84.35 | 82.26 | 76.75 | 76.99 | 77.02 |
| Ours-single | 88.12 | 72.31 | 89.24 | 85.19 | 81.25 | 80.59 | 83.62 |

## 4.6. Effect of Panoptic Segmentation

As a critical step in graph model construction, information of objects obtained through Detectron2 dramatically impacts the final performance. However, due to the lack of annotation with emotions and object categories, we adopt the panoptic segmentation model pre-trained on the COCO dataset, which contains a wide range of object categories. This situation leads to specific noise existing in the image information. As shown in Figure 6, the lefts are the original images from EmotionROI, and the detection results are on the right. In detail, there are omission cases (Figure 6d) and misclassification (Figure 6f) in detection results, which to a certain extent, affect the performance of the model, in the end, believe that if we can overcome this gap, our proposed method can obtain a better effect.

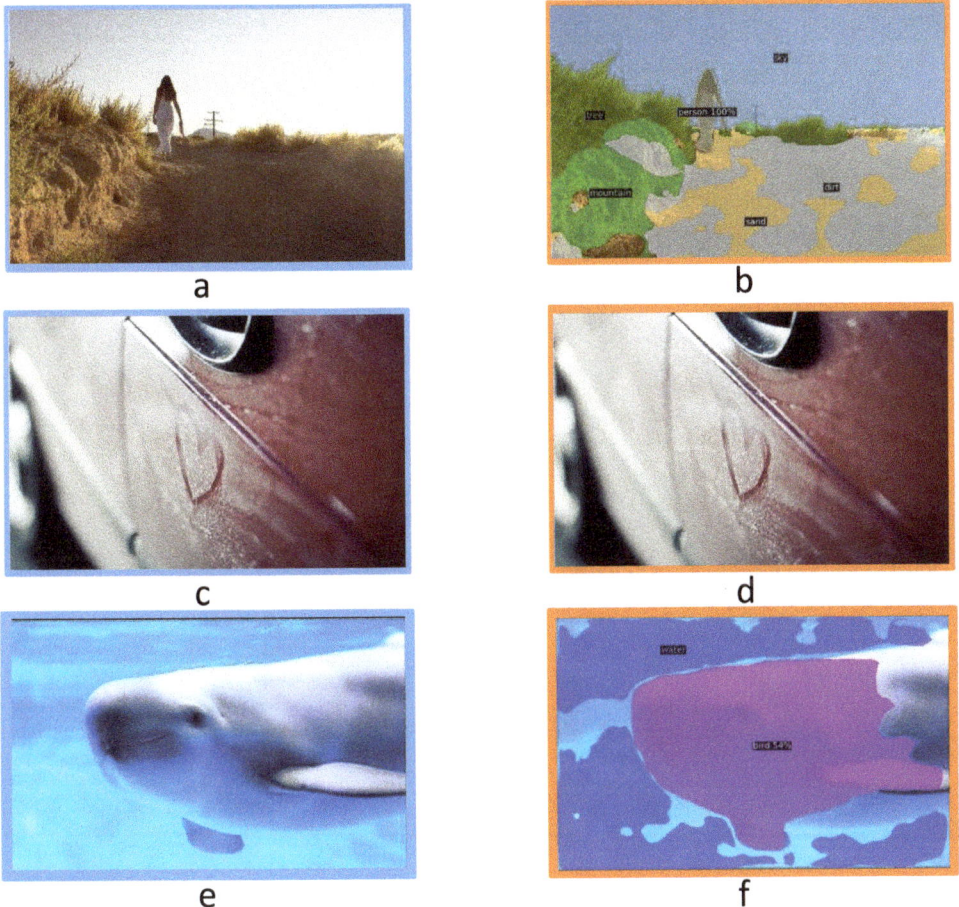

**Figure 6.** Example of panoptic segmentation. Given the raw images (**a**,**c**,**e**), panoptic segmentation generates the accurate result (**b**), category missing result (**d**) and misclassification result (**f**).

As stated above, some object information of images cannot be extracted by the panoptic segmentation model. So we further analyze the result on emotionROI, of which each image is annotated with emotion and attractive regions manually by 15 persons and forms with the Emotion Stimuli Map. By comparing them with the Emotion Stimuli Map, our method fails to detect the critical objects in 77 images of a total of 590 testing images, as shown in Figure 7 mainly caused by the inconsistent categories of the panoptic segmentation model. A part of the EmotionROI images and the corresponding stimuli map is shown in Figure 7a,b, these images in the process of classification using only a part or even no object interaction information, but our method still predicts their categories correctly, indicating that visual features still play an essential role in the classification, and the interaction feature generated by GCN branch further improve the accuracy of the model.

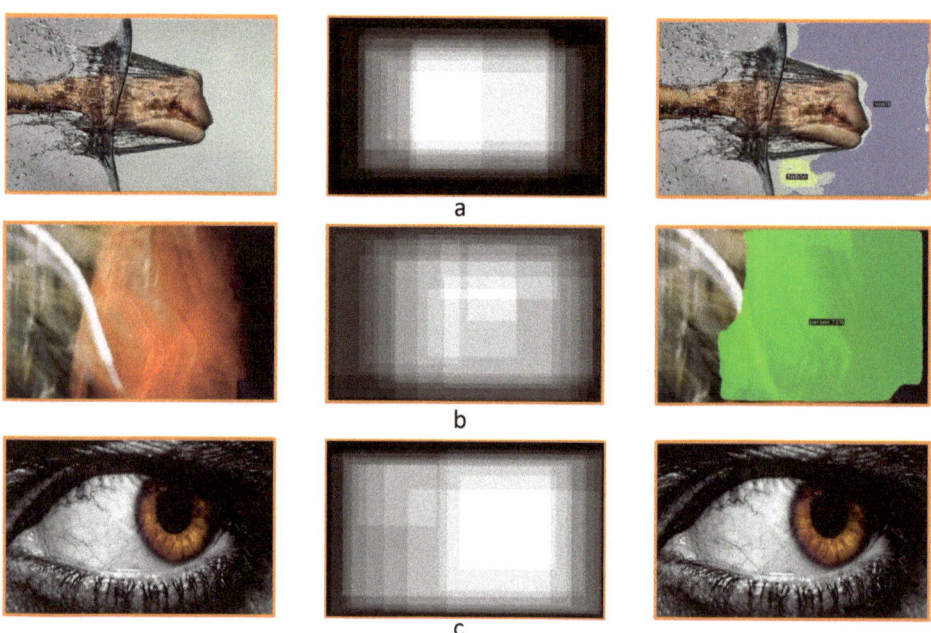

**Figure 7.** Some example images and corresponding Emotion Stimuli Maps whose object information is broken extracted by panoptic segmentation model, but correctly predicted by our method. The lefts of (**a**–**c**) are the raw images, the middles are the corresponding stimuli map and the rights are the visual results of segmentation.

## 5. Conclusions

This paper addresses the problem of visual sentiment analysis based on graph convolutional networks and convolutional neural networks. Inspired by the principles of human emotion and observation, we find that each type of interaction among objects in the image has an essential impact on sentiment. We present a framework that consists of two branches for sentimental interaction representations learning. First of all, we design an algorithm to build a graph model on popular affective datasets without category information annotated based on panoptic segmentation information. As an essential part of the graph model, we define the objects in the images as nodes and calculate the edges between nodes in the graph model according to sentimental value of each objects. According to the effect of brightness on sentiment, we select brightness and texture features as node features. A stacked GCN model is used to generate the relational features describing the interaction results of objects and integrate them with the visual features extracted by VGGNet to realize the classification of image sentiment. Experimental results show the effectiveness of our method on five popular datasets. Furthermore, making more effective utilizing of objects interaction information remains a challenging problem.

**Author Contributions:** Conceptualization, H.Z. and S.D.; methodology, H.Z.; software, H.Z. and S.D.; validation, S.D.; formal analysis, H.Z. and S.D.; investigation, H.Z. and S.D.; resources, H.Z. and X.L.; data curation, X.L. and S.D.; writing—original draft preparation, H.Z. and S.D.; writing—review and editing, L.W. and G.S.; visualization, H.Z.; supervision, L.W. All authors have read and agreed to the published version of the manuscript.

**Funding:** This work was supported by National Natural Science Foundation of China grant numbers 61802011, 61702022.

**Institutional Review Board Statement:** Not applicable.

**Informed Consent Statement:** Not applicable.

**Data Availability Statement:** Data sharing is not applicable to this paper as no new data were created or analyzed in this study.

**Conflicts of Interest:** The authors declare no conflict of interest.

# References

1. Zhao, S.; Gao, Y.; Ding, G.; Chua, T. Real-time Multimedia Social Event Detection in Microblog. *IEEE Trans. Cybern.* **2017**, *48*, 3218–3231. [CrossRef]
2. Peng, K.C.; Chen, T.; Sadovnik, A.; Gallagher, A.C. A Mixed Bag of Emotions: Model, Predict, and Transfer Emotion Distributions. In Proceedings of the IEEE Conference on Computer Vision and Pattern Recognition, Boston, MA, USA, 7–12 June 2015; pp. 860–868.
3. You, Q.; Luo, J.; Jin, H.; Yang, J. Building A Large Scale Dataset for Image Emotion Recognition: The Fine Print and the Benchmark. In Proceedings of the Thirtieth AAAI Conference on Artificial Intelligence, Phoenix, AZ, USA, 12–17 February 2016; pp. 308–314.
4. Zhu, X.; Li, L.; Zhang, W.; Rao, T.; Xu, M.; Huang, Q.; Xu, D. Dependency Exploitation: A Unified CNN-RNN Approach for Visual Emotion Recognition. In Proceedings of the 26th International Joint Conference on Artificial Intelligence, Melbourne, Australia, 19–25 August 2017; pp. 3595–3601.
5. Compton, R.J. The Interface Between Emotion and Attention: A Review of Evidence from Psychology and Neuroscience. *Behav. Cogn. Neurosci. Rev.* **2003**, *2*, 115–129. [CrossRef]
6. Zheng, H.; Chen, T.; You, Q.; Luo, J. When Saliency Meets Sentiment: Understanding How Image Content Invokes Emotion and Sentiment. In Proceedings of the 2017 IEEE International Conference on Image Processing, Beijing, China, 17–20 September 2017; pp. 630–634.
7. Fan, S.; Shen, Z.; Jiang, M.; Koenig, B.L.; Xu, J.; Kankanhalli, M.S.; Zhao, Q. Emotional Attention: A Study of Image Sentiment and Visual Attention. In Proceedings of the IEEE Conference on Computer Vision and Pattern Recognition, Salt Lake City, UT, USA, 18–22 June 2018; pp. 7521–7531.
8. Wu, L.; Qi, M.; Jian, M.; Zhang, H. Visual Sentiment Analysis by Combining Global and Local Information. *Neural Process. Lett.* **2019**, *51*, 1–13. [CrossRef]
9. Yang, J.; She, D.; Sun, M.; Cheng, M.M.; Rosin, P.L.; Wang, L. Visual Sentiment Prediction Based on Automatic Discovery of Affective Regions. *IEEE Trans. Multimed.* **2018**, *20*, 2513–2525. [CrossRef]
10. Esuli, A.; Sebastiani, F. Sentiwordnet: A Publicly Available Lexical Resource for Opinion Mining. In Proceedings of the Fifth International Conference on Language Resources and Evaluation, Genoa, Italy, 22–28 May 2006; pp. 417–422.
11. Nicolaou, M.A.; Gunes, H.; Pantic, M. A Multi-layer Hybrid Framework for Dimensional Emotion Classification. In Proceedings of the 19th International Conference on Multimedia, Scottsdale, AZ, USA, 28 November–1 December 2011; pp. 933–936.
12. Xu, M.; Jin, J.S.; Luo, S.; Duan, L. Hierarchical Movie Affective Content Analysis Based on Arousal and Valence Features. In Proceedings of the 16th International Conference on Multimedia, Vancouver, BC, Canada, 26–31 October 2008; pp. 677–680.
13. Zhao, S.; Gao, Y.; Jiang, X.; Yao, H.; Chua, T.S.; Sun, X. Exploring Principles-of-art Features for Image Emotion Recognition. In Proceedings of the 22nd International Conference on Multimedia, Orlando, FL, USA, 3–7 November 2014; pp. 47–56.
14. Machajdik, J.; Hanbury, A. Affective Image Classification Using Features Inspired by Psychology and Art Theory. In Proceedings of the 18th ACM international conference on Multimedia, Firenze, Italy, 25–29 October 2010; pp. 83–92.
15. Zhao, S.; Yao, H.; Yang, Y.; Zhang, Y. Affective Image Retrieval via Multi-graph Learning. In Proceedings of the 22nd International Conference on Multimedia, Orlando, FL, USA, 3–7 November 2014; pp. 1025–1028.
16. Hanjalic, A. Extracting Moods from Pictures and Sounds: Towards Truly Personalized TV. *IEEE Signal Process. Mag.* **2006**, *23*, 90–100. [CrossRef]
17. Zhao, S.; Yao, H.; Gao, Y.; Ding, G.; Chua, T.S. Predicting Personalized Image Emotion Perceptions in Social Networks. *IEEE Trans. Affect. Comput.* **2018**, *9*, 526–540. [CrossRef]
18. Yang, J.; She, D.; Lai, Y.K.; Yang, M.H. Retrieving and Classifying Affective Images via Deep Metric Learning. In Proceedings of the Thirty-Second AAAI Conference on Artificial Intelligence, New Orleans, LA, USA, 2–7 February 2018.
19. Sun, M.; Yang, J.; Wang, K.; Shen, H. Discovering Affective Regions in Deep Convolutional Neural Networks for Visual Sentiment Prediction. In Proceedings of the 2016 IEEE International Conference on Multimedia and Expo, Barcelona, Spain, 12–15 April 2016; pp. 1–6.
20. You, Q.; Jin, H.; Luo, J. Visual Sentiment Analysis by Attending on Local Image Regions. In Proceedings of the Thirty-First AAAI Conference on Artificial Intelligence, San Francisco, CA, USA, 4–9 February 2017; pp. 231–237.

21. Gori, M.; Monfardini, G.; Scarselli, F. A New Model for Learning in Graph Domains. In Proceedings of the 2005 IEEE International Joint Conference on Neural Networks, Montreal, QC, Canada, 31 July–4 August 2005; pp. 729–734.
22. Scarselli, F.; Gori, M,; Tsoi, A.C.; Hagenbuchner, M.; Monfardini, G. The Graph Neural Network Model. *IEEE Trans. Neural Netw.* **2008**, *20*, 61–80. [CrossRef] [PubMed]
23. Bruna, J.; Zaremba, W.; Szlam, A.; LeCun, Y. The Graph Neural Network Model. In Proceedings of the 2nd International Conference on Learning Representations, Banff, AB, Canada, 14–16 April 2014; pp. 61–80.
24. Chen, Z.M.; Wei, X.S.; Wang, P.; Guo, Y. Multi-label Image Recognition with Graph Convolutional Networks. In Proceedings of the 2016 IEEE International Conference on Multimedia and Expo, Long Beach, CA, USA, 16–20 June 2019; pp. 5177–5186.
25. Peng, K.C.; Sadovnik, A.; Gallagher, A.; Chen, T. Where Do Emotions Come From? Predicting the Emotion Stimuli Map. In Proceedings of the 2016 IEEE International Conference on Image Processing, Phoenix, AZ, USA, 25–28 September 2016; pp. 614–618.
26. Guo, D.; Wang, H.; Zhang, H.; Zha, Z.J.; Wang, M. Iterative Context-Aware Graph Inference for Visual Dialog. In Proceedings of the IEEE/CVF Conference on Computer Vision and Pattern Recognition, Seattle, WA, USA, 13–19 June 2020; pp. 10055–10064.
27. Simonyan, K.; Zisserman, A. Very Deep Convolutional Networks for Large-scale Image Recognition. *arXiv* **2014**, arXiv:1409.1556.
28. Borth, D.; Ji, R.; Chen, T.; Breuel, T.; Chang, S.F. Large-scale Visual Sentiment Ontology and Detectors Using Adjective Noun Pairs. In Proceedings of the 21st ACM International Conference on Multimedia, Barcelona, Spain, 21–25 October 2013; pp. 223–232.
29. You, Q.; Luo, J.; Jin, H.; Yang, J. Robust Image Sentiment Analysis Using Progressively Trained and Domain Transferred Deep Networks. In Proceedings of the Twenty-Ninth AAAI Conference on Artificial Intelligence, Austin, TX, USA, 26 January 2015; pp. 381–388.
30. Mikels, J.A.; Fredrickson, B.L.; Larkin, G.R.; Lindberg, C.M.;d Maglio, S.J.; Reuter-Lorenz, P.A. Emotional Category Data on Images from the International Affective Picture System. *Behav. Res. Methods* **2005**, *37*, 626–630. [CrossRef] [PubMed]
31. Siersdorfer, S,; Minack, E.; Deng, F.; Hare, J. Analyzing and Predicting Sentiment of Images on the Social Web. In Proceedings of the 18th ACM International Conference on Multimedia, Firenze, Italy, 25–29 October 2010; pp. 715–718.
32. Chen, T.; Borth, D.; Darrell, T.; Chang, S.F. Deep SentiBank: Visual Sentiment Concept Classification with Deep Convolutional Neural Networks. *arXiv* **2014**, arXiv:1410.8586.

*Article*

# Single Image Super-Resolution Method Using CNN-Based Lightweight Neural Networks

Seonjae Kim [1], Dongsan Jun [2,*], Byung-Gyu Kim [3], Hunjoo Lee [4] and Eunjun Rhee [4]

- [1] Department of Convergence IT Engineering, Kyungnam University, Changwon 51767, Korea; sjkim@kyungnam-ispl.kr
- [2] Department of Information and Communication Engineering, Kyungnam University, Changwon 51767, Korea
- [3] Department of IT Engineering, Sookmyung Women's University, Seoul 04310, Korea; bg.kim@sookmyung.ac.kr
- [4] Intelligent Convergence Research Lab., Electronics and Telecommunications Research Institute (ETRI), Daejeon 34129, Korea; hjoo@etri.re.kr (H.L.); ejrhee@etri.re.kr (E.R.)
- \* Correspondence: dsjun9643@kyungnam.ac.kr

**Abstract:** There are many studies that seek to enhance a low resolution image to a high resolution image in the area of super-resolution. As deep learning technologies have recently shown impressive results on the image interpolation and restoration field, recent studies are focusing on convolutional neural network (CNN)-based super-resolution schemes to surpass the conventional pixel-wise interpolation methods. In this paper, we propose two lightweight neural networks with a hybrid residual and dense connection structure to improve the super-resolution performance. In order to design the proposed networks, we extracted training images from the DIVerse 2K (DIV2K) image dataset and investigated the trade-off between the quality enhancement performance and network complexity under the proposed methods. The experimental results show that the proposed methods can significantly reduce both the inference speed and the memory required to store parameters and intermediate feature maps, while maintaining similar image quality compared to the previous methods.

**Keywords:** deep learning; convolutional neural networks; lightweight neural network; single image super-resolution; image enhancement; image restoration; residual dense networks

**Citation:** Kim, S.; Jun, D.; Kim, B.-G.; Lee, H.; Rhee, E. Single Image Super-Resolution Method Using CNN-Based Lightweight Neural Networks. *Appl. Sci.* **2021**, *11*, 1092. https://doi.org/10.3390/app11031092

Received: 10 December 2020
Accepted: 22 January 2021
Published: 25 January 2021

**Publisher's Note:** MDPI stays neutral with regard to jurisdictional claims in published maps and institutional affiliations.

**Copyright:** © 2021 by the authors. Licensee MDPI, Basel, Switzerland. This article is an open access article distributed under the terms and conditions of the Creative Commons Attribution (CC BY) license (https://creativecommons.org/licenses/by/4.0/).

## 1. Introduction

While the resolution of images has been rapidly increasing in recent years with the development of high-performance cameras, advanced image compression, and display panels, the demands to generate high resolution images from pre-existing low-resolution images are also increasing for rendering on high resolution displays. In the field of computer vision, single image super-resolution (SISR) methods aim at recovering a high-resolution image from a single low-resolution image. Since the low-resolution images cannot represent the high-frequency information properly, most super-resolution (SR) methods have focused on restoring high-frequency components. For this reason, SR methods are used to restore the high-frequency components from quantized images at the image and video post-processing stage [1–3].

Deep learning schemes such as convolutional neural network (CNN) and multi-layer perceptron (MLP) are a branch of machine learning which aims to learn the correlations between input and output data. In general, the output in the process of the convolution operations is one pixel, which is a weighted sum between an input image block and a filter, so an output image represents the spatial correlation of input image corresponding to the filters used. As CNN-based deep learning technologies have recently shown impressive results in the area of SISR, various CNN-based SR methods have been developed that surpass the conventional SR methods, such as image statistical methods and patch-based methods [4,5]. In order to improve the quality of low-resolution images, CNN-based SR networks tend to deploy more complicated schemes, which have deeper and denser CNN structures and cause increases in the computational complexity like the required memory to store network parameters, the number of convolution operations, and the inference speed. We propose two SR-based lightweight neural networks (LNNs) with hybrid residual and dense networks, which are the "inter-layered SR-LNN" and "simplified SR-LNN" respectively, which we denote in this paper as "SR-ILLNN" and "SR-SLNN", respectively. The proposed methods were designed to produce similar image quality while reducing the number of networks parameters, compared to previous methods. Those SR technologies can be applied to the pre-processing stages of face and gesture recognition [6–8].

The remainder of this paper is organized as follows: In Section 2, we review previous studies related to CNN-based SISR methods. In Section 3, we describe the frameworks of the proposed two SR-LNNs for SISR. Finally, experimental results and conclusions are given in Sections 4 and 5, respectively.

## 2. Related Works

Deep learning-based SR methods have shown high potential in the field of image interpolation and restoration, compared to the conventional pixel-wise interpolation algorithms. Dong et al. proposed a three-layer CNN structure called super-resolution convolutional neural network (SR-CNN) [9], which learns an end-to-end mapping from a bi-cubic interpolated low-resolution image to a high-resolution image. Since the advent of SR-CNN, a variety of CNN networks with deeper and denser network structure [10–13] have been developed to improve the accuracy of SR.

In particular, He et al. proposed a ResNet [11] for image classification. Its key idea is to learn residuals through global or local skip connection. It notes that ResNet can provide a high-speed training process and prevent the gradient vanishing effects. In addition to ResNet, Huang et al. proposed densely connected convolutional networks (DenseNet) [12] to combine hierarchical feature maps available along the network depth for more flexible and richer feature representations. Dong et al. proposed an artifacts reduction CNN (AR-CNN) [14], which effectively reduces the various compression artifacts such as block artifacts and ringing artifacts on Joint Photographic Experts Group (JPEG) compression images.

Kim et al. proposed a super-resolution scheme with very deep convolutional networks (VDSR) [15], which is connected with 20 convolutional layers and a global skip connection. In particular, the importance of receptive field size and the residual learning was verified by VDSR. Leding et al. proposed a SR-ResNet [16], which was designed with multiple residual blocks and generative adversarial network (GAN) for improving visually subjective quality. Here, a residual block is composed of multiple convolution layers, a batch normalization, and a local skip connection. Lim et al. exploited enhanced deep super-resolution (EDSR) and multi-scale deep super-resolution (MDSR) [17]. In particular, as these networks have been modified in a way of removing the batch normalization, it can reduce graphics processing unit (GPU) memory demand by about 40% compared with SR-ResNet.

Tong et al. proposed an image super-resolution using dense skip connections (SR-DenseNet) [18] as shown in Figure 1. Because SR-DenseNet consists of eight dense blocks and each dense block contains eight dense layers, this network has a total of 67 convolution layers and two deconvolution layers. Because the feature maps of the previous convolutional layer are concatenated with those of the current convolutional layer within a dense block, total number of the feature map from the last dense block reaches up to 1040 and it requires more memory capacity to store the massive network parameters and intermediate feature maps.

On the other hand, the aforementioned deep learning-based SR methods are also applied to compress raw video data. For example, Joint Video Experts Team (JVET) formed the Adhoc Group (AhG) for deep neural networks based video coding (DNNVC) [19] in 2020, which aims at exploring the coding efficiency using the deep learning schemes. Several studies [20–22] have shown better coding performance than the-state-of-the-art video coding technologies.

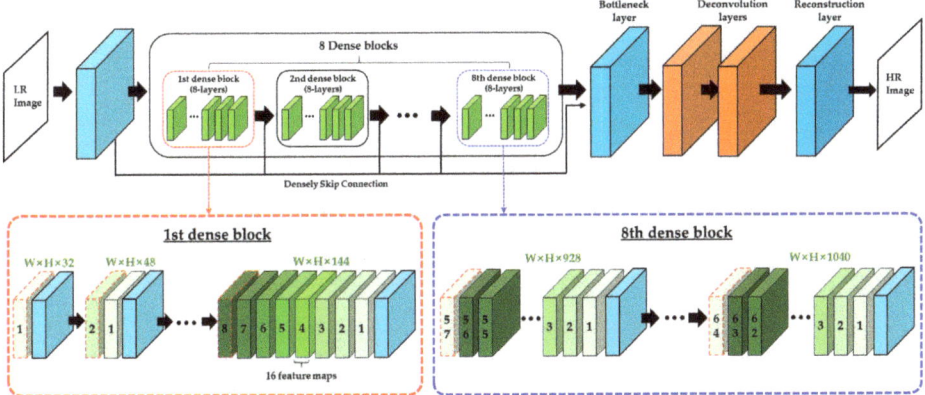

**Figure 1.** The framework of SR-DenseNet [18].

## 3. Proposed Method

Although more complicated deep neural network models have demonstrated better SR performance than conventional methods, it is difficult to implement them on low-complexity, low-power, and low-memory devices, due to the massive network parameters and convolution operations of deeper and denser networks. In case of SR-DenseNet, it is difficult to implement this model to the applications for real-time processing even though its SR performance is superior to that of other neural network models. To address this issue, we considered two lightweight network structures at the expense of unnoticeable quality degradation, compared to SR-DenseNet. The purpose of the proposed two lightweight neural networks for SISR is to quadruple the input images the same as SR-DenseNet. Firstly, SR-ILLNN learns the feature maps, which are derived from both low-resolution and interpolated low-resolution images. Secondly, SR-SLNN is designed to use only low-resolution feature maps of the SR-ILLNN for a few more reducing the network complexity.

### 3.1. Architecture of SR-ILLNN

Figure 2 shows the proposed SR-ILLNN, which consists of two inputs, 15 convolution layers and two deconvolution layers. The two inputs are denoted as a low-resolution (LR) image $X_{LR}$ and a bi-cubic interpolated low-resolution (ILR) image $X_{ILR}$ where N and M denote the width and height of the input image $X_{LR}$, respectively. The reason why we deployed the two inputs is to compensate the dense LR features of SR-DenseNet with high-resolution (HR) features of $X_{ILR}$, while reducing the number of convolutional layers as many as possible. As depicted in Figure 2, it consists of three parts, which are LR feature layers from convolutional layer 1 (Conv1) to Conv8, HR feature layers from Conv9 to Conv12, and shared feature layers from Conv13 to Conv15.

Each convolution is operated as in (1), where $W_i$, $B_i$, and '⊗' represent the kernels, biases, and convolution operation of the $i$th layer, respectively. In this paper, we notate a kernel as $[F_w \times F_h \times F_c]$, where $F_w \times F_h$ and $F_c$ are the spatial size of filter and the number of channels, respectively:

$$F_i(X_{LR}) = max(0, W_i \otimes F_{i-1}(X_{LR}) + B_i), \tag{1}$$

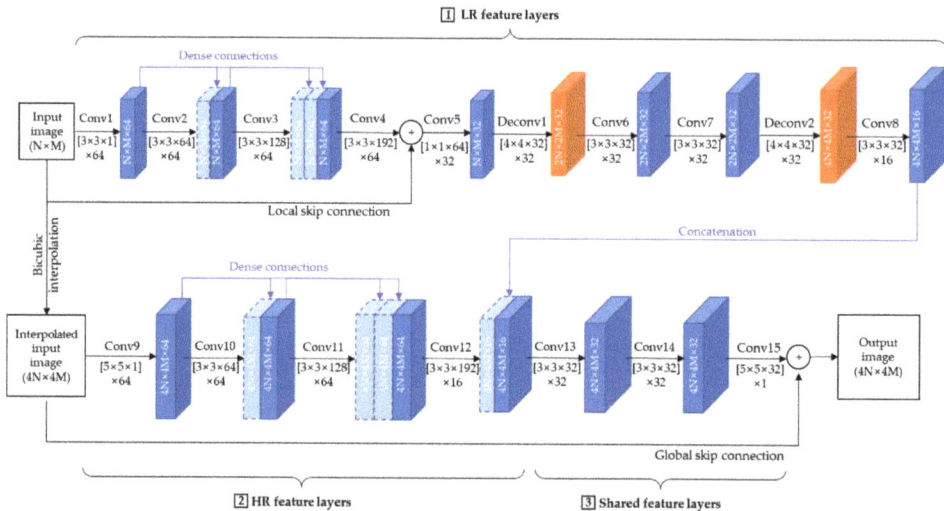

**Figure 2.** The framework of the proposed SR-ILLNN.

In the process of convolution operation, we applied rectified linear unit (ReLU, $max(0, x)$) on the filter responses and used a partial convolution-based padding scheme [23] to avoid the loss of boundary information. The padding sizes is defined so that the feature maps between different convolution layers can have the same spatial resolution as follows:

$$Padding\ Size = Floor(F_w/2), \quad (2)$$

where Floor(x) means the rounding down operation. Note that the convolutional layers of Conv1–4 and Conv9–12 of Figure 2 are conducted to generate output feature maps with dense connections for more flexible and richer feature representations, which are concatenated the feature maps of the previous layer with those of the current layer. So, convolution operations with dense connections are calculated as in (3):

$$\begin{aligned} F_i(X_{LR}) &= max(0, W_i \otimes [F_1(X_{LR}), \ldots, F_{i-1}(X_{LR})] + B_i), \\ F_j(X_{ILR}) &= max(0, W_j \otimes [F_9(X_{ILR}), \ldots, F_{j-1}(X_{ILR})] + B_j) \end{aligned} \quad (3)$$

A ResNet scheme [11] with skip connections can provide a high-speed training and prevent the gradient vanishing effect, so we deployed a local and a global skip connection to train the residual at the output feature maps of Conv4 and Conv15. Because the output feature maps $F_4$ and $X_{LR}$ have the different number of channels in local skip connection, $X_{LR}$ is copied up to the number of channels of $F_4$ before operating Conv5.

It should be noted that the number of feature maps has a strong effect on the inference speed. Therefore, the proposed SR-LNNs is designed to reduce the number of feature maps from 192 to 32, before deconvolution operation. Then, Deconv1 and Deconv2 are operated for image up-sampling as follows:

$$F_{deconv}(X_{LR}) = max(0, W_{deconv} \odot F_{i-1}(X_{LR}) + B_{deconv}), \quad (4)$$

where $W_{deconv}$, $B_{deconv}$ are the kernels and biases of the deconvolution layer, respectively, and the symbol '$\odot$' denotes the deconvolution operation. As each deconvolution layer has different kernel weights and biases, it is superior to the conventional SR methods like pixel-wise interpolation methods.

In the stage of the shared feature layers, the output feature maps of the LR feature layers $F_8(X_{LR})$ are concatenated with those of HR feature layers $F_{12}(X_{ILR})$. Then, the concatenated feature maps $[F_8(X_{LR}), F_{12}(X_{ILR})]$ are inputted to the shared feature layers as in (5):

$$F_{13}(X) = max(0, W_{13} \otimes [F_8(X_{LR}), F_{12}(X_{ILR})] + B_{13}) \quad (5)$$

Note that the activation function (ReLU) is not applied to the last feature map when the convolution operation is conducted in Conv15. Table 1 presents the structural analysis of the network parameters in SR-ILLNN.

Table 1. Analysis of network parameters in SR-ILLNN.

| Layer Name | Kernel Size | Num. of Kernels | Padding Size | Output Feature Map (W × H × C) | Num. of Parameters |
|---|---|---|---|---|---|
| Conv1 | 3 × 3 × 1 | 64 | 1 | N × M × 64 | 640 |
| Conv2 | 3 × 3 × 64 | 64 | 1 | N × M × 64 | 36,928 |
| Conv3 | 3 × 3 × 128 | 64 | 1 | N × M × 64 | 73,792 |
| Conv4 | 3 × 3 × 192 | 64 | 1 | N × M × 64 | 110,656 |
| Conv5 | 1 × 1 × 64 | 32 | 0 | N × M × 32 | 2080 |
| Deconv1 | 4 × 4 × 32 | 32 | 1 | 2N × 2M × 32 | 16,416 |
| Conv6, 7 | 3 × 3 × 32 | 32 | 1 | 2N × 2M × 32 | 9248 |
| Deconv2 | 4 × 4 × 32 | 32 | 1 | 4N × 4M × 32 | 16,416 |
| Conv8 | 3 × 3 × 32 | 16 | 1 | 4N × 4M × 16 | 4624 |
| Conv9 | 5 × 5 × 1 | 64 | 2 | 4N × 4M × 64 | 1664 |
| Conv10 | 3 × 3 × 64 | 64 | 1 | 4N × 4M × 64 | 36,928 |
| Conv11 | 3 × 3 × 128 | 64 | 1 | 4N × 4M × 64 | 73,792 |
| Conv12 | 3 × 3 × 192 | 16 | 1 | 4N × 4M × 16 | 27,664 |
| Conv13, 14 | 3 × 3 × 32 | 32 | 1 | 4N × 4M × 32 | 9248 |
| Conv15 | 5 × 5 × 32 | 1 | 2 | 4N × 4M × 1 | 801 |

### 3.2. Architecture of SR-SLNN

Because SR-ILLNN has hierarchical network structure due to the two inputs, we propose a SR-SLNN to reduce the network complexity of SR-ILLNN. As depicted in Figure 3, the SR-SLNN was modified to remove the HR feature layers and the shared feature layers of SR-ILLNN. In addition, it has seven convolution layers and two deconvolution layers, where two convolution layers between deconvolution layers are also removed. Table 2 presents the structural analysis of network parameters in SR-SLNN.

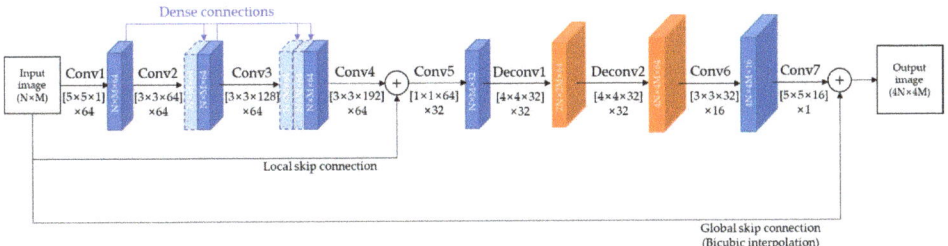

Figure 3. The framework of the proposed SR-SLNN.

Table 2. Analysis of network parameters in SR-SLNN.

| Layer Name | Kernel Size | Num. of Kernels | Padding Size | Output Feature Map (W × H × C) | Num. of Parameters |
|---|---|---|---|---|---|
| Conv1 | 5 × 5 × 1 | 64 | 2 | N × M × 64 | 1664 |
| Conv2 | 3 × 3 × 64 | 64 | 1 | N × M × 64 | 36,928 |
| Conv3 | 3 × 3 × 128 | 64 | 1 | N × M × 64 | 73,792 |
| Conv4 | 3 × 3 × 192 | 64 | 1 | N × M × 64 | 110,656 |
| Conv5 | 1 × 1 × 64 | 32 | 0 | N × M × 32 | 2080 |
| Deconv1 | 4 × 4 × 32 | 32 | 1 | 2N × 2M × 32 | 16,416 |
| Deconv2 | 4 × 4 × 32 | 32 | 1 | 4N × 4M × 32 | 16,416 |
| Conv6 | 3 × 3 × 32 | 16 | 1 | 4N × 4M × 16 | 4624 |
| Conv7 | 5 × 5 × 16 | 1 | 2 | 4N × 4M × 1 | 401 |

### 3.3. Loss Function and Hyper-Parameters

We set hyper-parameters as presented in Table 3. In order to find the optimal parameter $\theta$ = {Filter weights, Biases}, we defined mean square error (MSE) as the loss function (6) where $X_{HR}$, $Y$, and $N$ are the final output image of SR-LNN, the corresponding label image and the batch size. Here, $L(\theta)$ is minimized by Adam optimizer using the back-propagation. In particular, the number of epochs were set to 50 according to the Peak Signal-to-Nosie Ratio (PSNR) variations of the validation set (Set5) and the learning rates were empirically assigned to the intervals of epoch.

Since it is important to set the optimal number of network parameters in the design of lightweight neural network, we investigated the relation between the number of parameters and PSNR according to the various filter sizes. As measured in Table 4, we implemented the most of convolution layers with 3x3 filter size, except for deconvolution layers to generate the interpolated feature map that accurately corresponds to the scaling factor:

$$L(\theta) = \frac{1}{N}\sum_{i=0}^{N-1} \|X_{HR}(X^i) - Y^i\|_2^2 \quad (6)$$

Table 3. Hyper-parameters of the proposed methods.

| | |
|---|---|
| Optimizer | Adam |
| Learning Rate | $10^{-3}$ to $10^{-5}$ |
| Activation function | ReLU |
| Padding Mode | Partial convolutional based padding [23] |
| Num. of epochs | 50 |
| Batch size | 128 |
| Initial weight | Xavier |

Table 4. Relation between the number of parameters and PSNR according to the various filter sizes.

| Networks | Filter Size (Width × Height) | Num. of Parameters | PSNR (dB) |
|---|---|---|---|
| SR-ILLNN | 3 × 3 | 439,393 | 31.41 |
|  | 5 × 5 | 1,153,121 | 31.38 |
|  | 7 × 7 | 2,223,713 | 31.29 |
|  | 9 × 9 | 3,651,169 | 28.44 |
| SR-SLNN | 3 × 3 | 262,977 | 31.29 |
|  | 5 × 5 | 664,385 | 31.19 |
|  | 7 × 7 | 1,266,497 | 31.16 |
|  | 9 × 9 | 2,069,313 | 31.15 |

## 4. Experimental Results

As shown in Figure 4, we used the DIVerse 2K (DIV2K) dataset [24] whose total number is 800 images to train the proposed methods. In order to design SR-LNN capable of up-sampling input images four times, all training images with RGB components are converted into YUV components and extracted only Y component with the size of 100 × 100 patch without overlap. In order to generate interpolated input images, the patches are down-sampled and then up-sampled it again by bi-cubic interpolation.

Finally, we obtained three training datasets from DIV2K where the total number of each training dataset is 210,048 images for original images, low-resolution images, and interpolated low-resolution images, respectively. For testing our SR-LNN models, we used Set5, Set14, Berkeley Segmentation Dataset 100 (BSD100), and Urban100 as depicted in Figure 5, which are representatively used as testing datasets in most SR studies. For reference, Set5 was also used as a validation dataset.

All experiments were run on an Intel Xeon Skylake (eight cores @ 2.59 GHz) having 128 GB RAM and two NVIDIA Tesla V100 GPUs under the experimental environment described in Table 5. After setting a bicubic interpolation method as an anchor for performance comparison, we compared the proposed two SR-LNN models with SR-CNN [9], AR-CNN [14], and SR-DenseNet [18] in terms of image quality enhancement and network complexity.

**Figure 4.** Training dataset. (DIV2K).

**Figure 5.** Test datasets. (Set5, Set14, BSD100, and Urban100).

In order to evaluate the accuracy of SR, we used PSNR and the structural similarity index measure (SSIM) [25,26] on the Y component as shown in Tables 6 and 7, respectively. In general, PSNR has been commonly used as a fidelity measurement and it is the ratio between the maximum possible power of an original signal and the power of corrupting noise that affects the fidelity of its representation. In addition, SSIM is a measurement that calculates a score using structural information of images and is evaluated as similar to human perceptual scores. Compared with the anchor, the proposed SR-ILLNN and SR-SLNN enhance PSNR by as many as 1.81 decibel (dB) and 1.71 dB, respectively. Similarly, the proposed SR-LNNs show significant PSNR enhancement, compared to SR-CNN and AR-CNN. In contrast to the results of the anchor, the proposed SR-ILLNN has similar PSNR performance on most test datasets, compared with SR-DenseNet.

**Table 5.** Experimental environments.

| Num. of Training Samples | 210,048 |
|---|---|
| Input size ($X_{LR}$) | 25 × 25 × 1 |
| Interpolated input size ($X_{ILR}$) | 100 × 100 × 1 |
| Label size ($Y_{HR}$) | 100 × 100 × 1 |
| Linux version | Ubuntu 16.04 |
| CUDA version | 10.1 |
| Deep learning frameworks | Pytorch 1.4.0 |

**Table 6.** Average results of PSNR (dB) on the test dataset.

| Dataset | Bicubic | SR-CNN [9] | AR-CNN [14] | SR-DenseNet [18] | SR-ILLNN | SR-SLNN |
|---|---|---|---|---|---|---|
| Set5 | 28.44 | 30.30 | 30.35 | 31.43 | 31.41 | 31.29 |
| Set14 | 25.80 | 27.09 | 27.10 | 27.84 | 27.83 | 27.73 |
| BSD100 | 25.99 | 26.86 | 26.86 | 27.34 | 27.33 | 27.28 |
| Urban100 | 23.14 | 24.33 | 24.34 | 25.30 | 25.32 | 25.18 |
| Average | 24.73 | 25.80 | 25.81 | 26.53 | 26.54 | 26.44 |

**Table 7.** Average results of SSIM on the test dataset.

| Dataset | Bicubic | SR-CNN [9] | AR-CNN [14] | SR-DenseNet [18] | SR-ILLNN | SR-SLNN |
|---|---|---|---|---|---|---|
| Set5 | 0.8112 | 0.8599 | 0.8614 | 0.8844 | 0.8848 | 0.8827 |
| Set14 | 0.7033 | 0.7495 | 0.7511 | 0.7708 | 0.7709 | 0.7689 |
| BSD100 | 0.6699 | 0.7112 | 0.7126 | 0.7279 | 0.7275 | 0.7260 |
| Urban100 | 0.6589 | 0.7158 | 0.7177 | 0.7584 | 0.7583 | 0.7532 |
| Average | 0.6702 | 0.7192 | 0.7208 | 0.7481 | 0.7479 | 0.7447 |

In addition, we conducted an experiment to verify the effectiveness of skip connections and dense connections. In particular, the more dense connections are deployed in the between convolution layers, the more network parameters are required in the process of convolution operations. Table 8 shows the results of tool-off tests on the proposed methods. As both skip connections and dense connections contribute to improve PSNR in the test datasets, the proposed methods are deployed these schemes. Figure 6 shows MSE as well as PSNR corresponding to the number of epochs and these experiments were evaluated from all comparison methods (SR-CNN, AR-CNN, and SR-DenseNet), including the proposed methods. It is confirmed that although SR-DenseNet has the highest reduction-rate in terms of MSE, the proposed methods have an almost similar increase rate in terms of PSNR. Figure 7 shows the comparisons of subjective quality between the proposed methods and previous methods.

Table 8. The results of tool-off tests.

| Skip Connections | Dense Connections | Set5 (PSNR) | | Set14 (PSNR) | | BSD100 (PSNR) | | Urban100 (PSNR) | |
|---|---|---|---|---|---|---|---|---|---|
| | | SR-ILLNN | SR-SLNN | SR-ILLNN | SR-SLNN | SR-ILLNN | SR-SLNN | SR-ILLNN | SR-SLNN |
| Disable | Disable | 31.34 | 31.15 | 27.80 | 27.62 | 27.31 | 27.20 | 25.27 | 25.01 |
| Enable | Disable | 31.35 | 31.21 | 27.81 | 27.68 | 27.33 | 27.26 | 25.31 | 25.13 |
| Disable | Enable | 31.40 | 31.18 | 27.81 | 27.65 | 27.32 | 27.23 | 25.29 | 25.07 |
| Enable | Enable | 31.41 | 31.29 | 27.83 | 27.73 | 27.33 | 27.28 | 25.32 | 25.18 |

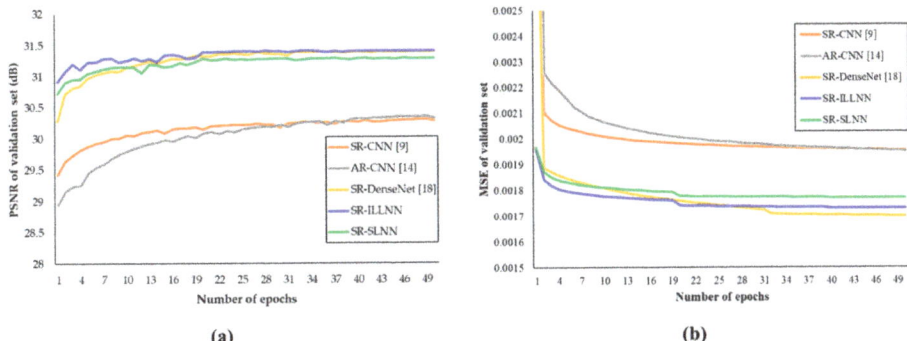

Figure 6. PSNR and MSE corresponding to the number of epochs. (a) PSNR per epoch. (b) MSE per epoch.

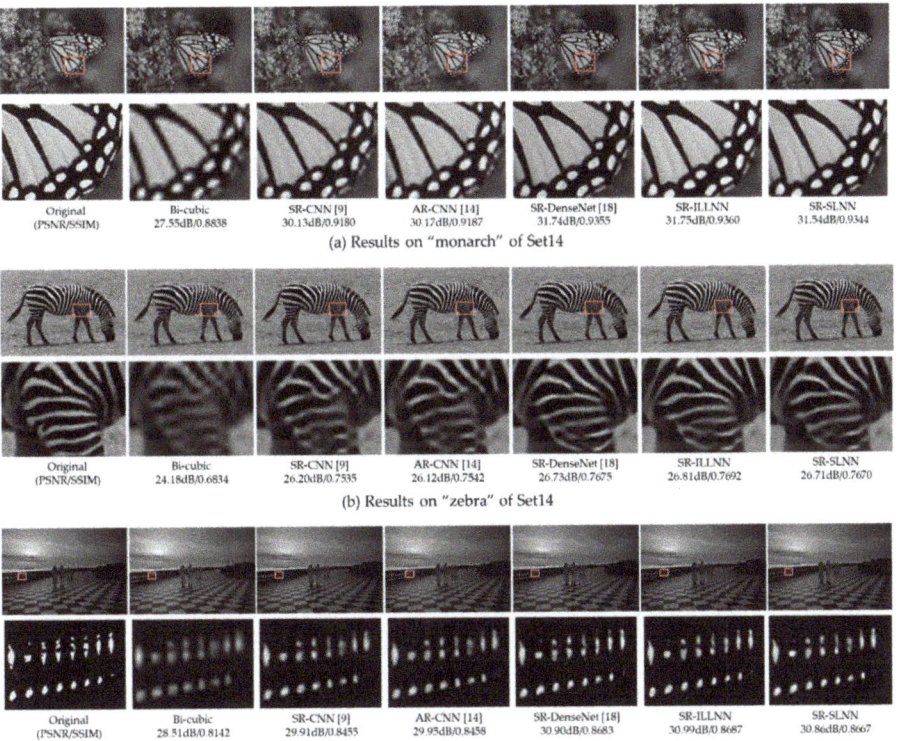

Figure 7. Comparisons of subjective quality on test dataset. (a) Results on "monarch" of Set14. (b) Results on "zebra" of Set14. (c) Results on "img028" of Urban100.

In terms of the network complexity, we analyzed the number of parameters, parameter size (MB), and total memory size (MB) where total memory size includes intermediate feature maps as well as the parameter size. In general, both the total memory size and the inference speed are proportional to the number of parameters. Table 9 presents the number of parameters and total memory size. Compared with SR-DenseNet, the proposed two SR-LNNs reduce the number of parameters by as low as 8.1% and 4.8%, respectively. Similarly, the proposed two SR-LNNs reduce total memory size by as low as 35.9% and 16.1%, respectively. In addition, we evaluated the inference speed on BSD100 test images. As shown in Figure 8, the inference speed of the proposed methods is much faster than that of SR-DenseNet. Even though the proposed SR-SLNN is slower than SR-CNN and AR-CNN, it is obviously superior to SR-CNN and AR-CNN in terms of PSNR improvements as measured in Tables 6 and 7.

Table 9. Analysis of the number of parameters and memory size.

|  | SR-CNN [9] | AR-CNN [14] | SR-DenseNet [18] | SR-ILLNN | SR-SLNN |
|---|---|---|---|---|---|
| Num. of parameters | 57,281 | 106,564 | 5,452,449 | 439,393 | 262,977 |
| Parameter size (MB) | 0.22 | 0.41 | 20.80 | 1.68 | 1.00 |
| Total memory size (MB) | 14.98 | 17.61 | 224.81 | 80.83 | 36.21 |

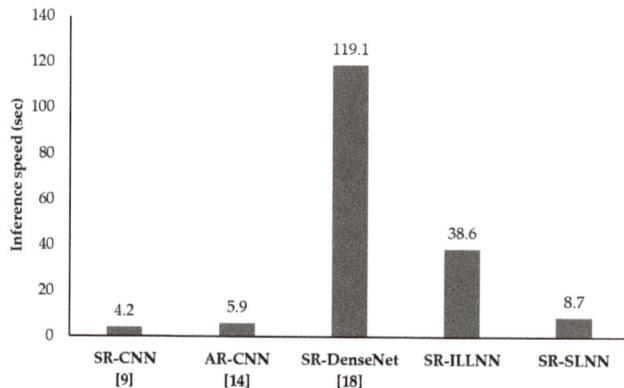

Figure 8. Inference speed on BSD100.

## 5. Conclusions and Future Work

In this paper, we have proposed two SR-based lightweight neural networks (SR-ILLNN and SR-SLNN) for single image super-resolution. We investigated the trade-offs between the accuracy of SR (PSNR and SSIM) and the network complexity, such as the number of parameters, memory capacity, and inference speed. Firstly, SR-ILLNN was trained on both low-resolution and high-resolution images. Secondly, SR-SLNN was designed to reduce the network complexity of SR-ILLNN. For training the proposed SR-LNNs, we used the DIV2K image dataset and evaluated both the accuracy of SR and the network complexity on Set5, Set14, BSD100, and Urban100 test image datasets. Our experimental results show that the SR-ILLNN and SR-SLNN can significantly reduce the number of parameters by 8.1% and 4.8%, respectively, while maintaining similar image quality compared to the previous methods. As future work, we plan to extend the proposed SR-LNNs to other color components as well as a luminance component for improving SR performance on color images.

**Author Contributions:** Conceptualization, S.K. and D.J.; methodology, S.K. and D.J.; software, S.K.; validation, D.J., B.-G.K., H.L., and E.R.; formal analysis, S.K. and D.J.; investigation, S.K. and D.J; resources, D.J.; data curation, S.K.; writing—original draft preparation, S.K.; Writing—Review and editing, D.J.; visualization, S.K.; supervision, D.J.; project administration, D.J.; funding acquisition, H.L. and E.R. All authors have read and agreed to the published version of the manuscript.

**Funding:** This research was funded by Ministry of Science and ICT(MSIT) of the Korea government, grant number20PQWO-B153369-02.

**Institutional Review Board Statement:** Not applicable.

**Informed Consent Statement:** Not applicable.

**Data Availability Statement:** Not applicable.

**Acknowledgments:** This research was supported by a grant (20PQWO-B153369-02) from Smart road lighting platform development and empirical study on testbed Program funded by Ministry of Science and ICT(MSIT) of the Korea government. This research was results of a study on the "HPC Support" Project, supported by the 'Ministry of Science and ICT' and NIPA.

**Conflicts of Interest:** The authors declare no conflict of interest.

# References

1. Yang, W.; Zhang, X.; Tian, Y.; Wang, W.; Xue, J.; Liao, Q. Deep Learning for Single Image Super-Resolution: A Brief Review. *IEEE Trans. Multimed.* **2019**, *21*, 3106–3121. [CrossRef]
2. Anwar, S.; Khan, S.; Barnes, N. A deep journey into super-resolution: A survey. *arXiv* **2019**, arXiv:1904.07523. [CrossRef]
3. Wang, Z.; Chen, J.; Hoi, S.C.H. Deep Learning for Image Super-resolution: A Survey. *IEEE Trans. Pattern. Anal. Mach. Intell.* **2020**. [CrossRef] [PubMed]
4. Keys, R. Cubic convolution interpolation for digital image processing. *IEEE Trans. Acoust.* **1981**, *29*, 1153–1160. [CrossRef]
5. Freeman, T.; Jones, R.; Pasztor, C. Example-based super-resolution. *IEEE Comput. Graph. Appl.* **2002**, *22*, 56–65. [CrossRef]
6. Kim, J.; Kim, B.; Roy, P.; Partha, P.; Jeong, D. Efficient facial expression recognition algorithm based on hierarchical deep neural network structure. *IEEE Access* **2019**, *7*, 41273–41285. [CrossRef]
7. Kim, J.; Hong, G.; Kim, B.; Dogra, D. deepGesture: Deep learning-based gesture recognition scheme using motion sensors. *Displays* **2018**, *55*, 38–45. [CrossRef]
8. Jeong, D.; Kim, B.; Dong, S. Deep Joint Spatiotemporal Network (DJSTN) for Efficient Facial Expression Recognition. *Sensors* **2020**, *20*, 1936. [CrossRef] [PubMed]
9. Dong, C.; Loy, C.C.; He, K.; Tang, X. Image super-resolution using deep convolutional networks. *IEEE Trans. Pattern. Anal. Mach. Intell.* **2015**, *38*, 295–307. [CrossRef] [PubMed]
10. Simonyan, K.; Zisserman, A. Very deep convolution networks for large -scale image recognition. In Proceedings of the International Conference on Learning Representations, San Diego, CA, USA, 7–9 May 2015.
11. He, K.; Zhang, X.; Ren, S.; Sun, J. Deep residual learning for image recognition. In Proceedings of the Conference on Computer Vision and Pattern Recognition, Las Vegas, NY, USA, 27–30 June 2016; pp. 770–778.
12. Huang, G.; Liu, Z.; Van Der Maaten, L.; Weinberger, K.Q. Densely connected convolutional networks. In Proceedings of the Conference on Computer Vision and Pattern Recognition, Honolulu, HI, USA, 21–26 July 2017; pp. 4700–4708.
13. Ioffe, S.; Szegedy, C. Batch normalization: Accelerating deep network training by reducing internal covariate shift. In Proceedings of the Machine Learning Research, Lille, France, 7–9 July 2015; pp. 448–456.
14. Dong, C.; Deng, Y.; Change Loy, C.; Tang, X. Compression artifacts reduction by a deep convolutional network. In Proceedings of the International Conference on Computer Vision, Santiago, Chile, 7–13 December 2015; pp. 576–584.
15. Kim, J.; Lee, J.; Lee, K. Accurate image super-resolution using very deep convolutional networks. In Proceedings of the Conference on Computer Vision and Pattern Recognition, Las Vegas, NY, USA, 27–30 June 2016; pp. 1646–1654.
16. Ledig, C.; Theis, L.; Huszár, F.; Caballero, J.; Cunningham, A.; Acosta, A.; Aitken, A.; Tejani, A.; Totz, J.; Wang, Z.; et al. Photo-realistic single image super-resolution using a generative adversarial network. In Proceedings of the Conference on Computer Vision and Pattern Recognition, Honolulu, HI, USA, 21–26 July 2017; pp. 4681–4690.
17. Lim, B.; Son, S.; Kim, H.; Nah, S.; Lee, K. Enhanced deep residual networks for single image super-resolution. In Proceedings of the Conference on Computer Vision and Pattern Recognition Workshops, Honolulu, HI, USA, 21–26 July 2017; pp. 136–144.
18. Tong, T.; Li, G.; Liu, X.; Gao, Q. Image super-resolution using dense skip connections. In Proceedings of the Conference on Computer Vision and Pattern Recognition, Honolulu, HI, USA, 21–26 July 2017; pp. 4799–4807.
19. Ye, Y.; Alshina, E.; Chen, J.; Liu, S.; Pfaff, J.; Wang, S. [DNNVC] AhG on Deep neural networks based video coding, Joint Video Experts Team (JVET) of ITU-T SG 16 WP 3 and ISO/IEC JTC 1/SC29, Document JVET-T0121, Input Document to JVET Teleconference. 2020.
20. Cho, S.; Lee, J.; Kim, J.; Kim, Y.; Kim, D.; Chung, J.; Jung, S. Low Bit-rate Image Compression based on Post-processing with Grouped Residual Dense Network. In Proceedings of the Conference on Computer Vision and Pattern Recognition Workshops, Long Beach, CA, USA, 16–20 June 2019.
21. Yang, R.; Xu, M.; Liu, T.; Wang, Z.; Guan, Z. Enhancing quality for HEVC compressed videos. *IEEE Trans. Circuits Syst. Video Technol.* **2018**, *29*, 2039–2054. [CrossRef]

22. Zhang, S.; Fan, Z.; Ling, N.; Jiang, M. Recursive Residual Convolutional Neural Network-Based In-Loop Filtering for Intra Frames. *IEEE Trans. Circuits Syst. Video Technol.* **2019**, *30*, 1888–1900. [CrossRef]
23. Lui, G.; Shih, K.; Wang, T.; Reda, F.; Sapra, K.; Yu, Z.; Tao, A.; Catanzaro, B. Partial convolution based padding. *arXiv* **2018**, arXiv:1811.11718.
24. Agustsson, E.; Timofte, R. NTIRE 2017 Challenge on Single Image Super-Resolution: Dataset and Study. In Proceedings of the Conference on Computer Vision and Pattern Recognition Workshops, Honolulu, HI, USA, 21–26 July 2017.
25. Wang, Z.; Bovik, A.C.; Sheikh, H.R.; Simoncelli, E.P. Image quality assessment: From error visibility to structural similarity. *IEEE Trans. Image Processing* **2004**, *13*, 600–612. [CrossRef] [PubMed]
26. Yang, C.; Ma, C.; Yang, M. Single-image super-resolution: A benchmark. In Proceedings of the European Conference on Computer Vision, Zurich, Switzerland, 6–12 September 2014; pp. 372–386.

Article

# Place Classification Algorithm Based on Semantic Segmented Objects

Woon-Ha Yeo, Young-Jin Heo, Young-Ju Choi and Byung-Gyu Kim *

Department of IT Engineering, Sookmyung Women's University, 100 Chungpa-ro 47gil, Yongsan-gu, Seoul 04310, Korea; wh.yeo@ivpl.sookmyung.ac.kr (W.-H.Y.); yj.heo@ivpl.sookmyung.ac.kr (Y.-J.H.); yj.choi@ivpl.sookmyung.ac.kr (Y.-J.C.)
* Correspondence: bg.kim@sookmyung.ac.kr; Tel.: +82-2-2077-7293

Received: 17 October 2020; Accepted: 15 December 2020; Published: 18 December 2020

**Abstract:** Scene or place classification is one of the important problems in image and video search and recommendation systems. Humans can understand the scene they are located, but it is difficult for machines to do it. Considering a scene image which has several objects, humans recognize the scene based on these objects, especially background objects. According to this observation, we propose an efficient scene classification algorithm for three different classes by detecting objects in the scene. We use pre-trained semantic segmentation model to extract objects from an image. After that, we construct a weight matrix to determine a scene class better. Finally, we classify an image into one of three scene classes (i.e., indoor, nature, city) by using the designed weighting matrix. The performance of our scheme outperforms several classification methods using convolutional neural networks (CNNs), such as VGG, Inception, ResNet, ResNeXt, Wide-ResNet, DenseNet, and MnasNet. The proposed model achieves 90.8% of verification accuracy and improves over 2.8% of the accuracy when comparing to the existing CNN-based methods.

**Keywords:** scene/place classification; semantic segmentation; deep learning; weighting matrix; convolutional neural network

## 1. Introduction

The scene is an important information which can be used as a metadata in image and video search or recommendation systems. This scene information can provide more detailed situation information with time duration and character who appears in image and video contents.

While humans naturally perceive the scene they are located, it is a challenging work for machines to recognize it. If the machines could understand the scene they are looking, this technology can be used for robots to navigate, or searching a scene in video data. The main purpose of scene classification is to classify name of scenes of given images.

In the early days, scene or place classification was carried out through traditional methods such as Scale-Invariant Feature Transformation (SIFT) [1], Speed-Up Robust Features (SURF) [2], and Bag of Words (BoW) [3]. In recent years, deep learning with the convolutional neural networks (CNNs) has been widely used for image classification ever since AlexNet [4] won the ImageNet Large Scale Visual Recognition Competition (ILSVRC) in 2012.

There have been several approaches to classify scenes and dplaces. One approach is using classification method such as *k*-nearest neighbor (KNN) classifier and other is based on the convolutional neural networks (CNNs). Chowanda et al. [5] proposed a new dataset for image classification and experimented with their dataset with CNNs such as VGG, GoogLeNet to classify Indonesian regions. Raja et al. [6] proposed a method of classifying indoor and outdoor by using KNN classifier. Viswanathan et al. [7] suggested an object-based approach. However, their methods could classify only indoor scenes such as kitchen, bathroom, and so forth. Yiyi et al. [8] also proposed an object-based classification method combining CNN and semantic segmentation model and classified five indoor scenes. Zheng et al. [9] suggested a method for aerial scene classification by using pre-trained CNN and multi-scale pooling. Liu et al. [10] proposed a Siamese CNN for remote sensing scene classification, which combined the identification and verification models of CNNs. Pires et al. [11] analyzed a CNN model for aerial scene classification by transfer learning. These methods were focused on verifying the aerial scene.

In this paper, we propose a method for classifying a scene and place image into one of three major scene categories: indoor, city, and nature which is different from the previous works in that we classify outdoor as well. There are many objects in the scene and place. It means that we are able to classify the scene by utilizing the information of the existing objects. Also, when humans see a scene or place, they recognize the scene or place based on objects, especially background objects. If there are mountains and the sky in the scene, they would perceive it as a natural scene, and if the scene is full of buildings, they would consider it as an urban scene. If there are ceiling and walls, they would recognize it as an indoor environment.

In order to classify a scene or place image based on this human perception process, we first conduct object segmentation using the image segmentation model pre-trained with MS COCO-stuff dataset [12]. While MS COCO dataset [13] is for object detection, MS COCO-stuff dataset is for segmentation. In this paper, we used DeepLab v2 [14] model which is semantic segmentation model and can be trained with the background (stuff) objects. MS COCO-stuff dataset contains 171 kinds of object classes which are suitable for our solution. To classify an image, we construct a weight matrix of each object classes so that we can give more weight to objects that are more dominant on determining a scene. Finally, we classify the scene by combining the weight matrix and detected objects in the scene or place.

We organize the rest of this paper as follows: Section 2 highlights the details of the proposed classification method. Section 3 presents the implementation of the experiment as well as the experimental results. Finally, we draw the conclusions and suggest further research directions in Section 4.

## 2. Proposed Classification Algorithm

The overall process of the proposed classification method is summarized in Figure 1. It contains image segmentation stage and computing scene score of the image. Before carrying this process out, we design the weight matrix with size 171 × 3. This matrix consists of 171 object classes, and each object class has 3 scene labels. The details of constructing the weight matrix will be explained in the following Section 2.2.

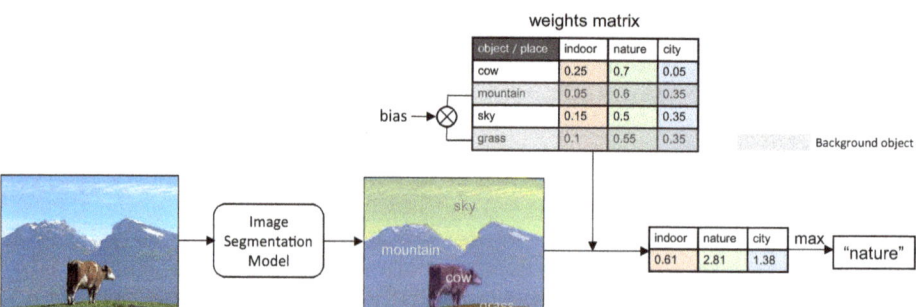

**Figure 1.** Overall structure of the proposed scheme.

## 2.1. Overall Structure

We define scene categories into three classes—indoor, city, and nature, which are the major categories for representing some scenes. To classify a scene image into a scene class, we first feed the image to the pre-trained image segmentation model. In this paper we used DeepLab v2 [14] to get the objects in the scene. For instance, if we get 'sky', 'mountain', 'cow', and 'grass' as a result of segmentation, we look for these object classes in the pre-constructed weight matrix. As humans perceive the scene in a scene image more by background objects rather than thing (foreground) objects, we want to focus more on the background classes by giving more weight on these classes. Therefore, if detected objects are part of background classes, we multiply *bias* value of 1.4 which is determined empirically.). We compute the scores of scene and place classes of the image by adding all weights of objects in the image. The scene class with the highest score is determined as the final scene and place class of the image. We determine objects in ground, solid, building in outdoor tree (Figure 2) as background objects.

## 2.2. Design of Weighting Factors

We build the weight matrix shown in Figure 3 by using 2017 validation images from the COCO-stuff dataset, including 5K images. These images are manually labeled as scene classes. After preparing dataset for constructing the matrix, the images are fed to the pre-trained image segmentation model one by one. We can get one or more objects as a result of segmentation.

The COCO-stuff dataset [12] includes 80 "thing" classes and 91 "stuff" classes, and stuff classes are divided into two wide categories—indoor and outdoor. The outdoor classes contain various classes representing background of an image such as building, mountain, and so forth (Figure 2).

The weight matrix $W$ has size of $M \times N$, $M$ is the number of object classes, and $N$ is the number of scene classes. Since we use COCO-stuff dataset and three scene labels, it turns out to $171 \times 3$. It is initialized with zeros at first.

As shown in Figure 3, assuming that we get classes of 'cow', 'mountain', 'sky', and 'grass' from an image. The image has a nature scene, so add 1 to nature column in weight matrix for each object class (cow, mountain, sky, and grass). After iteration of this process under 5 K images, the matrix would be constructed with various numbers. Therefore, we normalize it for each row. In the Equation (1), $W'$ denotes the normalized weight matrix. In addition, $m$ is $m$-th object class in the dataset and $n$ is $n$-th label in the place classes. The algorithm of constructing the weighting matrix is described in Algorithm 1 and the inference process with the model is shown in Algorithm 2.

In the inference process, we perform semantic segmentation for each test image, and we compute the scores of scene or place classes of the image by multiplying bias value to background object weights and adding all of them of each scene or place class by using pre-constructed weight matrix as:

$$W'_{mn} = \frac{W_{mn}}{\sum_{n'=1}^{N} W_{mn'}}. \tag{1}$$

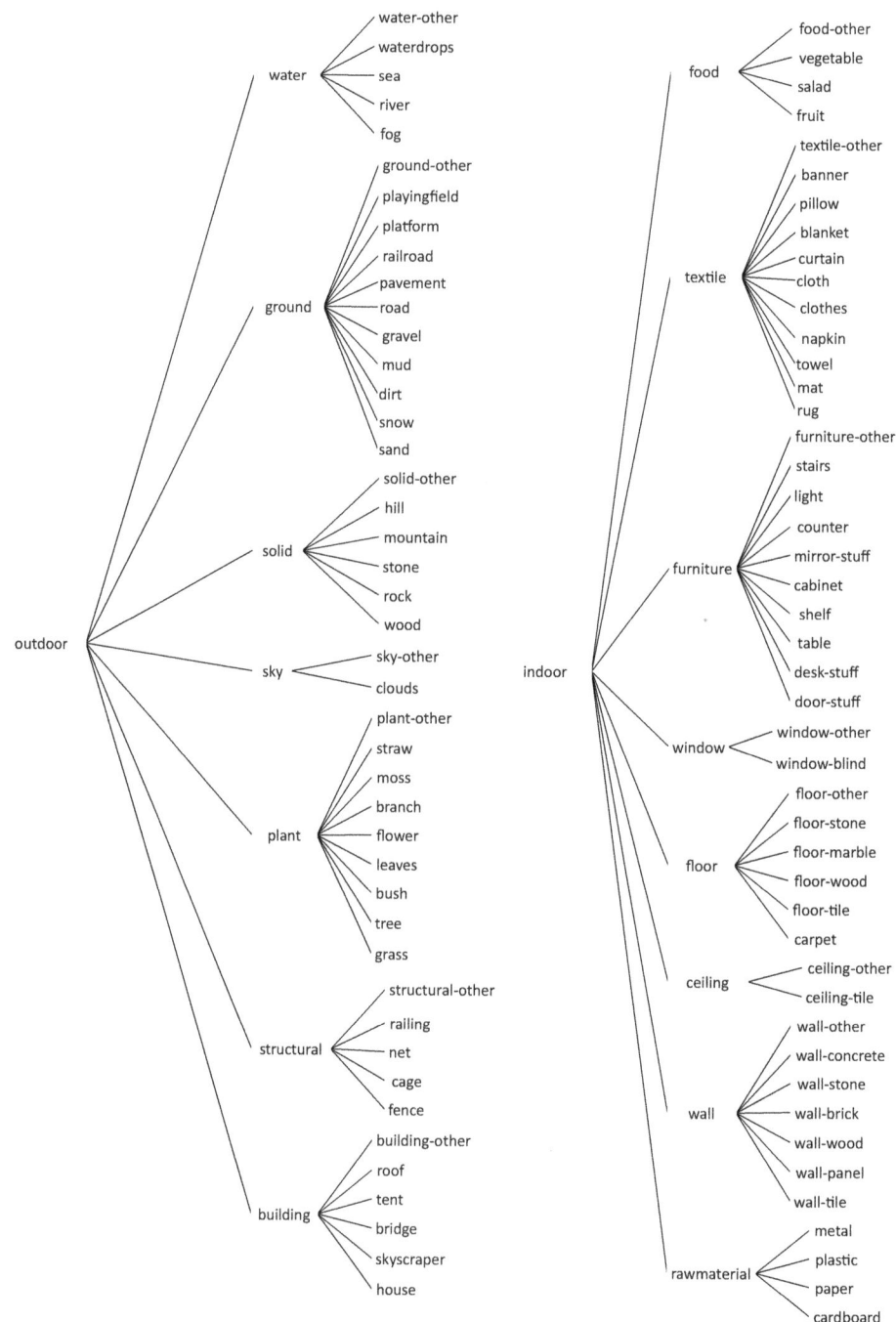

**Figure 2.** COCO-Stuff label hierarchy.

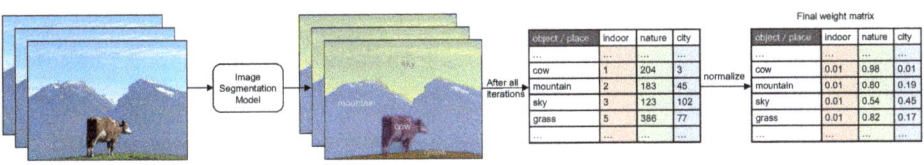

**Figure 3.** Construction of weight matrix W with a sample image.

---

**Algorithm 1** Weighting Matrix W

**Require:** $F$: pre-trained semantic segmentation model
1: **Inputs:**
   scene-labeled images $\{(x_1, y_1), (x_2, y_2), ..., (x_n, y_n)\}$
2: **Initialize:**
   $W[i][j] \leftarrow 0, i = 1, ..., o, j = 1, ..., p$
   $o$: # of object classes, $p$: # of scene classes

3: **for** t = 1 to n **do**
4:   $O_t \leftarrow F(x_t)$
5:   **for** $o \in O_t$ **do**
6:     $W[o][y_t] \leftarrow W[o][y_t] + 1$
7:   **end for**
8: **end for**
9: **for** i = 1 to o **do**
10:   **for** j = 1 to p **do**
11:     $W[i][j] \leftarrow W[i][j]/(W[i][1] + ... + W[i][p])$
12:   **end for**
13: **end for**

---

**Algorithm 2** Inference Process

**Require:** $F$: pre-trained semantic segmentation model and $W[i][j]$: weight matrix
1: **Inputs:**
   Test images $\{x_1, x_2, ..., x_n\}$
2: **Initialize:**
   $V[i] \leftarrow 0, i = 1, ..., p$
   $o$: # of object classes, $p$: # of scene classes

3: **for** t = 1 to n **do**
4:   $O_t \leftarrow F(x_t)$
5:   **for** $o \in O_t$ **do**
6:     $bias \leftarrow 1$
7:     **if** o is in Background **then**
8:       $bias \leftarrow 1.4$
9:     **end if**
10:    $V[t] \leftarrow V[t] + W[o] \times bias$
11:  **end for**
12: **end for**
13: $\hat{y} \leftarrow argmax(V[t])$

## 3. Experimental Results and Discussion

In this section, we will show the results of our classification model and well-renowned classification methods using CNNs. We implemented the proposed scheme by using PyTorch deep learning framework, and used single GPU for training. We trained DeepLab v2 model with COCO-stuff dataset which contains total 164k images. As we mentioned it in Section 2.2, the reason why we use COCO-stuff dataset is because its object classes are divided into indoor and outdoor categories.

In order to compare our method with CNN based classification models, we first built a custom test dataset that consists of 500 images as shown in Table 1. We extracted them from Korean movies, and the images were manually labeled into three scene classes (i.e., 0: indoor, 1: nature, 2: city) under certain criteria. We set some criteria according to the logic that humans more focus on the background objects rather than foreground objects. The criteria are described in Figure 4.

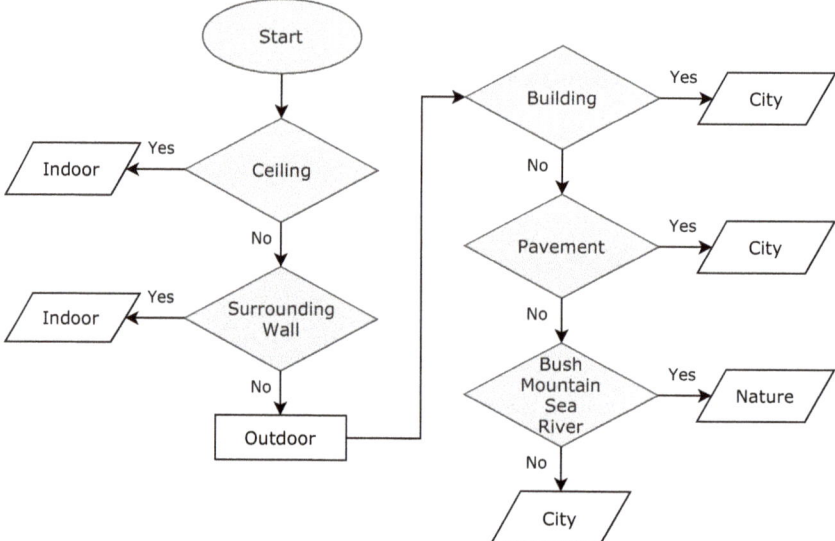

**Figure 4.** The criteria of labeling custom dataset. We first label the scene as either indoor or outdoor according to the existence of ceiling and surrounding wall. Then city images are labeled by buildings and pavements which are usually existing in urban area. Lastly, the scene is labeled as nature when there is nature stuff such as bush, mountain, sea, or river.

The sample test images are shown in Figure 5. We used COCO-stuff validation 2017 dataset for training the CNN models, which were also used for building the weight matrix. The test images were used for measuring accuracy of the classification. We experimented various CNNs, such as VGG [15], Inception [16,17], ResNet [18], ResNeXt [19], Wide-ResNet [20], DenseNet [21], and MnasNet [22] as shown in Table 2.

To be more specific, we trained each model by using transfer learning scheme [23], especially Feature Extraction [24,25]. The basic concept of feature extraction is represented in Figure 6. Typical CNNs have convolutional layers for extracting good features and fully connected layers to classify the feature. Feature extraction technique which trains only fully connected layers is used when there are insufficient data for training.

PyTorch Deep Learning Framework was used again to implement all structures for the experiment. The results of accuracy were computed after 200 iterations of training. We trained each model using cross entropy loss, and Adam optimizer with a batch size 64, learning rate 0.001. Learning rate was multiplied by 0.1 every 7 iteration.

In Table 2, we can observe the proposed scheme outperforms the existing well-known CNNs which were trained using transfer learning. In terms of the accuracy, the proposed scheme achieved 90.8% of the accuracy.

**Figure 5.** Samples of custom test dataset; each row represents each scene class.

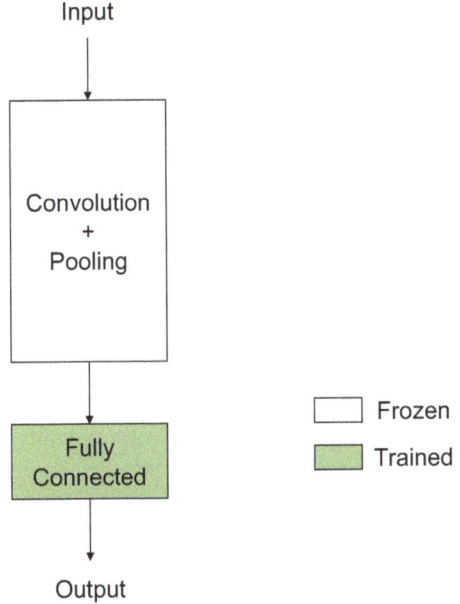

**Figure 6.** Basic Concept of Feature Extraction. The white box represents convolution and pooling layers in convolutional neural networks (CNNs), which are used for extracting features of an image. This part is not trained during transfer learning process. The green box represents fully connected layer in CNNs, which operates as classifier. This part is trained during whole training process.

**Table 1.** The number of images for each class in our custom dataset.

| Indoor | Nature | City | Total |
|---|---|---|---|
| 305 | 106 | 89 | 500 |

Table 2. Performance comparison with the existing CNNs on indoor/nature/city classes.

| Models | Accuracy (%) |
|---|---|
| VGG-19 [15] | 79.6 |
| VGG-19 (BN) [15] | 88.0 |
| GoogLeNet [16] | 84.6 |
| Inception-v3 [17] | 86.2 |
| ResNet-101 [18] | 84.0 |
| ResNeXt-101 [19] | 88.0 |
| Wide-ResNet [20] | 85.2 |
| DenseNet-121 [21] | 84.0 |
| DenseNet-161 [21] | 83.6 |
| DenseNet-201 [21] | 85.2 |
| MnasNet (0.5) [22] | 76.6 |
| MnasNet (1.0) [22] | 79.8 |
| Proposed Method | 90.8 |

When compared with VGG-19 (BN) [15] and ResNeXt-101 [19], the proposed method could improve 2.8% of the accuracy. Also, our scheme improved the performance over 13% comparing to MnasNet (0.5) [16]. VGG-19 was tested with batch normalization (BN) and without BN. The float values with MnasNet is the depth multiplier in Reference [22]. From this result, we can see that the proposed scheme is very reliable and better to classify the scene.

Figure 7 represents the graph of the experiment on determining optimal bias value in terms of test accuracy. It shows that the highest test accuracy when the bias is 1.4. This value is used in the inference process when multiplying weights of background objects.

In addition, we measured test accuracy on COCO-Stuff test dataset. We adopted first 100 images of the test images and labeled the images according to the criteria in Figure 4. We used same parameters as the previous experiment while training CNNs and building weight matrix. The result is shown in Table 3. The result shows that the proposed method outperforms the conventional CNNs and also indicates that it achieves better performance when test images are taken in the same domain with train images.

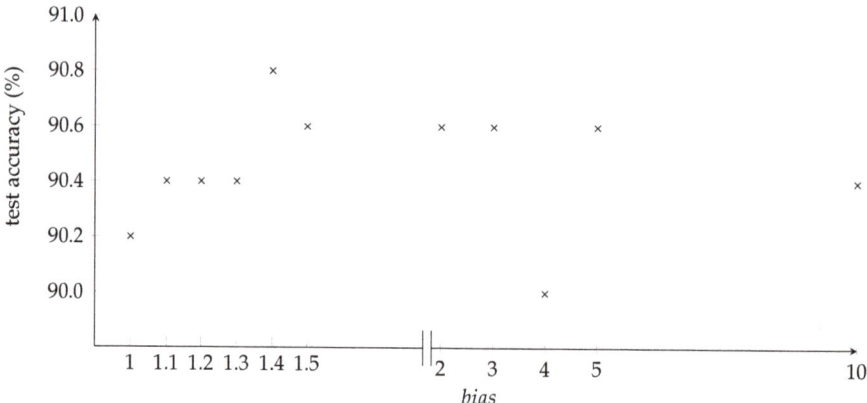

**Figure 7.** Experiment on different bias values for multiplying weights of background objects when inferencing.

Table 3. Performance comparison with the existing CNNs using COCO-Stuff test set.

| Models | Accuracy (%) |
| --- | --- |
| VGG-19 [15] | 91.0 |
| VGG-19 (BN) [15] | 85.0 |
| GoogLeNet [16] | 88.0 |
| Inception-v3 [17] | 86.0 |
| ResNet-101 [18] | 89.0 |
| ResNeXt-101 [19] | 89.0 |
| Wide-ResNet [20] | 88.0 |
| DenseNet-121 [21] | 82.0 |
| DenseNet-161 [21] | 86.0 |
| DenseNet-201 [21] | 90.0 |
| MnasNet (0.5) [22] | 87.0 |
| MnasNet (1.0) [22] | 85.0 |
| Proposed Method | 92.0 |

Lastly, we experimented on indoor classes and Table 4 shows the results. We used the subset of MITPlaces dataset [26]. It contains categories of 'library', 'bedroom' and 'kitchen' and 900 train images and 100 test images per class. As previous experiment, train images are used for building weight matrix and test images are used for measuring test accuracy in the proposed method. Classifying indoor categories must be treated differently from classifying indoor and outdoor. Since all indoor scenes have ceilings and walls, the bias value in Algorithm 2 must be given not by background objects, but by foreground objects. In this experiment, we defined foreground objects as furniture categories in Figure 2 and determined the value to be 3 empirically. Although the results shows that 7 of CNNs outperforms our method by less than 4.6%, it shows that our method can be extendable to indoor categorization problem.

Table 4. Performance comparison with the existing CNNs on library/bedroom/kitchen classes.

| Models | Accuracy (%) |
| --- | --- |
| VGG-19 [15] | 92.0 |
| VGG-19 (BN) [15] | 94.0 |
| GoogLeNet [16] | 88.0 |
| Inception-v3 [17] | 93.7 |
| ResNet-101 [18] | 91.0 |
| ResNeXt-101 [19] | 95.3 |
| Wide-ResNet [20] | 88.0 |
| DenseNet-121 [21] | 93.0 |
| DenseNet-161 [21] | 94.3 |
| DenseNet-201 [21] | 90.3 |
| MnasNet (0.5) [22] | 89.7 |
| MnasNet (1.0) [22] | 90.0 |
| Proposed Method | 90.7 |

CNNs which showed the best performance tested with our custom dataset are VGG-19 (BN) and ResNeXt-101. They both showed test accuracy of 88% and the proposed method showed 90.8%. Table 5 represents the performance of three models on each of three scene classes. VGG-19 (BN) predicted all images perfectly in indoor class and the proposed method is following. In nature class, the proposed method showed the best accuracy. When it comes to city class, ResNext-101 showed the best results. From this result, we can see that the proposed method is reliable for scene classification. Source code is available at https://github.com/woonhahaha/place-classification.

Table 5. Comparison between best CNNs and the proposed method. The correct number of images per class in our custom dataset.

|  | Indoor | Nature | City | Total |
|---|---|---|---|---|
| VGG-19 (BN) | 305 | 79 | 56 | 440 |
| ResNeXt-101 | 282 | 88 | 70 | 440 |
| Proposed Method | 304 | 89 | 61 | 454 |

## 4. Conclusions

In this work, we have proposed an efficient scene and place classification scheme using background objects and the designed weighting matrix. We designed this weighting matrix based on the open dataset which is widely used in the scene and object classifications. Also, we evaluated the proposed classification scheme which was based on semantic segmentation comparing to the existing image classification methods such as VGG [15], Inception [16,17], ResNet [18], ResNeXt [19], Wide-ResNet [20], DenseNet [21], and MnasNet [22]. The proposed scheme is the first approach of object-based classification that can classify outdoor categories as well. We have built a custom dataset of 500 images for testing which can help researchers who are dealing with scene classification. We crawled frames from Korean movies and labeled each image manually. The images were labeled as three major scene categories (i.e., indoor, nature, and city).

Experimental results showed that the proposed classification model outperformed several well-known CNNs mainly used for image classification. In the experiment, our model achieved 90.8% of verification accuracy and improved over 2.8% of the accuracy when comparing to the existing CNNs.

The Future work is to widen the scene classes to classify not just indoor (library, bedroom, kit) and outdoor (city, nature), but also more subcategories. It would be helpful for searching in videos with such semantic information.

**Author Contributions:** Conceptualization, B.-G.K.; methodology and formal analysis, W.-H.Y.; validation, Y.-J.C. and Y.-J.H.; writing–original draft preparation, W.-H.Y. and Y.-J.C.; writing–review and editing, B.-G.K.; supervision, B.-G.K. All authors have read and agreed to the published version of the manuscript.

**Funding:** This research project was supported by Ministry of Culture, Sports and Tourism (MCST) and from Korea Copyright Commission in 2020.

**Acknowledgments:** Authors thank for all reviewers who helped to improve this manuscript.

**Conflicts of Interest:** The authors declare no conflict of interest.

## Abbreviations

The following abbreviations are used in this manuscript:

CNN  Convolutional Neural Network
BN   Batch Normalization

## References

1. Lowe, D.G. Distinctive image features from scale-invariant keypoints. *Int. J. Comput. Vis.* **2004**, *60*, 91–110. [CrossRef]
2. Bay, H.; Ess, A.; Tuytelaars, T.; Van Gool, L. Speeded-up robust features (SURF). *Comput. Vis. Image Underst.* **2008**, *110*, 346–359. [CrossRef]
3. Philbin, J.; Chum, O.; Isard, M.; Sivic, J.; Zisserman, A. Object retrieval with large vocabularies and fast spatial matching. In Proceedings of the 2007 IEEE Conference on Computer Vision and Pattern Recognition, Minneapolis, MN, USA, 17–22 June 2007; pp. 1–8.

4. Krizhevsky, A.; Sutskever, I.; Hinton, G.E. Imagenet classification with deep convolutional neural networks. In Proceedings of the Advances in Neural Information Processing Systems, Lake Tahoe, NV, USA, 3–6 December 2012; pp. 1097–1105.
5. Chowanda, A.; Sutoyo, R. Deep Learning for Visual Indonesian Place Classification with Convolutional Neural Networks. *Procedia Comput. Sci.* **2019**, *157*, 436–443. [CrossRef]
6. Raja, R.; Roomi, S.M.M.; Dharmalakshmi, D.; Rohini, S. Classification of indoor/outdoor scene. In Proceedings of the 2013 IEEE International Conference on Computational Intelligence and Computing Research, Enathi, India, 26–28 December 2013; pp. 1–4.
7. Viswanathan, P.; Southey, T.; Little, J.; Mackworth, A. Place classification using visual object categorization and global information. In Proceedings of the 2011 Canadian Conference on Computer and Robot Vision, St. Johns, NL, Canada, 25–27 May 2011; pp. 1–7.
8. Liao, Y.; Kodagoda, S.; Wang, Y.; Shi, L.; Liu, Y. Understand scene categories by objects: A semantic regularized scene classifier using convolutional neural networks. In Proceedings of the 2016 IEEE International Conference on Robotics and Automation (ICRA), Stockholm, Sweden, 16–21 May 2016; pp. 2318–2325.
9. Zheng, X.; Yuan, Y.; Lu, X. A Deep Scene Representation for Aerial Scene Classification. *IEEE Trans. Geosci. Remote Sens.* **2019**, *57*, 4799–4809. [CrossRef]
10. Liu, X.; Zhou, Y.; Zhao, J.; Yao, R.; Liu, B.; Zheng, Y. Siamese convolutional neural networks for remote sensing scene classification. *IEEE Geosci. Remote Sens. Lett.* **2019**, *16*, 1200–1204. [CrossRef]
11. Pires de Lima, R.; Marfurt, K. Convolutional Neural Network for Remote-Sensing Scene Classification: Transfer Learning Analysis. *Remote Sens.* **2020**, *12*, 86. [CrossRef]
12. Caesar, H.; Uijlings, J.; Ferrari, V. Coco-stuff: Thing and stuff classes in context. In Proceedings of the IEEE Conference on Computer Vision and Pattern Recognition, Salt Lake City, UT, USA, 18–23 June 2018; pp. 1209–1218.
13. Lin, T.Y.; Maire, M.; Belongie, S.; Hays, J.; Perona, P.; Ramanan, D.; Dollár, P.; Zitnick, C.L. Microsoft coco: Common objects in context. In Proceedings of the European Conference on Computer Vision, Zurich, Switzerland, 6–12 September 2014; pp. 740–755.
14. Chen, L.C.; Papandreou, G.; Kokkinos, I.; Murphy, K.; Yuille, A.L. Deeplab: Semantic image segmentation with deep convolutional nets, atrous convolution, and fully connected crfs. *IEEE Trans. Pattern Anal. Mach. Intell.* **2017**, *40*, 834–848. [CrossRef] [PubMed]
15. Simonyan, K.; Zisserman, A. Very deep convolutional networks for large-scale image recognition. *arXiv* **2014**, arXiv:1409.1556.
16. Szegedy, C.; Liu, W.; Jia, Y.; Sermanet, P.; Reed, S.; Anguelov, D.; Erhan, D.; Vanhoucke, V.; Rabinovich, A. Going deeper with convolutions. In Proceedings of the IEEE Conference on Computer Vision and Pattern Recognition, Boston, MA, USA, 7–12 June 2015; pp. 1–9.
17. Szegedy, C.; Vanhoucke, V.; Ioffe, S.; Shlens, J.; Wojna, Z. Rethinking the inception architecture for computer vision. In Proceedings of the IEEE Conference on Computer Vision and Pattern Recognition, Las Vegas, NV, USA, 27–30 June 2016; pp. 2818–2826.
18. He, K.; Zhang, X.; Ren, S.; Sun, J. Deep residual learning for image recognition. In Proceedings of the IEEE Conference on Computer Vision and Pattern Recognition, Las Vegas, NV, USA, 27–30 June 2016; pp. 770–778.
19. Xie, S.; Girshick, R.; Dollár, P.; Tu, Z.; He, K. Aggregated residual transformations for deep neural networks. In Proceedings of the IEEE Conference on Computer Vision and Pattern Recognition, Honolulu, HI, USA, 21–26 July 2017; pp. 1492–1500.
20. Zagoruyko, S.; Komodakis, N. Wide residual networks. *arXiv* **2016**, arXiv:1605.07146.
21. Huang, G.; Liu, Z.; Van Der Maaten, L.; Weinberger, K.Q. Densely connected convolutional networks. In Proceedings of the IEEE Conference on Computer Vision and Pattern Recognition, Honolulu, HI, USA, 21–26 July 2017; pp. 4700–4708.
22. Tan, M.; Chen, B.; Pang, R.; Vasudevan, V.; Sandler, M.; Howard, A.; Le, Q.V. Mnasnet: Platform-aware neural architecture search for mobile. In Proceedings of the IEEE Conference on Computer Vision and Pattern Recognition, Long Beach, CA, USA, 15–21 June 2019; pp. 2820–2828.
23. Yosinski, J.; Clune, J.; Bengio, Y.; Lipson, H. How transferable are features in deep neural networks? In Proceedings of the Advances in Neural Information Processing Systems, Montreal, QC, Canada, 8–13 December 2014; pp. 3320–3328.

24. Sharif Razavian, A.; Azizpour, H.; Sullivan, J.; Carlsson, S. CNN features off-the-shelf: An astounding baseline for recognition. In Proceedings of the IEEE Conference on Computer Vision and Pattern Recognition Workshops, Columbus, OH, USA, 23–28 June 2014; pp. 806–813.
25. Donahue, J.; Jia, Y.; Vinyals, O.; Hoffman, J.; Zhang, N.; Tzeng, E.; Darrell, T. Decaf: A deep convolutional activation feature for generic visual recognition. In Proceedings of the International Conference on Machine Learning, Beijing, China, 21–26 June 2014; pp. 647–655.
26. Zhou, B.; Lapedriza, A.; Xiao, J.; Torralba, A.; Oliva, A. Learning deep features for scene recognition using places database. In Proceedings of the Advances in Neural Information Processing Systems, Montreal, QC, Canada, 8–13 December 2014; pp. 487–495.

**Publisher's Note:** MDPI stays neutral with regard to jurisdictional claims in published maps and institutional affiliations.

© 2020 by the authors. Licensee MDPI, Basel, Switzerland. This article is an open access article distributed under the terms and conditions of the Creative Commons Attribution (CC BY) license (http://creativecommons.org/licenses/by/4.0/).

Article

# Human Height Estimation by Color Deep Learning and Depth 3D Conversion

Dong-seok Lee [1], Jong-soo Kim [2], Seok Chan Jeong [3] and Soon-kak Kwon [1],*

[1] Department of Computer Software Engineering, Dong-eui University, Busan 47340, Korea; ulsan333@gmail.com
[2] Software Convergence Center, Dong-eui University, Busan 47340, Korea; avantas@naver.com
[3] Department of e-Business, Convergence of IT Devices Institute, AI Grand ICT Research Center, Dong-eui University, Busan 47340, Korea; scjeong@deu.ac.kr
* Correspondence: skkwon@deu.ac.kr; Tel.: +82-51-897-1727

Received: 7 July 2020; Accepted: 9 August 2020; Published: 10 August 2020

**Abstract:** In this study, an estimation method for human height is proposed using color and depth information. Color images are used for deep learning by mask R-CNN to detect a human body and a human head separately. If color images are not available for extracting the human body region due to low light environment, then the human body region is extracted by comparing between current frame in depth video and a pre-stored background depth image. The topmost point of the human head region is extracted as the top of the head and the bottommost point of the human body region as the bottom of the foot. The depth value of the head top-point is corrected to a pixel value that has high similarity to a neighboring pixel. The position of the body bottom-point is corrected by calculating a depth gradient between vertically adjacent pixels. Two head-top and foot-bottom points are converted into 3D real-world coordinates using depth information. Two real-world coordinates estimate human height by measuring a Euclidean distance. Estimation errors for human height are corrected as the average of accumulated heights. In experiment results, we achieve that the estimated errors of human height with a standing state are 0.7% and 2.2% when the human body region is extracted by mask R-CNN and the background depth image, respectively.

**Keywords:** human-height estimation; depth video; depth 3D conversion; artificial intelligence; convolutional neural networks

---

## 1. Introduction

The physical measurements of a person such as human height, body width and stride length are important bases for identifying a person from video. For example, the height of the person captured by a surveillance video is important evidence for identifying a suspect. Physical quantities are also used as important information for continuously tracking a specific person in video surveillance system consisting of multiple cameras [1]. A specific behavior such as falling down can be recognized by detecting changes in human height. Various studies have been conducted to estimate human height from color video. Human height is estimated by obtaining 3D information of the human body from color video [2–6]. Both the position and the pose of the camera are required in order to obtain 3D information of human body. Human height can also be estimated by calculating the ratio of the length between human body and a reference object whose length is already known [7–12]. The estimation methods of human height based on color video have a disadvantage in that the camera parameters or information about a reference object are required.

Depth video stores depth values, meaning the distances between subjects and the camera. The pixels of depth video are converted to 3D coordinates by the depth values. Object detection [13–15] and behavior recognition [16–18] by depth video are possible by extracting the 3D features of the objects.

Recently, smartphones recognize a human face through an equipped TOF sensor for recognizing the identity of a person. Object lengths can be also measured from depth video without the additional information, so the problems of the human-height estimation based on color video can be solved by using depth video.

The field of artificial intelligence has made significant progress by researching neural network structures which consist of multilayers. In particular, convolutional neural network (CNN) [19] respectably improves object detection that categorizes the object and detects the boundary boxes and pixels of the objects [20–26].

In this study, a human-height estimation method is proposed using depth and color information. The human-height estimation is improved by extracting a human body and a human head from color information and by measuring human height from depth information. The human body and the human head of current frame in color video are extracted through mask R-CNN [26]. If color images are not available due to a low light environment, then the human body region is extracted by comparing between current frame in depth video and a pre-stored background depth image. The topmost point of the human head region is extracted as a head-top and bottommost point of the human body region as a foot-bottom. Two top head and foot-bottom points are converted to 3D real-world coordinates by these image coordinates and depth pixel values. Human height is estimated by calculating a Euclidean distance between two real-world coordinates.

The proposed method improves the human-height estimation by using both color and depth information and by applying mask R-CNN which is an art-of-state algorithm for object detection. In addition, the proposed method removes the need for the camera parameters or the length of other object in the human-height estimation using depth information.

This study is organized as follows: In Section 2, the related works for object detection by CNN and for the human-height estimation based on color or depth video are described. In Section 3, the human-height estimation by depth and color videos is proposed. The experimental results of the proposed method are presented in Section 4. Finally, a conclusion for this study is described in Section 5.

## 2. Related Works

### 2.1. Object Detection from Color Information by Convolutional Neural Network

Object detection problems in color image can generally be classified into four categories: classification, localization, detection and object segmentation. First, the classification determines an object category for single object in the image. Second, the localization finds the boundary box for single object in the image. Third, the detection finds the boundary boxes and determines object categories for multiple objects. Finally, the object segmentation finds pixels where each object is. CNN can solve whole categories of object detection problems. CNN replaces the weights of the neural networks with kernels which are rectangular filters. Generally, object detection through CNN are classified as 1-stage and 2-stage methods [27]. The 1-stage method performs both the location and the classification at once. The 2-stage method performs the classification after the location. The 1-stage method is faster than the 2-stage method but is less accurate. R-CNN [20] is a first proposed method for the detection through CNN. R-CNN applies a selective search algorithm to find the boundary box with a high probability where an object exists. The selective search algorithm is the method of constructing the boundary box by connecting adjacent pixels with similar texture, color and intensity. The object is classified through SVM (support vector machine). The feature map of the boundary boxes is extracted through AlexNet. R-CNN has disadvantages that the object detection is seriously slow and SVM should be trained separately from CNN. Fast R-CNN [21] has higher performance than R-CNN. Fast R-CNN applies a RoIPool algorithm and introduces a softmax classifier instead of SVM, so the feature map extraction and the classification are integrated into one neural network. faster R-CNN [22] replaces the selective search algorithm into a region proposal network (RPN) so whole processes of object detection are performed in one CNN. YOLO [23,24] is the 1-stage method for object detection. YOLO defines the object detection problem as a regression problem. YOLO divides an input image into the grid cells of a certain size. The boundary boxes and the reliability of the object are predicted for each cell at same time. YOLO detects the objects more quickly than the 2-stage methods but is less accurate. SSD [25] allows the various sizes of the grid cells in order to increase the accuracy of object detection. mask R-CNN [26] is proposed for the object segmentation unlike other R-CNNs.

*2.2. Object Length Measurement from Color or Depth Information*

Length estimation methods based on color video are classified into length estimations by camera parameters [2–6], by vanishing points [7–12], by prior statistical knowledge [28,29], by gaits [30,31] and by neural networks [32,33]. The length estimation methods by the camera parameters generate an image projection model into an color image by the focal length, the height and the poses of a camera. The object length is estimated by converting the 2D coordinates in the pixels of the image into 3D coordinates through the projection model. The length estimation methods by the camera parameters have a disadvantage that the accurate camera parameters should be provided in advance. In order to overcome this disadvantage, Liu [2] introduces an estimation method for the camera parameters using prior knowledge about the distribution of relative human heights. Cho [6] proposes an estimation method for the camera parameters by tracking the poses of human body from a sequence of frames. The length estimation methods by the vanishing points use a principle that several parallel lines in 3D space meet at one point in a 2D image. The vanishing point is found by detecting the straight lines in the image. The length ratio between two objects can be calculated using the vanishing points. If the length of one object is known in advance, then the length of another object can be calculated by the length ratio. Criminisi [7] introduces a length estimation method by the given vanishing points of the ground. Fernanda [8] proposes the detection method of the vanishing points by clustering the straight lines iteratively without camera calibration. Jung [9] proposes the method of detecting the vanishing points for color videos captured by multiple cameras. Viswanath [10] proposes an error model for the human-height estimation by the vanishing points. The error of the human-height estimation is corrected by the error model. Rother [11] detects the vanishing points by tracking specific object such as traffic signs from a sequence of frames. Pribyl [12] estimates the object length by detecting the specific objects. Human height can be also estimated by the prior statistical knowledge of human anthropometry [28,29] or by the gaits [30,31]. In recent years, estimation studies in various fields achieve great success by applying neural networks. The neural networks are also applied to the human-height estimation. Gunel [32] proposes a neural network for predicting a relationship between each proportion of human joints and human height. Sayed [33] estimates human height by CNN using a length ratio between a human body width and a human head size.

Since depth video has distance information from the depth camera, the distance between two points in a depth image can be measured without the camera parameters or the vanishing points. Many studies [34–36] extract a skeleton, which is the connection structure of human body parts, from the depth image for the human-height estimation. However, the human body region extraction is some inaccurate due to noises in depth video.

## 3. Proposed Method

In this study, we propose a human-height estimation method using color and depth information. It is assumed that a depth camera is fixed in a certain position. Color and depth videos are captured by the depth camera. Then, a human body and a human head are extracted from current frame in color or depth video. A head-top and a foot-bottom are found in the human head and the human body, respectively. Two head-top and foot-bottom points are converted into 3D real-world coordinates by the corresponding pixel values of the frame in depth video. Human height is estimated by calculating a distance between two real-world coordinates. The flow of the proposed methods is shown in Figure 1.

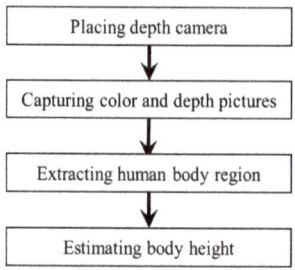

**Figure 1.** Flow of proposed method.

## 3.1. Human Body Region Extraction

It is important to accurately detect the human body region for estimating human height precisely. Most frames in depth video have many noises, so it is difficult to extract the accurate human body region from the depth frame. In contrast, color video allows to detect the human body region accurately by CNN. In the proposed method, mask R-CNN [26] is applied to extract the accurate human body region from current frame in color video. Then, the human body region is mapped to current frame in depth video. If color video is not available for extracting the human body region, then the human body region is extracted from current frame in depth video directly. In this case, the human body region is extracted by comparing with current depth frame with a pre-captured background depth image. Figure 2 shows the flow of extracting the human body region in the proposed method.

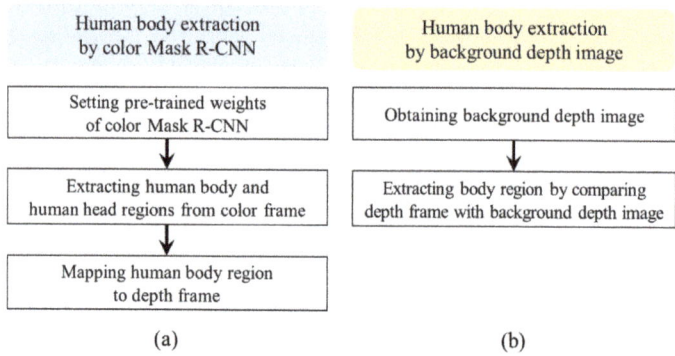

**Figure 2.** Flow of human body region extraction. (**a**) From color information; (**b**) from depth information.

### 3.1.1. Human Body Region Extraction Using Color Frames

Mask R-CNN [26] consists of three parts: a feature pyramid network (FPN) [37], a residual network (ResNet) [38,39] and a RPN. A FPN detects the categories and the boundary boxes of objects in color video. ResNet extracts an additional feature map from each boundary box. Figure 3 shows the processes of extracting the human body and the human head using mask R-CNN.

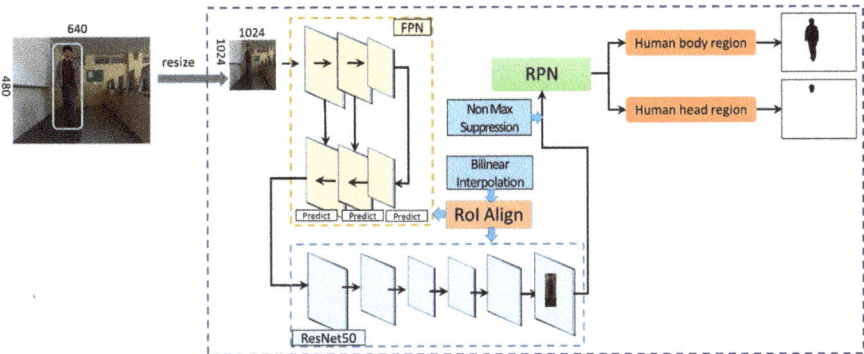

**Figure 3.** Processes of extracting human body and human head regions using mask R-CNN.

FPN condenses the scale of the input frame through bottom–up layers and expands the scale through top–down layers. Various scale objects can be detected through FPN. ResNet introduces a skip connection algorithm that the output value of each layer feeds into the next layer and directly into the layers about more than 2 hops away. The skip connection algorithm reduces the amount of values to be learned for weights of layers, so the learning efficiency of ResNet is improved. The feature map is extracted from the frame through FPN and ResNet. RPN extracts the boundary boxes and the masks which are object area in rectangle and in pixels, respectively. Comparing with faster R-CNN which applies RoIPool, mask R-CNN extends RPN to extract not only the boundary box, but also the masks. RoIPool rounds off the coordinates of the boundary box to integer. In contrast, RoIAlign allows the floating coordinates. Therefore, the detection of the object areas of mask R-CNN is more precisely than of faster R-CNN. Figure 4 shows an example for calculating the coordinates of the regions of interest (RoIs) for detecting the boundary box and the masks by RoIPool and RoIAlign. Non-max-suppression (NMS) removes overlapping areas between the boundary boxes. The size of the overlapping area is calculated for each boundary box. Two boundary boxes are merged if the size of the overlapping area is more than 70%.

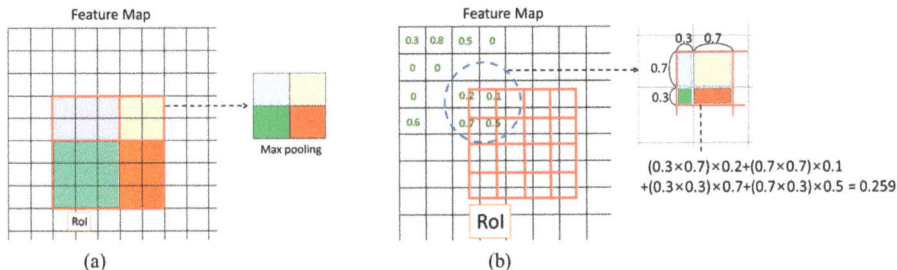

**Figure 4.** Example of region of interest (RoI) coordinate calculation by RoIPool and RoIAlign. (a) RoIPool; (b) RoIAlign.

Table 1 shows the performances of mask R-CNN when various types of backbones are applied to mask R-CNN. When X-101-FPN is applied as the backbone of mask R-CNN, the average precision of the boundary boxes (Box AP), which is a metric for detecting the boundary box, is the highest. However, the times for a train and a detection are slowest. In consideration of a tradeoff between the accuracy and the time for the detection, the proposed method applies ResNet-50 FPN which consists of 50 CNNs as the backbone.

Table 1. Performance of backbones for mask R-CNN. Bold represents the best performance.

| Backbone | Learn Rate Schedules | Train Time (s/iter) | Inference Time (s/im) | Box AP (%) |
| --- | --- | --- | --- | --- |
| R50-C4 | 1x | 0.584 | 0.110 | 35.7 |
| R50-DC5 | 1x | 0.471 | 0.076 | 37.3 |
| ResNet-50 FPN | 1x | **0.261** | **0.043** | 37.9 |
| R50-C4 | 3x | 0.575 | 0.111 | 38.4 |
| R50-DC5 | 3x | 0.470 | 0.076 | 39.0 |
| ResNet-50 FPN | 3x | **0.261** | **0.043** | 40.2 |
| R101-C4 | 3x | 0.652 | 0.145 | 41.1 |
| R101-DC5 | 3x | 0.545 | 0.092 | 40.6 |
| ResNet-101 FPN | 3x | 0.340 | 0.056 | 42.0 |
| X-101-FPN | 3x | 0.690 | 0.103 | **43.0** |

The human body and the human head are detected by mask R-CNN. Mask R-CNN is trained using 3000 images of COCO dataset [40] with information about the human body and the human head. In the training mask R-CNN, a learn rate and epochs are set to 0.001 and 1000, respectively. A threshold for detection of the human body and the human head is set to 0.7. If a detection accuracy for RoI is more than the threshold, then corresponding RoI is detected as the human body or the human head. The process of extracting the human body and human head regions through mask R-CNN is as follows.

1. Resizing a color image to a certain size
2. Extracting a feature map through FPN and ResNet50
3. Extracting RoI boundary boxes from feature map by RoIAlign
4. Boundary box regression and classification for boundary boxes through RPN
5. Generating boundary box candidates by projecting the boundary box regression results onto the color frame
6. Detecting a boundary box for each object by non-max-suppression
7. Adjusting the boundary box area through RoIAlign
8. Finding pixels in boundary boxes to obtain a mask for each boundary box

3.1.2. Human Body Region Extraction Using Depth Frames

If the depth camera is fixed in a certain position, then the pixels of the human body region in the depth frame have different values from the depth pixels of a background. Therefore, the body region can be extracted by comparing depth pixels between the depth frame and the background depth image which has depth information about background. In order to extract the human body region accurately, the background depth image should be generated from several depth frames that capture the background because the depth video includes temporary noises. A depth value at the certain position of the background depth image is determined as a minimum value among pixels in the corresponding position of the depth frames capturing the background.

The human body region is extracted by comparing the pixels between the depth frame and the background depth image. A binarization image B is generated for the human body region extraction as follows:

$$B(x,y) = \begin{cases} 1, & d_b(x,y) - d(x,y) > T_b \\ 0, & \text{otherwise} \end{cases}, \quad (1)$$

where $d_b(x,y)$ and $d(x,y)$ are the depth pixel values of the background depth image and the depth frame at position of $(x,y)$, respectively and $T_b$ is a threshold for the binarization.

3.2. Extraction of Head Top and Foot Bottom Points

The topmost point of the human head region is extracted as the head-top, $(x_h, y_h)$ and the bottommost point of the human body region as the foot-bottom, $(x_f, y_f)$. If horizontal continuous pixels exist as shown in Figure 5, then the head-top or the foot-bottom is extracted as a center point among these pixels. If human stands with legs apart as shown in Figure 6, then two separate regions may be found from the bottommost of the human body region. In this case, the center points of two regions are the candidates of the foot-bottom. One candidate which has a depth value closer to the depth pixel value of the head-top point is selected as the foot-bottom point.

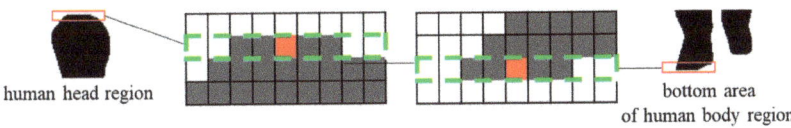

**Figure 5.** Extracting the head-top and foot-bottom points.

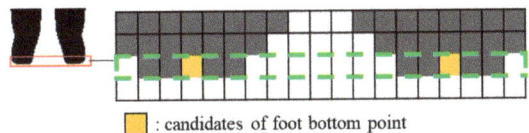

**Figure 6.** Extracting the foot-bottom point in case of apart human legs.

### 3.3. Human Height Estimation

Human height is estimated by measuring a length in the 3D real world between the head-top and foot-bottom points. In order to measure the length on the real world, 2D image coordinates of two head-top and the foot-bottom are converted into 3D real-world coordinates by applying a pinhole camera model [41] as follows:

$$\begin{aligned} X &= \frac{(x-W/2)}{f} d(x,y) \\ Y &= \frac{(y-H/2)}{f} d(x,y) \\ Z &= d(x,y), \end{aligned} \quad (2)$$

where $X, Y, Z$ are the real-world coordinates, $f$ is a focal length of the depth camera, which means the parameter of the depth camera, and $W$ and $H$ are the horizontal and vertical resolutions of the depth image, respectively. In (2), the origin of the image coordinate system is the top–left of the image, but the origin of 3D camera coordinate system is the camera center. In order to compensate for the difference in the position of the origin between two coordinate systems, the coordinates of the image center are subtracted from the image coordinates. the real-world coordinates of the head-top $(X_h, Y_h, Z_h)$ and of the foot-bottom $(X_f, Y_f, Z_f)$ are calculated by substituting the real-world coordinates and the depth values of the head-top and the foot-bottom for (2), respectively, as follows:

$$\begin{aligned} X_h &= \frac{(x_h - W/2)}{f} d_h \\ Y_h &= \frac{(y_h - H/2)}{f} d_h \\ Z_h &= d_h, \end{aligned} \quad (3)$$

$$\begin{aligned} X_f &= \frac{(x_f - W/2)}{f} d_f \\ Y_f &= \frac{(y_f - H/2)}{f} d_f \\ Z_f &= d_f, \end{aligned} \quad (4)$$

where $d_h$ and $df$ are the depth values of the head-top and the foot-bottom, respectively. Human height is estimated by calculating an Euclidean distance between the real-world coordinates of the head-top and the foot-bottom as follows:

$$\begin{aligned} H &= \sqrt{(X_h - X_f)^2 + (Y_h - Y_f)^2 + (Z_h - Z_f)^2} \\ &= \sqrt{((x_h d_h - x_f d_f)/f)^2 + ((y_h d_h - y_f d_f)/f)^2 + (d_h - d_f)^2}. \end{aligned} \quad (5)$$

The unit of the estimated human height by (5) is same as the unit of the depth pixels. If the pixels of the depth video store the distance as millimeters, then the unit of $H$ is millimeter.

The estimated human height by (5) may have an error. One reason of the error is the noise of $d_h$. Generally, $(x_h, y_h)$ may be in a hair area. The depth values in the hair area have large noises because the hair causes the diffuse reflection of an infrared ray emitted by the depth camera. Therefore, $d_h$ should be corrected as the depth value of a point which is close to the head top but is not on the hair area. The depth value of the point which is not on the hair area has a high similarity to the depth values of neighboring pixels. The similarity is obtained by calculating the variance of the pixels located within $r$ pixels to the left, right and bottom including the corresponding pixel as follows:

$$\sigma_r^2 = \frac{1}{(r+1)(2r+1)} \sum_{i=0}^{r} \sum_{j=-r}^{r} \left(d(x+i, y+j)^2\right) - \left(\frac{1}{(r+1)(2r+1)} \sum_{i=0}^{r} \sum_{j=-r}^{r} d(x+i, y+j)\right)^2. \quad (6)$$

If $\sigma_r^2$ is smaller than $T_\sigma$, then the $d_h$ is corrected as the depth value of the corresponding pixel as shown in Figure 7. Otherwise, the point is found between pixels below one pixel and the similarity of the found point is calculated by (6). In (6), $r$ is smaller as $d_h$ is larger because the width of an object is larger as the distance from the camera is closer as follows [42]:

$$\frac{P_1}{P_2} = \frac{d_2}{d_1}, \quad (7)$$

where $P_1$ and $P_2$ are the pixel lengths of the object widths when the depth values are $d_1$ and $d_2$, respectively. Therefore, $r$ depended on the depth value is determined as follows:

$$r = \frac{d_0}{d_h} r_0. \quad (8)$$

In (8), $d_0$ and $r_0$ are constants so $d_0 \times r_0$ can be regarded as a parameter. If $d_0 \times r_0$ is represented as $\gamma$, (6) is modified as follows:

$$\begin{aligned} \sigma_r^2 = &\frac{1}{(\gamma/d_h+1)(2\gamma/d_h+1)} \sum_{i=0}^{\gamma/d_h} \sum_{j=-\gamma/d_h}^{\gamma/d_h} \left(d(x+i, y+j)^2\right) \\ &- \left(\frac{1}{(\gamma/d_h+1)(2\gamma/d_h+1)} \sum_{i=0}^{2\gamma/d_h} \sum_{j=-2\gamma/d_h}^{2\gamma/d_h} d(x+i, y+j)\right)^2. \end{aligned} \quad (9)$$

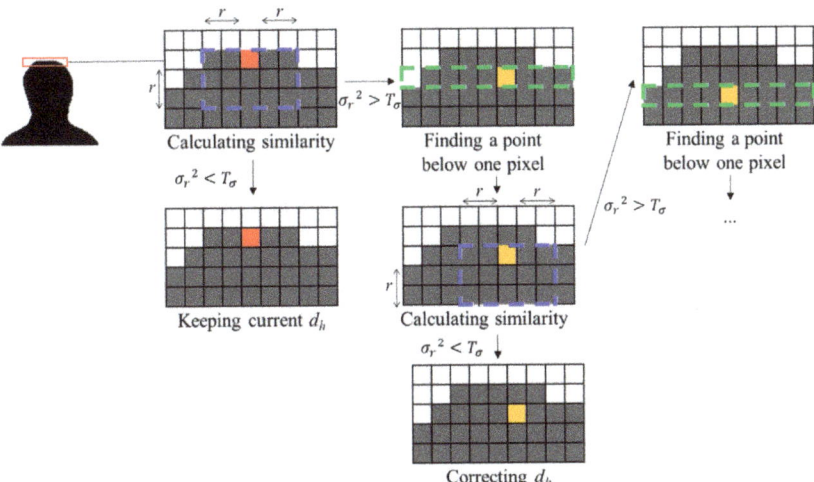

**Figure 7.** Flow of correcting $d_h$ by calculating similarity to neighboring pixels.

Mask R-CNN occasionally detects slightly wider human body region than the actual region. In particular, the region detection error in the lower part of human body may causes that a point on the ground is extracted as the foot-bottom. Assuming the ground is flat, the difference in a depth gradient is little between two vertically adjacent pixels that is in the ground area. The depth gradient of certain pixel $(x, y)$ is defined as follows:

$$g(x, y) = d(x + 1, y) - d(x, y). \tag{10}$$

If certain point is on the ground, then the difference in the depth gradients is the same between the point and one pixel down. In order to determine whether the extraction of the foot-bottom is correct, two depth gradients are compared as follows:

$$\begin{aligned} D &= g(x_f - 1, y_f) - g(x_f, y_f) \\ &= (d(x_f, y_f) - d(x_f - 1, y_f)) - (d(x_f + 1, y_f) - d(x_f, y_f)) \\ &= 2d(x_f, y_f) - d(x_f - 1, y_f) - d(x_f + 1, y_f). \end{aligned} \tag{11}$$

If $D$ is 0, then the point is removed from the human body region. The comparison of the depth gradients by (11) is applied to the bottommost pixels of the human body region in order to correct the foot-bottom. If all of the bottommost pixels are removed, then this process is repeated for the points of the human body region where is one pixel up. The foot-bottom is extracted as a center pixel among the bottommost pixels which are not removed. Figure 8 shows correcting the position of the foot-bottom point.

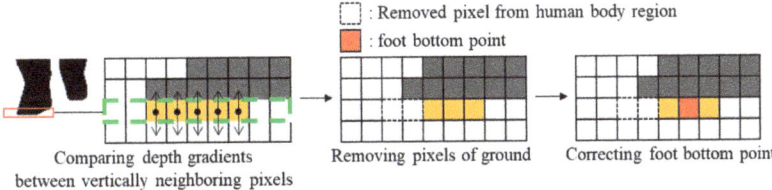

**Figure 8.** Foot bottom point correction.

The noises of depth video may cause the temporary error of the human-height estimation. The temporary error of the human height is corrected by the average of the estimated human heights among a sequence of depth frames as follows:

$$\overline{H}(n+1) = \frac{1}{n+1}\left(\sum_{i}^{n+1} H(i)\right)$$
$$= \frac{1}{n}\left(\sum_{i}^{n} H(i)\right) \times \frac{n}{n+1} + \frac{1}{n+1}H(n+1) \quad (12)$$
$$= \frac{n}{n+1}\overline{H}(n) + \frac{1}{n+1}H(n+1),$$

where $n$ is the order of the captured depth frames and $H(n)$ and $\overline{H}(n)$ are the estimated and corrected human heights in the $n$th frame order, respectively.

## 4. Experiment Results

Intel Realsense D435 is used as a depth camera for the experiments of the proposed method. A focal length $f$ and a frame rate of the depth camera are 325.8 mm and 30 Hz, respectively. The resolutions of depth video are specified as 640 × 480. The threshold $T_b$ for (1) and $T_\sigma$ for (6) are set to 100 and 50, respectively. The parameter $\gamma$ in (9) is 4000, which means $r$ is 2 when $d_h$ is 2000 mm.

Figures 9 and 10 show the extractions of the human body region through by mask R-CNN and by the background depth image, respectively. Both methods of the human body region extraction accurately extract the human body region at not only a standing state, but also a walking state. In addition, the human body region is accurately extracted regardless of the states of the human body. In Figure 9, areas painted in green and red are the human body and human head regions, respectively. The human head regions are accurately found even though the position of the hand is above the head. The human body region extraction by the background depth image extracts larger regions than by mask R-CNN, so some part of the background is included in the human body region. In addition, the bottom area of the human body is not included in the human body region because the depth values of these area are similar to the depth value of the ground.

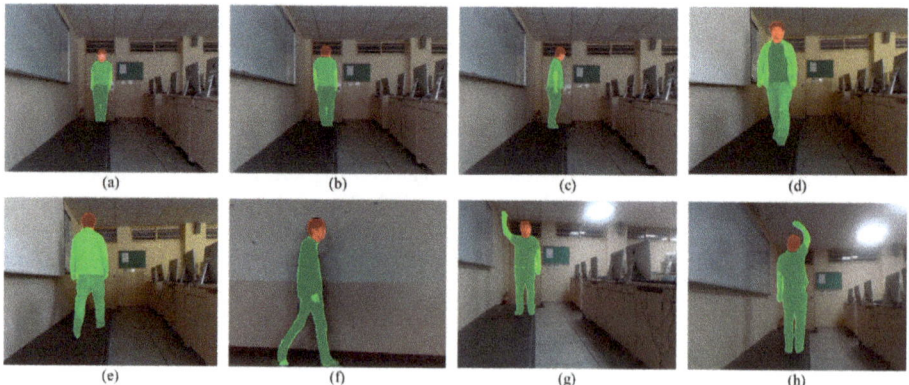

**Figure 9.** Extraction of human body region by mask R-CNN. (**a**) Standing toward front; (**b**) standing backward; (**c**) standing sideways; (**d**) walking toward camera; (**e**) walking opposite to camera; (**f**) lateral walking, (**g**) standing toward front and waving hand; (**h**) standing backward and waving hand.

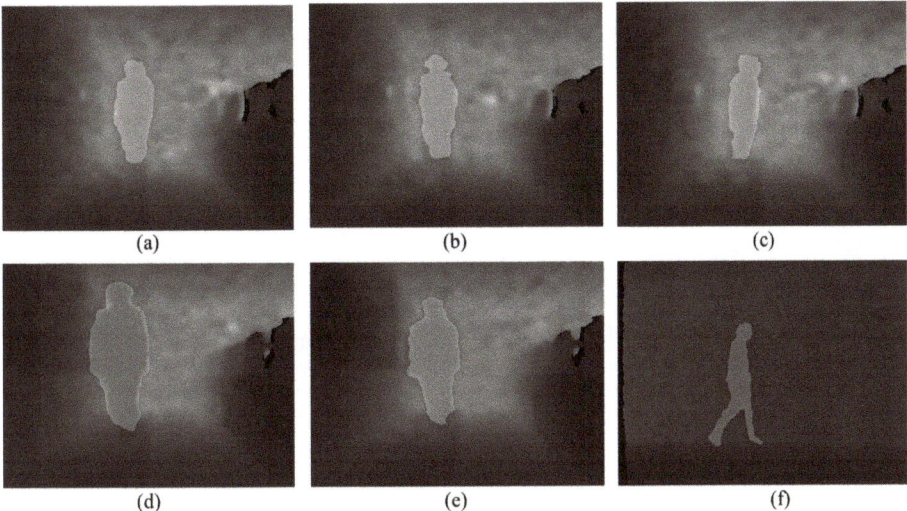

**Figure 10.** Extraction of human body region by background depth image. (**a**) Standing toward front; (**b**) standing backward; (**c**) standing sideways; (**d**) walking toward camera; (**e**) walking opposite to camera; (**f**) lateral walking.

Figure 11 shows the correction of human height by (12) when the body region is extracted by mask R-CNN. The distributions of the estimated human height are large because of the noises of the depth frame when the correction of human height is not applied. After applying the correction of the human-height estimation by (12), the human heights are estimated as certain heights after about 20 frames.

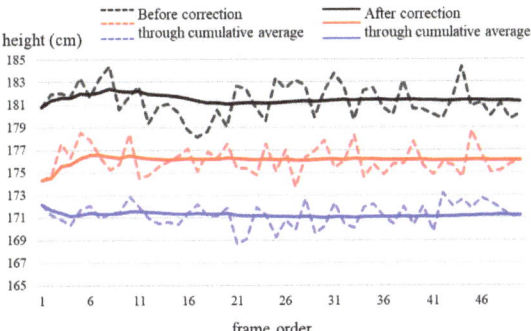

**Figure 11.** Results of height estimation after correction through cumulative average for each three persons.

Figure 12 shows the result of the human-height estimation depending on the methods of the human body region extraction. The actual height of a person is 177 cm. In the human body region extraction through the background depth image, first 50 frames are accumulated to generate the background depth image. The human body keeps at a distance of 3.5 m from the camera. The body height is estimated as 176.2 cm when the human body region is extracted by mask R-CNN. The body height is estimated as 172.9 cm when the human body region is extracted by the background depth image.

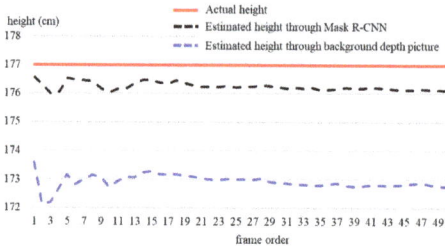

**Figure 12.** Results of height estimation depending on methods of human body region extraction.

Figure 13 shows the result of the human-height estimation according to the distance between human body and the camera. The actual height of a person is 182 cm. The distances from the camera are 2.5 m, 3 m, 3.5 m, 4 m and 4.5 m. The averages of the human heights are estimated as 181.5 cm, 181.1 cm, 181.2 cm, 179.7 cm and 179.8 cm when the distances are 2.5 m, 3 m, 3.5 m, 4 m and 4.5 m, respectively.

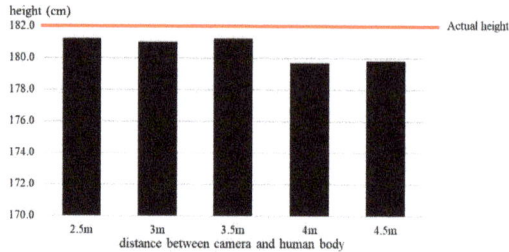

**Figure 13.** Results of height estimation according to distance between camera and human body.

Figure 14 shows the result of the human-height estimation when a person whose height is 180 cm is standing, walking toward the camera and lateral walking. Human body keeps at a distance of 2.5 m from the camera when the human is standing and lateral walking. When the human is walking toward the camera, the distance from the camera is in range of 2.5 m to 4 m. In the standing state, the human height is estimated as 178.9 m. The human height is 177.1 cm and 174.9 cm in the lateral walking and the walking toward the camera, respectively. The magnitude of the estimated error in the lateral walking state is similar to in the standing state. The estimated error in walking toward the camera is larger than the others. The reason is that the vertical length of the human body is reduced because human knees are bent to a certain degree while walking.

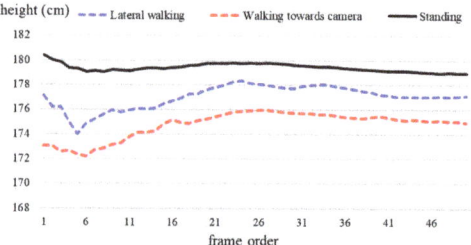

**Figure 14.** Results of height estimation in standing and walking states.

Figure 15 shows the positions of $d_h$ and $(x_f, y_f)$ before and after the correction of the head-top and the foot-bottom, respectively. The green and red points in Figure 15 represent the head-top and foot-bottom points, respectively. In Figure 15a, the position of $d_h$ is on the hair area, so $d_h$ has some error and the changes in $d_h$ are large as shown in Figure 16. After correcting $d_h$, the changes is smaller. The changes of the estimated human body height are reduced after correcting of the foot-bottom point as shown in Figure 17. Two persons whose actual heights are 182 cm and 165 cm are estimated as 188.6 cm and 181.5 cm before correcting $d_h$, respectively, as 172.2 cm and 163.3 cm after the correction of the head-top point, respectively and as 181.5 cm and 163.1 cm after the correction of both head-top and foot-bottom points, respectively.

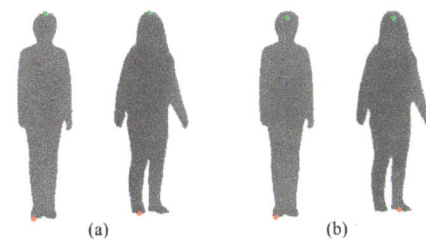

**Figure 15.** Positions of $d_h$ and $(x_f, y_f)$. (**a**) Before correction; (**b**) after correction.

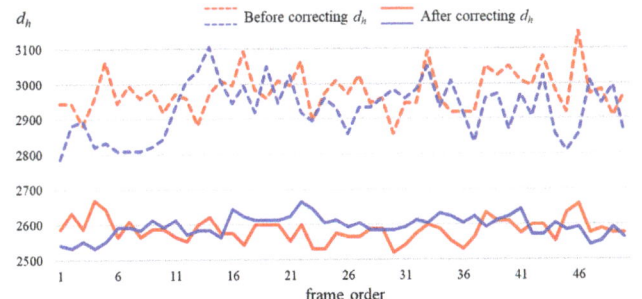

**Figure 16.** Changes in $d_h$ according to frame order.

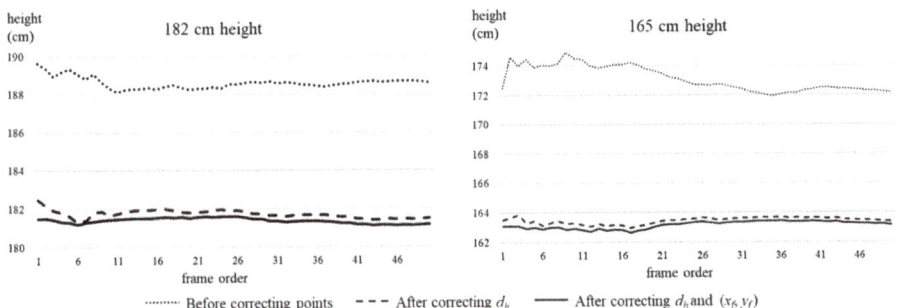

**Figure 17.** Result of height estimation after correction of $d_h$ and $(x_f, y_f)$.

Figure 18 shows the results of the human-height estimation depending on $r_0$ and $T_\sigma$, which are the parameters for (8) and (9) when $d_0$ is 2000. The estimated height drops sharply when $r_0$ is less than or equal to 2 and decreases smoothly when $r_0$ is larger than 2. In addition, the estimated height linearly increases when $T_\sigma$ is less than or equal to 250 and slowly increases when $T_\sigma$ is larger than 250. Body height is estimated most accurately when $r_0$ is 2 and $T_\sigma$ is 125.

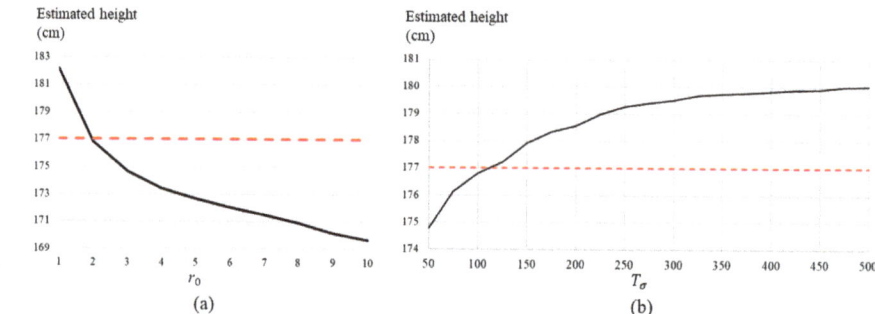

**Figure 18.** Results of human-height estimation depending on parameters for (8) and (9). (**a**) $r$; (**b**) $T_\sigma$.

Tables 2–4 show the results of the human-height estimation depending on human body postures for 10 persons. All of persons are captured within a range of 2.5 m to 4 m from the camera. Each person is captured with 150 frames. The error of the human-height estimation is calculated as follows:

$$\frac{|\overline{H} - H_{actual}|}{H_{actual}}, \qquad (13)$$

where $\overline{H}$ and $H_{actual}$ are the average of the corrected human heights by (12) and an actual height for a person, respectively. When the human body region is extracted by mask R-CNN, the errors of the human-height estimation with standing, lateral walking and walking towards the camera are 0.7%, 1.3% and 1.8%, respectively. The accurate foot-bottom for the human-height estimation is the point of a foot heel which is on the ground. However, the bottommost pixel of the body region which is extracted as the foot-bottom point is usually a foot toe point in the proposed method. The position difference between the foot heel and foot toe points may make the error of the human-height estimation. The human-height estimation errors with standing, lateral walking, and walking towards the camera are 2.2%, 2.9%, 4.6%, respectively, when the body region is extracted by the background depth image. The human-height estimation using only depth frames has more error than using both color and depth frames.

**Table 2.** Results of human-height estimation by proposed method while standing.

| Person No. | Actual Height (cm) | Extracting Human Body by Mask R-CNN | | Extracting Human Body by Background Depth Image | |
|---|---|---|---|---|---|
| | | Estimated Height (cm) | Estimation Error (%) | Estimated Height (cm) | Estimation Error (%) |
| 1 | 177 | 179.6 | 1.5 | 172.1 | 2.8 |
| 2 | 183 | 182.1 | 0.5 | 176.5 | 3.6 |
| 3 | 165 | 164.2 | 0.5 | 161.3 | 2.2 |
| 4 | 178 | 176.5 | 0.8 | 173.9 | 2.3 |
| 5 | 182 | 180.9 | 0.6 | 177.7 | 2.4 |
| 6 | 173 | 174.6 | 0.9 | 175.2 | 1.3 |
| 7 | 175 | 174.4 | 0.3 | 171.3 | 2.1 |
| 8 | 170 | 169.2 | 0.5 | 167.1 | 1.7 |
| 9 | 168 | 167.3 | 0.4 | 165.6 | 1.4 |
| 10 | 181 | 178.9 | 0.6 | 177.4 | 1.4 |
| Average error (%) | | | 0.7 | | 2.2 |

Table 3. Results of human-height estimation by proposed method while lateral walking.

| Person No. | Actual Height (cm) | Extracting Human Body by Mask R-CNN | | Extracting Human Body by Background Depth Image | |
|---|---|---|---|---|---|
| | | Estimated Height (cm) | Estimation Error (%) | Estimated Height (cm) | Estimation Error (%) |
| 1 | 177 | 176.1 | 0.5 | 171.5 | 3.1 |
| 2 | 183 | 180.7 | 1.3 | 175.9 | 3.9 |
| 3 | 165 | 163.0 | 1.2 | 160.1 | 3.0 |
| 4 | 178 | 175.2 | 1.6 | 174.1 | 2.2 |
| 5 | 182 | 179.6 | 1.3 | 176.6 | 3.0 |
| 6 | 173 | 170.7 | 1.3 | 168.1 | 2.8 |
| 7 | 175 | 172.6 | 1.4 | 170.7 | 2.5 |
| 8 | 170 | 167.5 | 1.5 | 166.1 | 2.3 |
| 9 | 168 | 166.2 | 1.1 | 162.8 | 3.1 |
| 10 | 181 | 177.1 | 1.6 | 174.8 | 2.9 |
| Average error (%) | | | 1.3 | | 2.9 |

Table 4. Results of human-height estimation by proposed method while walking towards camera.

| Person No. | Actual Height (cm) | Extracting Human Body by Mask R-CNN | | Extracting Human Body by Background Depth Image | |
|---|---|---|---|---|---|
| | | Estimated Height (cm) | Estimation Error (%) | Estimated Height (cm) | Estimation Error (%) |
| 1 | 177 | 174.1 | 1.6 | 167.5 | 5.4 |
| 2 | 183 | 178.7 | 2.3 | 172.9 | 5.5 |
| 3 | 165 | 161.0 | 2.4 | 157.1 | 4.8 |
| 4 | 178 | 173.8 | 2.4 | 172.1 | 3.3 |
| 5 | 182 | 178.4 | 2.0 | 173.6 | 4.6 |
| 6 | 173 | 169.7 | 1.9 | 164.1 | 5.1 |
| 7 | 175 | 171.2 | 2.2 | 168.7 | 3.6 |
| 8 | 170 | 166.4 | 2.1 | 163.1 | 4.1 |
| 9 | 168 | 165.2 | 1.7 | 158.8 | 5.5 |
| 10 | 181 | 174.9 | 2.8 | 171.8 | 4.6 |
| Average error (%) | | | 2.1 | | 4.6 |

## 5. Conclusions

In this study, a human-height estimation method using color and depth information was proposed. The human body region was extracted through the pre-trained mask R-CNN to color video. The human body region extraction from depth video was also proposed by comparing with the background depth image. Human height was estimated from depth information by converting two points of head-top and foot-bottom into two 3D real-world coordinates and by measuring the Euclidean distance between two 3D coordinates. Human height was accurately estimated even if the person is not in front or a walking state. In the experiment results, the errors of the human-height estimation by the proposed method with the standing state were 0.7% and 2.2% when the human body region was extracted by mask-R CNN and by the background depth image, respectively. The proposed method significantly improves the human-height estimation by combining color and depth information. The proposed method can be applied to estimate not only the body height, but also the height of other object types such as animals. The proposed method can also be applied to gesture recognition and body posture estimation which require the types and the 3D information of objects.

**Author Contributions:** Conceptualization, D.-s.L., J.-s.K., S.C.J. and S.-k.K.; software, D.-s.L. and J.-s.K.; writing—original draft preparation, D.-s.L., J.-s.K., S.C.J. and S.-k.K.; supervision, S.-k.K. All authors have read and agreed to the published version of the manuscript.

**Funding:** This research was supported by the BB21+ Project in 2020 and was supported by Dong-eui University Foundation Grant (2020).

**Conflicts of Interest:** The authors declare no conflict of interest.

## References

1. Tsakanikas, V.; Dagiuklas, T. Video surveillance systems-current status and future trends. *Comput. Electr. Eng.* **2018**, *70*, 736–753. [CrossRef]
2. Momeni, M.; Diamantas, S.; Ruggiero, F.; Siciliano, B. Height estimation from a single camera view. In Proceedings of the International Conference on Computer Vision Theory and Applications, Rome, Italy, 24–26 February 2012; pp. 358–364.
3. Gallagher, A.C.; Blose, A.C.; Chen, T. Jointly estimating demographics and height with a calibrated camera. In Proceedings of the International Conference on Computer Vision, Kyoto, Japan, 29 September–2 October 2009; pp. 1187–1194.
4. Zhou, X.; Jiang, P.; Zhang, X.; Zhang, B.; Wang, F. The measurement of human height based on coordinate transformation. In Proceedings of the International Conference on Intelligent Computing, Lanzhou, China, 2–5 August 2016; pp. 717–725.
5. Liu, J.; Collins, R.; Liu, Y. Surveillance camera auto calibration based on pedestrian height distributions. In Proceedings of the British Machine Vision Conference, Dundee, UK, 29 August–2 September 2011; pp. 1–11.
6. Cho, W.; Shin, M.; Jang, J.; Paik, J. Robust pedestrian height estimation using principal component analysis and its application to automatic camera calibration. In Proceedings of the International Conference on Electronics, Information, and Communication, Honolulu, HI, USA, 24–27 January 2018; pp. 1–2.
7. Criminisi, A.; Reid, I.; Zisserman, A. Single view metrology. *Int. J. Comput. Vis.* **2000**, *40*, 123–148. [CrossRef]
8. Andaló, F.A.; Taubin, G.; Goldenstein, G. Efficient height measurements in single images based on the detection of vanishing points. *Comput. Vis. Image Underst.* **2015**, *138*, 51–60. [CrossRef]
9. Jung, J.; Kim, H.; Yoon, I.; Paik, J. Human height analysis using multiple uncalibrated cameras. In Proceedings of the International Conference on Consumer Electronics, Las Vegas, NV, USA, 7–11 January 2016; pp. 213–214.
10. Viswanath, P.; Kakadiaris, I.A.; Shah, S.K. A simplified error model for height estimation using a single camera. In Proceedings of the International Conference on Computer Vision, Kyoto, Japan, 29 September–2 October 2009; pp. 1259–1266.
11. Rother, D.; Patwardhan, K.A.; Sapiro, G. What can casual walkers tell us about a 3d scene? In Proceedings of the International Conference on Computer Vision, Rio de Janeiro, Brazil, 14–21 October 2007; pp. 1–8.
12. Pribyl, B.; Zemcik, P. Simple single view scene calibration. In *Lecture Notes in Computer Science*; Springer: Berlin/Heidelberg, Germany, 2011; Volume 6915, pp. 748–759.
13. Zhao, Y.; Carraro, M.; Munaro, M.; Menegatti, E. Robust multiple object tracking in rgb-d camera networks. In Proceedings of the International Conference on Intelligent Robots and Systems, Vancouver, BC, Canada, 24–28 September 2017; pp. 6625–6632.
14. Ren, C.Y.; Prisacariu, V.A.; Kahler, O.; Reid, I.D.; Murray, D.W. Real-time tracking of single and multiple objects from depth-colour imagery using 3d signed distance functions. *Int. J. Comput. Vis.* **2017**, *124*, 80–95. [CrossRef] [PubMed]
15. Jiang, M.X.; Luo, X.X.; Hai, T.; Wang, H.Y.; Yang, S.; Abdalla, A.N. Visual Object Tracking in RGB-D Data via Genetic Feature Learning. *Complexity* **2019**, *2019*, 1–8. [CrossRef]
16. Ren, Z.; Yuan, J.; Meng, J.; Zhang, Z. Robust part-based hand gesture recognition using kinect sensor. *IEEE Trans. Multimed.* **2013**, *15*, 1110–1120. [CrossRef]
17. Li, Y.; Miao, Q.; Tian, K.; Fan, Y.; Xu, X.; Li, R.; Song, J. Large-scale gesture recognition with a fusion of rgb-d data based on the c3d model. In Proceedings of the International Conference on Pattern Recognition, Cancun, Mexico, 4–8 December 2016; pp. 25–30.
18. Li, Y. Hand gesture recognition using Kinect. In Proceedings of the 2012 IEEE International Conference on Computer Science and Automation Engineering, Beijing, China, 22–24 June 2012; pp. 196–199.
19. Lecun, Y.; Bottou, L.; Bengio, Y.; Haffner, P. Gradient-based learning applied to document recognition. *Proc. IEEE* **1998**, *86*, 2278–2324. [CrossRef]

20. Girshick, R.; Donahue, J.; Darrell, T.; Malik, J. Rich feature hierarchies for accurate object detection and semantic segmentation. In Proceedings of the Conference on Computer Vision and Pattern Recognition, Columbus, OH, USA, 23–28 June 2014; pp. 580–587.
21. Girshick, R. Fast R-CNN. In Proceedings of the International Conference on Computer Vision, Santiago, Chile, 7–13 December 2015; pp. 1440–1448.
22. Shaoqing, R.; Kaiming, H.; Girshick, R.; Sun, J. Faster R-CNN: Towards real-time object detection with region proposal networks. *IEEE Trans. Pattern Anal. Mach. Intell.* **2017**, *39*, 1137–1149. [CrossRef]
23. Redmon, J.; Divvala, S.; Girshick, R.; Farhadi, A. You only look once: Unified, real-time object detection. In Proceedings of the Conference on Computer Vision and Pattern Recognition, Las Vegas, NV, USA, 27–30 June 2016; pp. 779–788.
24. Redmon, J.; Farhadi, A. YOLO9000: Better, faster, stronger. In Proceedings of the Conference on Computer Vision and Pattern Recognition, Honolulu, HI, USA, 21–26 July 2017; pp. 6517–6525.
25. Liu, W.; Anguelov, D.; Erhan, D.; Szegedy, C.; Reed, S.; Fu, C.Y.; Berg, A.C. SSD: Single shot multibox detector. In Proceedings of the European Conference on Computer Vision, Amsterdam, The Netherlands, 11–14 October 2016; pp. 21–37.
26. He, K.; Gkioxari, G.; Dollár, P.; Girshick, R. Mask R-CNN. *IEEE Trans. Pattern Anal. Mach. Intell.* **2020**, *42*, 386–397. [CrossRef] [PubMed]
27. Soviany, P.; Ionescu, R.T. Optimizing the trade-off between single-stage and two-stage deep object detectors using image difficulty prediction. In Proceedings of the International Symposium on Symbolic and Numeric Algorithms for Scientific Computing, Timisoara, Romania, 20–23 September 2018; pp. 209–214.
28. BenAbdelkader, C.; Yacoob, Y. Statistical body height estimation from a single image. In Proceedings of the International Conference on Automatic Face & Gesture Recognition, Amsterdam, The Netherlands, 17–19 September 2008; pp. 1–7.
29. Guan, Y. Unsupervised human height estimation from a single image. *J. Biomed. Eng.* **2009**, *2*, 425–430. [CrossRef]
30. BenAbdelkader, C.; Cutler, R.; Davis, L. Stride and cadence as a biometric in automatic person identification and verification. In Proceedings of the International Conference on Automatic Face Gesture Recognition, Washington, DC, USA, 21 May 2002; pp. 372–377.
31. Koide, K.; Miura, J. Identification of a specific person using color, height, and gait features for a person following robot. *Robot. Auton. Syst.* **2016**, *84*, 76–87. [CrossRef]
32. Günel, S.; Rhodin, H.; Fua, P. What face and body shapes can tell us about height. In Proceedings of the International Conference on Computer Vision, Seoul, Korea, 27–28 October 2019; pp. 1819–1827.
33. Sayed, M.R.; Sim, T.; Lim, J.; Ma, K.T. Which Body Is Mine? In Proceedings of the Winter Conference on Applications of Computer Vision, Waikoloa Village, HI, USA, 8–10 January 2019; pp. 829–838.
34. Contreras, A.E.; Caiman, P.S.; Quevedo, A.U. Development of a kinect-based anthropometric measurement application. In Proceedings of the IEEE Virtual Reality, Minneapolis, MN, USA, 29 March–2 April 2014; pp. 71–72.
35. Robinson, M.; Parkinson, M. Estimating anthropometry with microsoft Kinect. In Proceedings of the International Digital Human Modeling Symposium, Ann Arbor, MI, USA, 11–13 June 2013; pp. 1–7.
36. Samejima, I.; Maki, K.; Kagami, S.; Kouchi, M.; Mizoguchi, H. A body dimensions estimation method of subject from a few measurement items using KINECT. In Proceedings of the International Conference on Systems, Man, and Cybernetics, Seoul, Korea, 14–17 October 2012; pp. 3384–3389.
37. Lin, T.; Dollár, P.; Girshick, R.; He, K.; Hariharan, B.; Belongie, S. Feature pyramid networks for object detection. In Proceedings of the Conference on Computer Vision and Pattern Recognition, Honolulu, HI, USA, 21–26 July 2017; pp. 936–944.
38. He, K.; Zhang, X.; Ren, S.; Sun, J. Deep residual learning for image recognition. In Proceedings of the IEEE Conference on Computer Vision and Pattern Recognition, Las Vegas, NV, USA, 27–30 June 2016; pp. 770–778.
39. He, K.; Zhang, X.; Ren, S.; Sun, J. Identity Mappings in Deep Residual Networks. In Proceedings of the European Conference on Computer Vision, Amsterdam, The Netherlands, 8–16 October 2016; pp. 630–645.
40. Lin, T.Y.; Maire, M.; Belongie, S.; Hays, J.; Perona, P.; Ramanan, D.; Dollar, P.; Zitnick, C.L. Microsoft COCO: Common objects in context. In Proceedings of the European Conference on Computer Vision, Zurich, Switzerland, 6–12 September 2014; pp. 740–755.

41. Hartley, R.; Zisserman, A. *Camera Models, in Multiple View Geometry in Computer Vision*, 2nd ed.; Cambridge University Press: New York, NY, USA, 2000; pp. 153–177.
42. Kwon, S.K.; Lee, D.S. Zoom motion estimation for color and depth videos using depth information. *EURASIP J. Image Video Process.* **2020**, *2020*, 1–13. [CrossRef]

© 2020 by the authors. Licensee MDPI, Basel, Switzerland. This article is an open access article distributed under the terms and conditions of the Creative Commons Attribution (CC BY) license (http://creativecommons.org/licenses/by/4.0/).

*Article*

# The Application and Improvement of Deep Neural Networks in Environmental Sound Recognition

**Yu-Kai Lin [1], Mu-Chun Su [1,*] and Yi-Zeng Hsieh [2,3,4,*]**

1. Department of Computer Science & Information Engineering, National Central University, Taoyuan City 32001, Taiwan; stan.yk.lin@g.ncu.edu.tw
2. Department of Electrical Engineering, National Taiwan Ocean University, Keelung City 20224, Taiwan
3. Institute of Food Safety and Risk Management, National Taiwan Ocean University, Keelung City 20224, Taiwan
4. Center of Excellence for Ocean Engineering, National Taiwan Ocean University, Keelung City 20224, Taiwan
* Correspondence: muchun@csie.ncu.edu.tw (M.-C.S.); yzhsieh@mail.ntou.edu.tw (Y.-Z.H.)

Received: 28 July 2020; Accepted: 21 August 2020; Published: 28 August 2020

**Featured Application: Authors are encouraged to provide a concise description of the specific application or a potential application of the work. This section is not mandatory.**

**Abstract:** Neural networks have achieved great results in sound recognition, and many different kinds of acoustic features have been tried as the training input for the network. However, there is still doubt about whether a neural network can efficiently extract features from the raw audio signal input. This study improved the raw-signal-input network from other researches using deeper network architectures. The raw signals could be better analyzed in the proposed network. We also presented a discussion of several kinds of network settings, and with the spectrogram-like conversion, our network could reach an accuracy of 73.55% in the open-audio-dataset "Dataset for Environmental Sound Classification 50" (ESC50). This study also proposed a network architecture that could combine different kinds of network feeds with different features. With the help of global pooling, a flexible fusion way was integrated into the network. Our experiment successfully combined two different networks with different audio feature inputs (a raw audio signal and the log-mel spectrum). Using the above settings, the proposed ParallelNet finally reached the accuracy of 81.55% in ESC50, which also reached the recognition level of human beings.

**Keywords:** deep neural network; convolutional neural network; environmental sound recognition; feature combination

## 1. Introduction

We live in a world surrounded by various acoustic signals. People react from their sense of hearing in situations like passing streets, finding someone in a building, or communicating with others. The development of computer vision has given machines the ability to support our lives in many ways. Hearing sense, as another important factor of our lives, is also an appropriate target to develop with artificial intelligence. A machine assistance acoustic detection system could be applied in several aspects, such as healthcare [1], monitoring [2], security [3] and multi-media applications [4].

In the artificial intelligence domain, neural networks have been a popular research field in recent years. Many acoustic topics have been researched with this technique, such as speech recognition [5,6] and music information retrieval (MIR) [7,8]. However, this kind of acoustic research only work for a certain purpose. Unlike this kind of content, the general acoustic events in our lives might not have periodicity or clear rhythms that can be detected, and the non-stationary properties of environmental sound make this problem difficult and complex. To achieve a system that can deal with general acoustic cases, the first step might be to recognize the current environmental scene. Scenes such as coffee shops, streets, and offices all have a unique event set; by adding the scene information into the detection system, the system complexity could be reduced. This is why environmental sound recognition techniques are important and essential.

This study attempted to provide an end-to-end solution for an environmental sound recognition system. There were two major contributions from this research. First, we improved the performance of the network feed with a raw audio signal. Second, we proposed a more flexible parallel network that could combine several kinds of features together. The result showed that this kind of network could combine raw audio signals and the log-mel spectrum efficiently.

The rest of this paper is organized as follows: In Section 2, we introduce the background of this research, including the current research on environmental sound recognition and the fundamental knowledgement of neural networks. In Section 3, a detailed description of our network and development methods is introduced. In Section 4, we perform experiments to examine our network architecture and the proposed development method, and we compare our results with those of other research, using a number of public datasets. In Section 5, we present a conclusion of our work and provide suggestions for further research.

## 2. Related Works

### 2.1. Environmental Sound Recognition

The intention of the study is to resolve the conditions around Environmental sound recognition (ESR), which is also known as environmental sound classification. The study is not specifically intended to detect the event trigger time precisely, but more important to understand what the acoustic scene is. In past years, numerous methods, such as the Gaussian Mixture Model (GMM) [9], Hidden Markov model (HMM) [10,11], random forest [12], and support vector machine [13], etc., have been used to solve the ESR problem. However, none of these methods can reach the level of human beings. Since 2012, neural networks have shown the great potential in computer vision [14]. Increasingly, researchers have begun to apply neural networks in the ESR field.

For a neural network, it is important to choose a suitable feature to be the input value. In 2014, Piczak [15] proposed a usable network structure using the log-mel spectrum and delta as the input features, which was once considered state-of-art in the ESR field. The log-mel spectrum has been a popular feature used in the ESR field in recent years. In Challenge on Detection and Classification of Acoustic Scenes and Events (DCASE challenges) [16,17], most of the researchers still choose to take the log-mel spectrum as one of the network inputs in acoustic scene classification tasks.

In 2015, Sainath et al. [6] used a raw audio wave as the network input to train for speech recognition and had promising results. Raw signals seem to be one of the choices in the ESR field.

In 2016, Aytar et al. [18] proposed SoundNet, which is trained using both images and raw audio. The image part is used to assist in the training, but the scene is still recognized according to the raw audio signals. The result of the network was impressive. Although, the performance might drop considerably, the network structure can still be trainable using the raw audio signal only. In the same year, Dai et al. [19] proposed an 18-layer network that could also work with raw audio signals, and the larger number of filters and a deeper structure provided a much better result using raw audio signals. We could clearly see that network architecture has a huge effect with the raw signal input when comparing these two works [18,19]. The depth and the filter numbers are obviously worth further discussions. On the other hand, both the two works use the global pooling strategy [20] to integrate the network output information, which has shown an outstanding effect on dimension reduction. Global pooling has other benefits in structure integration, which is explained in our method development.

In 2017, Tokozume and Harada [21] proposed EnvNet, which transforms a signal from a raw 1d signal to a 2d spectrogram-like graph through the network. This is an interesting idea, because training with a spectrum might also be adapted to this kind of graph. In the same year, Tokozume et al. [22] proposed another augmentation method that could be applied in the same kind of network, and the results could even reach the level of human beings.

These related works reveal that the input features greatly influence the performance of a network. Although many features have been tried on the network, a proper way to combine individual acoustic features are lacking. Moreover, network architectures that use raw signals as the input also require further discussion. Therefore, based on the existing research, this study focuses on improving the above-mentioned aspects.

## 2.2. Review of Neural Networks

The concept of neural networks has been proposed for a long time [23]. However, it was not considered useful due to the enormous computation requirements. The recent development of computer hardware has given researchers new opportunities to apply the technique in various problems, such computer vision [14] and speech recognition [5], etc. Neural networks show great potential in these aspects. In the following sections, we introduce the fundamental concepts of a neural network, as well as some techniques to tune up a network.

## 2.3. Feed-Forward Neural Network

The simplest feed-forward neural networks would be Single-layer perceptrons, which can be built up to do a regression. Assume we would like to project a $\mathcal{X} \in \mathbb{R}^n$ to $\mathcal{Y} \in \mathbb{R}^m$, the two variables could be rewrite as two vectors $\underline{\mathcal{X}} = [s_1 \cdots s_n]^T$ and $\underline{\mathcal{Y}} = [t_1 \dots t_m]^T$, so we could simply try the formula below:

$$\underline{\mathcal{Y}} = W\underline{\mathcal{X}} + b. \tag{1}$$

In (1), $W \in \mathbb{R}^{m \times n}$ and $b \in \mathbb{R}^{m \times 1}$, therefore, the main purpose to solve the equation is to find the suitable w and b. If we already had a certain sample $S_i = (\mathcal{X}_i, \mathcal{Y}_i)$, obviously, we make the result of input $\mathcal{X}_i$ could be as close to $\mathcal{Y}_i$ as possible. There are several methods that can be used to retrieve the correct value of w and b, such as the stochastic gradient descent (SGD) or Newton's method. No matter which method is applied, the equation will have a good result when $\underline{\mathcal{X}}$ and $\underline{\mathcal{Y}}$ among all the samples are linearly dependent. Inspired by the animal neuron system, the activation function $\varphi$ was added to improve Equation (1), and the new equations are listed as (2) and (3):

$$\underline{v} = W\underline{\mathcal{X}} + b \tag{2}$$

$$\underline{\mathcal{Y}} = \varphi(\underline{v}). \tag{3}$$

The activation function provides a non-linear transform to filter out the weaker signal. For example, the classic activation function sigmoid is:

$$\text{sigmoid}(x) = \frac{1}{1 + e^{-x}}. \tag{4}$$

After passing Equation (4), every output value is straightly normalized to a range between 0 and 1, which is a superior non-linear transform. Equation (3) can now have the ability to make a regression to the non-linear equation.

From Equation (3), it can be clearly seen that each element in $\underline{\mathcal{Y}}$ is actually composed of every element of $\underline{\mathcal{X}}$ in different weights. Here, $\underline{\mathcal{X}}$ form an input layer, and each node create an element of $\underline{\mathcal{Y}}$ is called a neuron in the network.

To enhance the network structure, a hidden layer can be added to improve the network performance. The neuron numbers in the hidden layer needs to be decided by users, it usually setup to the value bigger than both input and output dimensions. The hidden layer is used to project the input vector into another dimension, resulting in a greater chance of finding a linear way to transform from a higher dimension to the output layers. Several non-linear transform also make the network hold greater power to complete complex regression.

To obtain the correct weights of the feed-forward neural networks, the backpropagation method [24] (BP) is widely used. By calculating the gradient from the loss function, the gradients can be backward propagated to each of the weights.

It seems that the network would better be design deeper, more layers, or wider, more neurons per layer, but actually both of the two methods all get some issues need to be deal with. The weight number grows exponentially with the width of the network, which also leads to a large growth of the computation times and also causes the network to face a serious overfitting condition. This also means that the network might easily fit the training data but still result in poor performance while testing. Deeper networks need to solve the gradient decent problem. When performing BP, the gradient travels from the end of the network and gets thinner and thinner while arriving at the front, and it can even vanish directly. A number of methods have been proposed to improve the vanishing gradient problem, of which a deeper network is recommended to be built as a solution.

### 2.4. Convolutional Neural Networks

The convolutional neural network (CNN) is a special type of neural network used in image recognition. LeNet [25] is considered to be the first complete CNN. It is composed of convolution layers, activation layers, pooling layers, and fully connected layers. These layers all have special usages, which are introduced later in the paper. CNN resembles the original input image into a series of feature maps, by which each pixel in the feature map is actually a neuron. Unlike the way in which a normal neural network acts, each neuron does not connect to all the neurons in the previous layer; the connections only build up when these two neurons have a certain locality relationship. It makes sense because the information revealed in a certain location intuitively has a little chance to be related to another distant location. In this way, the total weight is reduced, which helps to improve the over-fitting condition.

### 2.5. Convolutional Layers

Each convolutional layer is composed of several graphic filters called kernels, it works just like the way in image processing does. Through convolutions, the kernels enhance part of the image's characteristic and turn the image into an individual feature map. The feature maps are all the same size and are bundled together to become a brand-new image. The convolutional layer provides an example regarding what the new image will look like. Each map in the same image is called a *channel*, and the number of channels becomes the depth of the image. When working through the convolution layers, the kernel actually processes all the channels once at a time. Another important aspect of the convolutional layer is *parameter sharing*. If we look back to the processing method of MLP, we can discover that each pixel in the same image needs to be applied to different kernels. However, in convolutional layers, the whole image shares the same kernel to create a feature map, which gives CNN an important *shift invariant* characteristic. As the kernel can move all around the image, the features correlated with the kernel can be detected anywhere, which gives CNN superior performance in image recognition.

### 2.6. Activation Layers

As mentioned in the previous section, the main purpose of the activation layer is to provide a non-linear transform. These functions need to be derivative. There are several types of activation functions, including sigmoid (4), tanh (5) and rectified linear unit (ReLU) (6):

$$\tanh(x) = \frac{e^x - e^{-x}}{e^x + e^{-x}} \qquad (5)$$

$$\text{ReLU}(x) = max(x, 0). \qquad (6)$$

Unfortunately, these activation functions all have some flaws. When using the gradient decent methods, ReLU can be affected by gradient exploding, because ReLU does not change the range of the output value from the input. Another problem cause by ReLU is the dead ReLU problem. When a relu has a negative input value, it will give an output of 0, which will cause the whole chain of the input to not update at that time, or even worse, never update until training is finished. On the other hand, sigmoid and tanh are affected by the vanishing gradient problem, because the gradients BP from these functions will at most only have 1/4 left. Comparing with these two groups of activation functions, we observe that the problem of ReLU can be solved by adding a normalization layer, which also results in a faster processing speed. For these reasons, ReLU is now the most commonly-used activation function.

### 2.7. Pooling Layers

Even though parameter sharing reduces the large number of parameters for CNN, for a large-scale picture, it is still necessary to find a reasonable way to perform subsampling. Pooling layers can be used to finish this job.

For a continuous signal (like an image), it is intuitive to perform downsampling by grouping a fixed number of adjacent values and then picking up an output value from each group. The pickup method could be based on the average, maximum, or minimum. Among these methods, maximum pooling has shown the best result and is commonly used now.

However, care must be taken, as not all feature maps can take pooling as the down sampling method. According to the previous description, each value in the same group needs to be adjacent, which means these values actually have some spatial relationships, and each group also needs to have the same spatial meaning. Therefore, pooling layers might not be suitable to in some cases using CNN, such as in game maps [26].

### 2.8. Fully Connected Layers

Fully connected (FC) layers are similar to the typical MLP. The processing feature maps are flattened before entering this layer and transform from several dimensions to a single vector. Most of the parameters in a CNN are set in FC layers, and the size of the FC layer determines the capacity of the network.

### 2.9. Loss Function

A neural network can be used for classification and regression, each of which needs a different loss function, and these functions all need to be derivate:

$$Loss_{mse}(\underline{y}, \underline{t}) = \frac{1}{2}(\underline{y} - \underline{t})^2. \qquad (7)$$

Equation (7) is the mean square loss (MSE) function, which is often used in regression tasks. It directly shows the difference between the output value and the target value. Another loss function often used in classification is cross entropy, which usually works with the softmax logistic function. In Equation (7), $\underline{y} = [y_1 \cdots y_J]^T$ is the output vector coming from the FC layer and the $J$ is the final class number. Softmax tends to find the probability distribution of the classification result. After passing through the function, the sum of output vector $\underline{S}$ becomes 1, and $S_j$ represents the probability of the input being classified as class $j$:

$$S_j(\underline{y}) = \frac{e^{y_j}}{\sum_{k=1}^{J} e^{y_k}} \qquad (8)$$

$$Loss_{cross\ entropy}(\underline{y}, \underline{t}) = -\sum_{j=1}^{J} t_j \log S_j(\underline{y}) \qquad (9)$$

$$Loss_{cross\ entropy} = -\log S_j(\underline{y}). \qquad (10)$$

The purpose of cross entropy is to estimate the difference between two vectors by calculating the log likely-hood function. The result is the same as that shown by Equation (9). In most classification cases, the final result will be a one-hot vector, in which target j has a value of one and the other element is zero, that is, only $S_j$ has the value. Therefore, the loss function then be simplified to (10).

## 2.10. Model Initialization

In a network, there are numerous hyper parameters that need to be decided, it is normally to consider a way to do the initialize. An ideal properly-initialized network could have the following property: if we take a series of random inputs into the network, the output should be fairly distributed in each of the classes, and there should not be any particular trends at the beginning. Obviously, randomly initializing the parameters will not have this effect. Glorot and Bengio proposed normalized initialization [27] to keep the various from the layer input to output.

$$W \sim U\left[-\frac{\sqrt{6}}{\sqrt{n_j+n_{j+1}}}, \frac{\sqrt{6}}{\sqrt{n_j+n_{j+1}}}\right] \tag{11}$$

$n_j$ in Equation (11) means the number of inputs in layer $j$. Equation (11) performs well for linear layers, but for nonlinear layers like ReLU, the equation needs to be adjusted.

He et al. proposed another method [28] to fix the formula, in which $n_{j+1}$ in (11) can be simply ignored. Our experiment used He's method to initialize the network.

## 2.11. Batch Normalization

In the previous section, it was mentioned that ReLU needs a method to support it in arranging the output value. The distribution of the output value also needs to be controlled. Sergey et al. proposed a method called *batch normalization* [29]. The main concept of this method is to force the addition of a linear transform before the nonlinear layer to make the variance and mean of the nonlinear layer input X, $X \in \mathbb{R}^{i \times j \times k}$, $i+j+k = m$ be in a certain range:

$$\mu_\beta \leftarrow \frac{1}{m}\sum_{i=1}^{m} x_i \tag{12}$$

$$\sigma_\beta^2 \leftarrow \frac{1}{m}\sum_{i=1}^{m}(x_i - \mu_\beta)^2 \tag{13}$$

$$\hat{x}_i \leftarrow \frac{x_i - \mu_\beta}{\sqrt{\sigma_\beta^2 + \epsilon}} \tag{14}$$

In Equations (12) and (13), the value of m is the total number of elements in the mini-batch and channels. After Equation (14), the purpose is to find the current mean value $\mu_\beta$ and current variance $\sigma_\beta$, and then adjust them to become 0 and 1:

$$y_i \leftarrow \gamma \hat{x}_i + \beta \equiv BN_{\gamma,\beta}(x_i). \tag{15}$$

Other learnable transform parameters can be added into the formula, and the final result will be similar to Equation (15). These two variables help the input values to do a little bit adjustment, which helps to solve the dead ReLU problem. It is essential to take batch normalization in a deep network.

## 3. Method Development

### 3.1. Data Sets

In our experiments, we took two kinds of public data sets to evaluate our network structure improvements: ESC50 [30] and ESC10 (Warsaw University of Technology, Warsaw, Poland).

ESC50 is a collection of environmental sound recordings that contains 50 classes, such as airplanes, car horns, cats, humans laughing, and rain, etc. There are 2000 short clips in total, and each class has 40 files. Each clip is five seconds long, and there is a total length of 2.78 h. It was recommended to test with the official 5-fold setting, as some of the files in the same class are extracted from the same source, using the official fold could avoid some problems.

ESC10 is a subset of ESC50 that takes out 10 classes from ESC50, while other configurations remain the same. It was beneficial to do a small-scale test in this dataset first.

### 3.2. Data Preprocessing

There are three kind of data put into our CNN such as the raw signal, the mel-spetrogram, and the output of 1D network. The output of 1D network is that we input signal into the 2D network. For the preprocessing, we first down-sampled the audio files to a sample rate of 16,000, averaged the stereo audio to mono, and eliminated the empty segments at the front and the tail of the files. If the resulting file was less than 1.5 s, we equally filled up the length with the 0 value at the beginning and end of the files. In the training phase, based on the method in [22], we appended 750 ms of 0 to both sides of the audio and then randomly cropped a 1.5 s clip, while the variance of the clip was 0. We then continued to repeat the procedure. After cropping the file, the mean and variance of the clip were normalized to 0 and 1. In the testing phase, we sequentially cropped 10 clips of 1.5 s each from the test audio. Each clip overlapped for about 390 ms. We chose the majority of probability scheme to do the final classification for each test file.

For the log spectrum, we transferred from the normalized clip with a sample rate of 16,000. The frame size was set to 512 (about 30 ms) with a 50% overlap, and the resulting values were then put through the log operation and mel-filters. This finally resulted in a 128-bin mel-spectrum. We did not make further normalizations to the spectrum graph, and they were fed into the network directly.

### 3.3. Data Augmentation

Compared to image datasets [31,32], acoustic datasets are not very popular; the number of files is insufficient, and there is a lack of diversity. Some researches [22,33] have revealed that data augmentation can help to enhance the result of classification. Common acoustic augmentation methods include pitch shifting and time stretching. Although CNN is shift invariant, these augmentation methods still have an effect on network training, therefore we chose both of them to be our augmentation methods.

We performed another augmentation method, known as wave overlapping, which was inspired by the study in [22] and their use of between class learning. We simplified the method to perform augmentation for just for a signal class. We, first, randomly cropped two segments of the same size from a single file, and then multiplied each of them by an individual random ratio from 0 to 1. These two crops were then summed up together, and the mean and variance were normalized to 0 and 1. The difference of volume we create for the new segment riches the diversity of the data. It is a simple method to enhance the dataset, and keeps the labels unchanged. The result shows that it is even better than just provide two of the individual crops. The experiment is described in the following chapter.

### 3.4. Network Customization

CNN provides a flexible method to extract features from any kind of input. Many researches [18,19,21,22] have shown that raw signals can be the input of a network. Inspired by [21], we assumed that the concatenation of a 1d feature map would form a spectrum-like graph. In fact, the 1d convolution along the time axis could actually fetch the frequency information from the raw signal. Each channel represents a set of frequencies, and the Y axis of the concatenation map means the frequency sets the response at a certain moment. We believed that more features could be extracted from this kind of map. Therefore one of our purposes was to optimize the extraction network. As shown in Figure 1, we proposed a network structure feed with raw signals and output a feature vector to entering a full connected layer to do the classification.

The network was composed of a 1D and 2D network. Just like the description above, the 1D network was used to extract a spectrum-like map, and the 2D network was used to find detailed features from the map.

Furthermore, the 2D network could not only be used in our network-organized map but could also be applied in the mel filter bank spectrum. In the next chapter. We would show the result of our network processing these two kinds of feature maps. Our network architectures are listed in Tables 1 and 2.

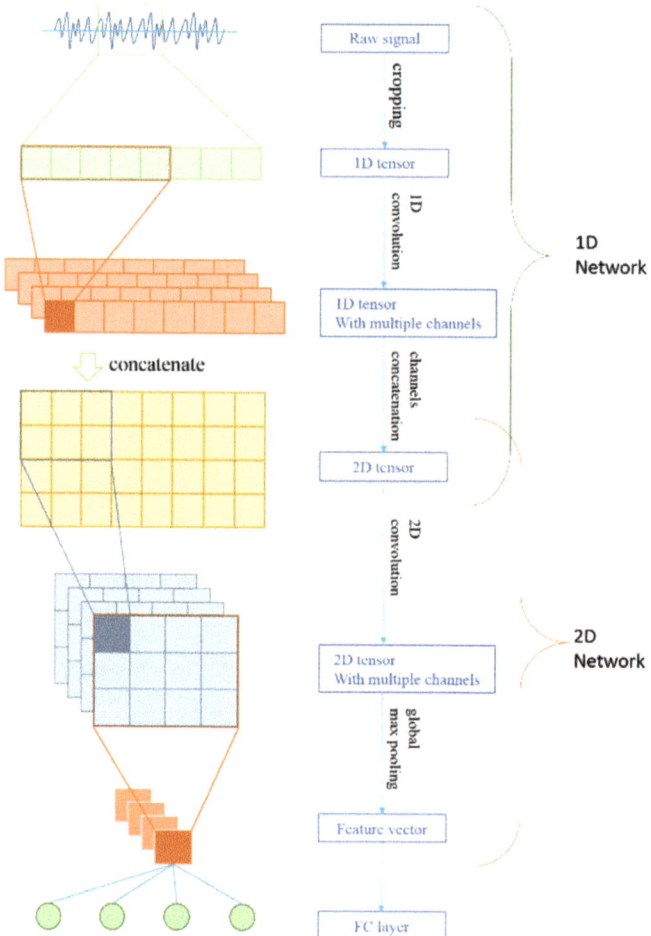

**Figure 1.** Network structure for ESR classification. The 1D and 2D networks are serialized together but could also work alone to be fit with different situations.

In Table 1, Conv refers to the convolutional layers and Pool refers to the max pooling layers. All the Conv layers were appended with a batch normalization layer and a ReLU layer. The input tensor of the network was a 1d tensor of 1 × 24,000 (a 1.5 s clip under a sample rate of 16,000), and the output tensor was a 2d tensor of 1 × 128 × 120.

Table 1. Architecture of the 1d network.

| Layer | Ksize | Stride | Padding | #Filters |
|---|---|---|---|---|
| Input (1 × 24,000) | | | | |
| Conv1 | 13 | 1 | same | 32 |
| Pool | 2 | 2 | 0 | – |
| Conv2 | 13 | 1 | same | 64 |
| Conv3 | 15 | 3 | same | 128 |
| Conv4 | 7 | 5 | same | 128 |
| Conv5 | 11 | 8 | same | 128 |
| Channel concatenation (1 × 128 × 100) | | | | |

Table 2. Architecture of the 2D network and full connections.

| Layer | Ksize | Stride | #Filters |
|---|---|---|---|
| Input (1 × M × N) | | | |
| 3 × Conv (1~3) | (15, 1) | (1, 1) (1, 1) (1, 1) | 32 |
| 3 × Conv (4~6) | (15, 1) | (1, 1) (1, 1) (2, 1) | 64 |
| 3 × Conv (7~9) | (15, 1) | (1, 1) (1, 1) (2, 1) | 128 |
| 3 × Conv (10~12) | (5, 5) | (1, 1) (1, 2) (2, 2) | 256 |
| 3 × Conv (13~15) | (3, 3) | (1, 1) (1, 1) (1, 1) | 256 |
| 3 × Conv (16~18) | (3, 3) | (1, 1) (1, 1) (1, 1) | 512 |
| Conv19 | (3, 3) | (2, 2) | 512 |
| Conv20 | (3, 3) | (1, 1) | 1024 |
| Global max pooling (1024) | | | |
| FC1 (2048) | | | |
| FC2 (# classes) | | | |

In Table 2, the first six blocks contained three Conv layers each. These three Conv layers had the same kernel size and filter number, but were constructed with different stride settings. All the Conv layers were appended with a batch normalization layer and a ReLU layer. FC1 was also followed by a batch normalization layer, a ReLU layer, and a drop out [34] for 50%. Padding was always applied on Conv layers, and if there was no stride, the size of the output would be the same as the input of each layer. The input size of this network was adjustable, but due to global pooling, the output size of the max pooling layers could be controlled as the last channel number, which was 1024 in Conv 20.

### 3.5. Network Parallelization

One of the main purposes of our work was to find a suitable method to combine several features in the network, we desired these features could eventually help adjust other networks during the training procedure. Applying the idea to the features with high homogeneity is intuitive. Based on our 1D and 2D networks, we proposed a feature parallel network and took raw signals and the mel-spectrum as examples. Figure 2 represents the concept of our method. In the last layer of the two-dimensional (2D) network, we used the global max pooling [20] to extract the feature vector from different kinds of feature maps. These extracted vectors could easily connect along the same axis whether their length being the same or not. In our experiment, we tested the parallel features using the same vector size of 1024; therefore, the length of the 1d tensor entering the FC layer shown in Figure 2 was 2048.

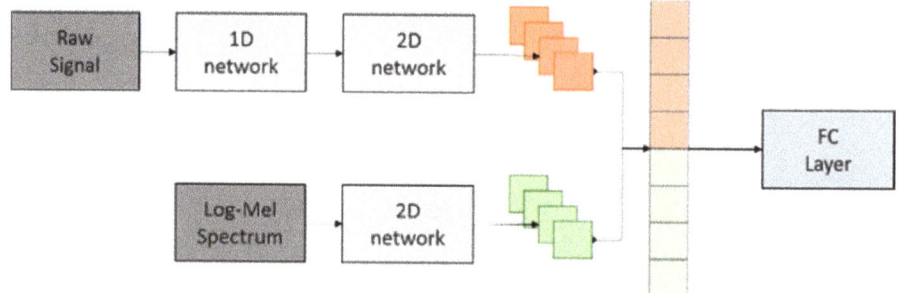

**Figure 2.** Parallel architecture of the network.

## 4. Results and Discussion

### 4.1. Experiment Setup

The neural network could be highly complex, but have only a few non-linear layers. The increasing of the depth makes the network easier to fetch the abstract expression form the source. Our intention of increasing the depth of the network allows the network to generate an effective acoustic filter, like mel-filter from signal processing.

We were interested in some particular setting within the network, so we would like to try modify some of these settings to examine what would they take effect. Our experiment focused on the following topics:

- Signal length per frame in the 1d network
- Depth of the 1d network
- Channel numbers of the 1D network (the height of the generated map)
- Kernel shape in the 2d network
- The effect of pre-training before parallel
- The effect of the augmentation

We took ESC50 and ESC10 as our datasets, and used the official fold settings. The experiments only consisted of training and testing, and we did not perform additional validations. Each experiment ran for 350 epochs, and we used Nesterov's accelerated gradient [35] as our network optimization method. The momentum factor was set to 0.8. The initial learning rate was set to 0.005 and decayed to 1/5 in the following epochs: 150 and 200. L2 regulation was also used and was set to $5 \times 10^{-6}$. Our models were built using PyTorch 0.2.0 and were trained on the machine with GTX1080Ti (NVIDIA, Santa Clara, CA, USA). The audio wav files were extracted using PySoundFile 0.9, and LibROSA 0.6 was used to create the mel-spectrums.

### 4.2. Architecture of the 1D Network

To test the influence of the filter size in the 1D network, it was necessary to slightly modify the network structure shown in Table 1.

The kernel sizes of Conv 4 and Conv 5 in the 1D network affected the frame length the most, therefore we tried three kinds of combinations to reach 25 ms, 35 ms, and 45 ms per frame. Test accuracy with the different frame lengths setting: the frame indicates the output unit of the 1D network. The dataset used in this test was ESC50, as shown in Table 3.

**Table 3.** Test accuracy with different frame length settings.

| Frame Length  | ~25 ms | ~35 ms | ~45 ms |
|---|---|---|---|
| ksize of Conv4 | 7 | 15 | 21 |
| ksize of Conv5 | 11 | 15 | 19 |
| Avg. | 73.55% | 72.45% | 72.55% |
| Std. dev. | 2.89% | 2.57% | 1.54% |

Clearly, the most stable result was found at a frame size of 25 ms; as the frame size increased, the accuracy worsened. The result found at 25 ms showed the most generalizability.

*4.3. Network Depth*

In the depth test, the epochs times is 350, but the result always converged before the 350 epochs. Increasing the depth of the network could enhance the non-linear transform ability of the network, and it is known as to enrich the abstract expressiveness. Also, the non-linear transform could consider as the processes to form the acoustic filter just like mel-filter, gammatone-filter, etc.

In this test, we inserted a certain number of layers before Conv 2–5. The settings of these layers were ksize = 3 and strides = 1. Padding was introduced to maintain the input length, the filter numbers were equal to the former Conv layer, and each of the insertion layers was followed by a batch normalization layer and a ReLU layer. The distribution of these layers was considered to not significantly affect the frame length. We get two kinds of setting, 12 or 24 additional layers, the location means in front of where these layers would be put. The configuration is shown in Table 4, and the result in Table 5.

**Table 4.** Configuration of the depth test.

| Location \ #Layers | Distribution of 12 Layers | Distribution of 24 Layers |
|---|---|---|
| Conv2 | 6 | 10 |
| Conv3 | 3 | 8 |
| Conv4 | 2 | 5 |
| Conv5 | 1 | 1 |

**Table 5.** Test accuracy with different depth settings of the 1d network.

| Depth | 12 + 5 | 24 + 5 |
|---|---|---|
| Avg. | 72.00% | 66.95% |
| Std. | 1.28% | 2.62% |

Although parameter numbers of the network increased slightly, we found that the network could still converge within 350 epochs, so we kept it the same.

It was surprising that the depth of the 1D network did not significantly affect the result, or even worse, the results going down while the network becomes deeper. The results were not consistent with that of Dai et al. [19]. The main reason for this discrepancy may have been due to the frame size in their experiment but not the depth of the network. More researches and experiments may be needed to prove this argument.

*4.4. Number of Filters*

A sufficient number of filters was necessary to provide sufficient capacity for the network to load the frequency information. Therefore, it could not be set too small. On the other hand, an excessively large setting would cause a large graph to pass into the 2D network, which would slow down the network processing but not provide a significant improvement in accuracy. We tried three different settings: 64, 128, and 256, and the result is shown in Table 6.

**Table 6.** Test accuracy using different filter numbers in the 1d network.

| #Filters | 64 | 128 | 256 |
|---|---|---|---|
| Avg. | 71.05% | 73.55% | 74.15% |
| Std. | 2.05% | 2.89% | 3.16% |

Although the setting of 256 had a slightly better result than 128, it required almost three times the amount of training compared with the 128 filters model, we chose 128 filters as our final decision.

### 4.5. Architecture of the 2D Network

The kernel shape could affect the invariant shifting of CNN, and it is not desired for this kind of invariant characteristic to show up in the frequency domain. In fact, a square kernel has been proven to not be suitable for spectrum content. We tried three different shape settings to see which would the best performer using our 1D-2D network by modified the size value of Conv (1~9). The test result is shown in Table 7.

**Table 7.** Test accuracy using different kernel shapes in the 2D network.

| Size | (15,1) | (15,15) | (1,15) |
|---|---|---|---|
| Avg. | 73.55% | 67.25% | 70.85% |
| Std. | 2.89% | 2.57% | 2.75% |

### 4.6. Parallel Network: The Effect of Pre-Training

To achieve the best performance of the parallel network, a pre-training procedure was required. Our network was composed of a raw-signal-1D-2D network, a spectrogram-2D network, and a set of fc layers. The pre-training procedure was built on the first two parts individually with their own fc layers (see Figure 1), trained the network alone, and then took over the essentials part and connected them into the parallel network. Likewise, we added an additional data source to improve the network analysis capability. These two data sources with high homogeneity are chosen. The neural network tends to ignore some information during the training, and our adding procedure is additional information, the feature vector from another network, back to it after training. We tried two kinds of the pre-training settings and compared them with the network before pre-training: Only pre-trained the raw-signal-1D-2D network. Both the upper reaches were pre-trained. The result is shown in Table 8.

**Table 8.** Test accuracy.

| Pre-Train Network | Without Pre-Train | Raw Signal Only | Both Network |
|---|---|---|---|
| Avg. | 78.20% | 75.25% | 81.55% |
| Std. | 2.96% | 2.18% | 2.79% |

The worst result occurred when the network was pre-trained only using the raw-signal-input network; however, if we pre-trained both networks, we could then get the best result. This revealed that the trained 1D-2D network could disturb the training procedure of the spectrogram network.

### 4.7. Data Augmentation

In this section, we test the augment method mentioned in the former paragraph. The *wave-overlapping* method was used to insert two different crops into a single clip, and then three different kinds of settings were applied. The original ESC50 had 1600 sets of training data in each fold. We randomly picked up certain sets of data to make the additional training clips. Setting 1 took 800 original clips, and Setting 2 took 400 clips. The wave overlapping created two crops from a single original source.

Setting 3 caused these two crops to be pitch shifted or time stretched. We tested the result using a partially pre-trained network and a fully pre-trained one. The results are shown in Tables 9 and 10.

Table 9. Test accuracy of augments applied in partial pre-pertained network.

| Aug Type | Original with Extra | Overlap with Raw | Overlap with PS/TS |
|---|---|---|---|
| Avg. | 74.05% | 76.95% | 75.20% |
| Std. | 3.55% | 3.59% | 2.65% |

Table 10. Test accuracy of augments applied in full pre-trained network.

| Aug Type | Original with Extra | Overlap with Raw |
|---|---|---|
| Avg. | 81.40% | 81.35% |
| Std. | 3.04% | 3.03% |

*4.8. Network Conclusion*

The previous experiments found a network architecture with the most efficient settings. We next compared our results with the networks with other researches based on raw signals or spectrograms without augmentations, as shown in Table 11.

Table 11. Test result of different kinds of models with open-datasets.

| Models | Features | Accuracy ESC50 | Accuracy ESC10 |
|---|---|---|---|
| Piczak's CNN [15] | log-mel spectrogram | 64.5% | 81.0% |
| m18 [19] | raw audio signal | 68.5% [22] | 81.8% [22] |
| EnvNet [21] * | raw audio signal ⊕ log-mel spectrogram | 70.8% | 87.2% |
| SoundNet (5 layers) [18] | raw audio signal | 65.0% | 82.3% |
| AlexNet [36] | spectrogram | 69% | 86% |
| GoogLeNet [36] | spectrogram | 73% | 91% |
| EnvNet with BC [22] * | raw audio signal ⊕ log-mel spectrogram | 75.9% | 88.7% |
| EnvNet–v2 [22] | raw audio signal | 74.4% | 85.8% |
| 1D-2D network (ours) | raw audio signal | 73.55% | 90.00% |
| ParallelNet (ours) | raw audio signal &log-mel spectrogram | 81.55% | 91.30% |
| Human accuracy [30] | | 81.3% | 95.7% |

* Result combined before softmax.

We summarized our experiment results as following:

- Our proposed method is an end to end system achieving 81.55% of accuracy in ESC50.
- Our proposed 1D-2D network could properly extract features from raw audio signal, compared with the old works.
- Our proposed ParallelNet could efficiently raising the performance with multiple types of input features.

## 5. Conclusions and Perspectives

This study proposed a 1D-2D network to perform audio classification, using only the raw signal input, as well as obtain the current best result in ESC50 among the networks using only the raw signal input. In the 1D network, we showed that the frame size had the largest effect, and that a deeper network might not be helpful when only using batch normalization. In addition, our parallel network showed great potential in combining different audio features, and the result was better than that for networks taking only one kind of feature individually. The final accuracy level corresponded to that of a human being.

Although we found that different frame size and network depth settings could affect the performance of a 1D network, the reasons causing these phenomena require more studies. Much research [37–39] has proposed methods to show the response area in the input graphics of CNN for the classification result or even for certain filters. The deep learning on sequential data processing with Kolmogorov's theorem is more and more important. Fine-grained visual classification tasks, Zheng et al. [40] proposed a novel probability fusion decision framework (named as PFDM-Net) for fine-grained visual classification. The authors in [41] proposed a novel device preprocessing of a speech recognizer, leveraging the online noise tracking and deep learning of nonlinear interactions between speech and noise. While, Osayamwen et al. [42] showed that such a supervisory loss of function is not optimal in human activity recognition, and they improved the feature discriminative power of the CNN models. These techniques could help find out hot spots in a spectrogram and could also help to generate a highly response audio clip for certain layers, which could provide a good direction for analyzing the true effect behind each kind of setting.

Most of our work focused on the arrangement of a 1D network, and there are still some topics for a 2D network that need to be discussed, such as network depth, filter size, and layer placement. These topics are all good targets for future work.

Our parallel network combined the raw audio input and log-mel spectrum successfully. More input features could also be tried in the future, such as mel-frequency cepstral coefficients (MFCC) and gammatone spectral coefficients (GTSC), etc. The feature vector ratio is also a good topic for future discussion. With the help of global pooling, we could even combine different kinds of network structures to perform the parallel test, just like the diverse fusion network structures used in computer visual. Our parallel methods have excellent research potential.

**Author Contributions:** Conceptualization, Y.-K.L. and Y.-Z.H.; Methodology, Y.-K.L., M.-C.S.; Software, L.-K.L.; Validation, Y.-K.L. and M.-C.S.; Formal Analysis, Y.-K.L.; Investigation, Y.-K.L.; Resources, M.-C.S.; Data Curation, Y.-K.L.; Writing Original Draft Preparation, Y.-K.L.; Writing Review & Editing, Y.-K.L., M.-C.S. and Y.-Z.H.; Visualization, Y.-K.L.; Supervision, M.-C.S.; Project Administration, M.-C.S., Y.-Z.H.; Funding Acquisition, Y.-Z.H. and M.-C.S. All authors have read and agreed to the published version of the manuscript.

**Funding:** This research was funded by [Ministry of Science and Technology, R.O.C] grant number [MOST 107-2221-E-019-039-MY2, MOST 109-2221-E-019-057-, MOST 109-2634-F-008-007-, 109-2634-F-019-001-, 109-2221-E-008-059-MY3, 107-2221-E-008-084-MY2 and 109-2622-E-008-018-CC2]. This research was funded by [University System of Taipei Joint Research Program] grant number [USTP-NTUT-NTOU-109-01] and NCU-LSH-109-B-010.

**Conflicts of Interest:** The authors declare no conflict of interest.

## References

1. Chen, J.; Cham, A.H.; Zhang, J.; Liu, N.; Shue, L. Bathroom Activity Monitoring Based on Sound. In Proceedings of the International Conference on Pervasive Computing, Munich, Germany, 8–13 May 2005.
2. Weninger, F.; Schuller, B. Audio Recognition in the Wild: Static and Dynamic Classification on a Real-World Database of Animal Vocalizations. In Proceedings of the Acoustics, Speech and Signal Processing (ICASSP) 2011 IEEE International Conference, Prague, Czech, 22–27 May 2011.
3. Clavel, C.; Ehrette, T.; Richard, G. Events detection for an audio-based Surveillance system. In Proceedings of the ICME 2005 IEEE International Conference Multimedia and Expo., Amsterdam, The Netherlands, 6–8 July 2005.

4. Bugalho, M.; Portelo, J.; Trancoso, I.; Pellegrini, T.; Abad, A. Detecting Audio Events for Semantic Video search. In Proceedings of the Tenth Annual Conference of the International Speech Communication Association, Bighton, UK, 6–9 September 2009.
5. Mohamed, A.-R.; Hinton, G.; Penn, G. Understanding how deep Belief Networks Perform Acoustic Modelling. In Proceedings of the Acoustics, Speech and Signal Processing (ICASSP), 2012 IEEE International Conference, Kyoto, Japan, 23 April 2012.
6. Sainath, T.N.; Weiss, R.J.; Senior, A.; Wilson, K.W.; Vinyals, O. Learning the speech front-end with raw waveform CLDNNs. In Proceedings of the Sixteenth Annual Conference of the International Speech Communication Association, Dresden, Germany, 6–10 September 2015.
7. Lee, H.; Pham, P.; Largman, Y.; Ng, A.Y. Unsupervised Feature Learning for Audio Classification Using Convolutional Deep Belief Networks. In Proceedings of the Advances in Neural Information Processing Systems, Vancouver, BC, Canada, 7–10 December 2009.
8. Van den Oord, A.; Dieleman, S.; Schrauwen, B. Deep Content-Based Music Recommendation. In Proceedings of the Advances in Neural Information Processing Systems, Lake Tahoe, NV, USA, 5–10 December 2013.
9. Peltonen, V.; Tuomi, J.; Klapuri, A.; Huopaniemi, J.; Sorsa, T. Computational Auditory Scene Recognition. In Proceedings of the Acoustics, Speech, and Signal Processing (ICASSP), 2002 IEEE International Conference, Orlando, FL, USA, 13–17 May 2002.
10. Rabiner, L. A tutorial on hidden Markov models and selected applications in speech recognition. *Proc. IEEE* **1989**, *77*, 257–286. [CrossRef]
11. Peng, Y.-T.; Lin, C.-Y.; Sun, M.-T.; Tsai, K.-T. Healthcare audio event classification using hidden markov models and hierarchical hidden markov models. In Proceedings of the ICME 2009 IEEE International Conference Multimedia and Expo, Cancun, Mexico, 28 June–3 July 2009.
12. Elizalde, B.; Kumar, A.; Shah, A.; Badlani, R.; Vincent, E.; Raj, B.; Lane, I. Experiments on the DCASE challenge 2016: Acoustic scene classification and sound event detection in real life recording. *arXiv* **2016**, arXiv:1607.06706.
13. Wang, J.-C.; Wang, J.-F.; He, K.W.; Hsu, C.-S. Environmental Sound Classification Using Hybrid SVM/KNN Classifier and MPEG-7 Audio Low-Level Descriptor. In Proceedings of the Neural Networks IJCNN'06 International Joint Conference, Vancouver, BC, Canada, 16–21 July 2006.
14. Krizhevsky, A.; Sutskever, I.; Hinton, G.E. Imagenet Classification with Deep Convolutional Neural Networks. In Proceedings of the Advances in Neural Information Processing Systems, Lake Tahoe, NV, USA, 3–8 December 2012.
15. Piczak, K.J. Environmental sound classification with convolutional neural networks. In Proceedings of the Machine Learning for Signal Processing (MLSP), 2015 IEEE 25th International Workshop, Boston, MA, USA, 17–20 September 2015.
16. Stowell, D.; Giannoulis, D.; Benetos, E.; Lagrange, M.; MPlumbley, D. Detection and Classification of Acoustic Scenes and Events. *IEEE Trans. Multimed.* **2015**, *17*, 1733–1746. [CrossRef]
17. DCASE 2017 Workshop. Available online: http://www.cs.tut.fi/sgn/arg/dcase2017/ (accessed on 30 June 2017).
18. Aytar, Y.; Vondrick, C.; Torralba, A. Soundnet: Learning Sound Representations from Unlabeled Video. In Proceedings of the Advances in Neural Information Processing Systems, Barcelona, Spain, 5–10 December 2016; pp. 892–900.
19. Dai, W.C.; Dai, S.; Qu, J.; Das, S. Very Deep Convolutional Neural Networks for Raw Waveforms. In Proceedings of the Acoustics, Speech and Signal Processing (ICASSP), 2017 IEEE International Conference, New Orleans, LA, USA, 5–9 March 2017.
20. Lin, M.; Chen, Q.; Yan, S. Network in Network. *arXiv* **2013**, arXiv:1312.4400.
21. Tokozume, Y.; Harada, T. Learning Environmental Sounds with End-to-End Convolutional Neural Network. In Proceedings of the Acoustics, Speech and Signal Processing (ICASSP), 2017 IEEE International Conference, New Orleans, LA, USA, 5–9 March 2017.
22. Tokozume, Y.; Ushiku, Y.; Harada, T. Learning from Between-class Examples for Deep Sound Recognition. In Proceedings of the ICLR 2018 Conference, Vancouver, BC, Canada, 30 April–3 May 2018.
23. Rosenblatt, F. The perceptron: A probabilistic model for information storage and organization in the brain. *Psychol. Rev.* **1958**, *65*, 386–408. [CrossRef] [PubMed]
24. Rumelhart, D.E.; Hinton, G.E.; Williams, R.J. Learning representations by back-propagating errors. *Nature* **1986**, *323*, 533. [CrossRef]

25. LeCun, Y.; Bottou, L.; Bengio, Y.; Haffner, P. Gradient-based learning applied to document recognition. *Proc. IEEE* **1998**, *86*, 2278–2324. [CrossRef]
26. Silver, D.; Huang, A.; Maddison, C.J.; Guez, A.; Sifre, L.; van den Driessche, G.; Schrittwieser, J.; Antonoglou, I.; Panneershelvam, V.; Lanctot, M. Mastering the game of Go with deep neural networks and tree search. *Nature* **2016**, *529*, 484. [CrossRef] [PubMed]
27. Glorot, X.; Bengio, Y. Understanding the Difficulty of Training Deep Feedforward Neural Networks. In Proceedings of the thirteenth international conference on artificial intelligence and statistics, Chia Laguna, Italy, 13–15 May 2010.
28. He, K.; Zhang, X.; Ren, S.; Sun, J. Delving Deep into Rectifiers: Surpassing Human-Level Performance on Imagenet Classification. In Proceedings of the IEEE International Conference on Computer Vision, Las Condes, Chile, 11–18 December 2015.
29. Ioffe, S.; Szegedy, C. Batch normalization: Accelerating deep network training by reducing internal covariate shift. *arXiv* **2015**, arXiv:1502.03167.
30. Piczak, K.J. ESC: Dataset for Environmental Sound Classification. In Proceedings of the 23rd ACM international conference on Multimedia, Brisbane, Australia, 26 October 2015.
31. Deng, J.; Dong, W.; Socher, R.; Li, L.-J.; Li, K.; Fei-Fei, L. Imagenet: A Large-Scale Hierarchical Image Database. In Proceedings of the IEEE Conference Computer Vision and Pattern Recognition CVPR, Miami, FL, USA, 20–25 June 2009.
32. Lin, T.-Y.; Maire, M.; Belongie, S.; Hays, J.; Perona, P.; Ramanan, D.; Doll'ar, P.; Zitnick, C.L. Microsoft Coco: Common Objects in Context. In Proceedings of the European Conference on Computer Vision, Zurich, Switzerland, 6–12 September 2014.
33. Salamon, J.; Bello, J.P. Deep convolutional neural networks and data augmentation for environmental sound classification. *IEEE Sign. Process. Lett.* **2017**, *24*, 279–283. [CrossRef]
34. Srivastava, N.; Hinton, G.; Krizhevsky, A.; Sutskever, I.; Salakhutdinov, R. Dropout: A simple way to prevent neural networks from overfitting. *J. Mach. Learn. Res.* **2014**, *15*, 1929–1958.
35. Nesterov, Y. *Gradient Methods for Minimizing Composite*; Springer: Berlin/Heidelberg, Germany, 2007.
36. Boddapati, V.; Petef, A.; Rasmusson, J.; Lundberg, L. Classifying environmental sounds using image recognition networks. *Proc. Comput. Sci.* **2017**, *112*, 2048–2056. [CrossRef]
37. Simonyan, K.; Vedaldi, A.; Zisserman, A. Deep inside convolutional networks: Visualising image classification models and saliency maps. *arXiv* **2013**, arXiv:1312.6034.
38. Zeiler, M.D.; Fergus, R. Visualizing and Understanding Convolutional Networks. In Proceedings of the European Conference on Computer Vision, Zurich, Switzerland, 6–12 September 2014.
39. Salamon, J.; Jacoby, C.; Bello, J.P. A Dataset and Taxonomy for Urban Sound Research. In Proceedings of the 22nd ACM international conference on Multimedia, Orlando, FL, USA, 3–7 November 2014.
40. Zheng, Y.-Y.; Kong, J.-L.; Jin, X.-B.; Wang, X.-Y.; Su, T.-L.; Wang, J.-L. Probability fusion decision framework of multiple deep neural networks for fine-grained visual classification. *IEEE Access* **2019**, *7*, 122740–122757. [CrossRef]
41. Tu, Y.; Du, J.; Lee, C. Speech enhancement based on teacher–student deep learning using improved speech presence probability for noise-robust speech recognition. *IEEE ACM Trans. Audio Speech Lang. Process.* **2019**, *27*, 2080–2091. [CrossRef]
42. Osayamwen, F.; Tapamo, J. Deep learning class discrimination based on prior probability for human activity recognition. *IEEE Access* **2019**, 14747–14756. [CrossRef]

© 2020 by the authors. Licensee MDPI, Basel, Switzerland. This article is an open access article distributed under the terms and conditions of the Creative Commons Attribution (CC BY) license (http://creativecommons.org/licenses/by/4.0/).

Article

# A Multi-Resolution Approach to GAN-Based Speech Enhancement

Hyung Yong Kim, Ji Won Yoon, Sung Jun Cheon, Woo Hyun Kang and Nam Soo Kim *

Department of Electrical and Computer Engineering and the Institute of New Media and Communications, Seoul National University, 1 Gwanak-ro, Gwanak-gu, Seoul 08826, Korea; hykim@hi.snu.ac.kr (H.Y.K.); jwyoon@hi.snu.ac.kr (J.W.Y.); sjcheon@hi.snu.ac.kr (S.J.C.); whkang@hi.snu.ac.kr (W.H.K.)
* Correspondence: nkim@snu.ac.kr; Tel.: +82-2-880-8419

**Abstract:** Recently, generative adversarial networks (GANs) have been successfully applied to speech enhancement. However, there still remain two issues that need to be addressed: (1) GAN-based training is typically unstable due to its non-convex property, and (2) most of the conventional methods do not fully take advantage of the speech characteristics, which could result in a sub-optimal solution. In order to deal with these problems, we propose a progressive generator that can handle the speech in a multi-resolution fashion. Additionally, we propose a multi-scale discriminator that discriminates the real and generated speech at various sampling rates to stabilize GAN training. The proposed structure was compared with the conventional GAN-based speech enhancement algorithms using the VoiceBank-DEMAND dataset. Experimental results showed that the proposed approach can make the training faster and more stable, which improves the performance on various metrics for speech enhancement.

**Keywords:** speech enhancement; generative adversarial network; relativistic GAN; convolutional neural network

Citation: Kim, H.Y.; Yoon, J.W.; Cheon, S.J.; Kang, W.H.; Kim, N.S. A Multi-Resolution Approach to GAN-Based Speech Enhancement. *Appl. Sci.* **2021**, *11*, 721. https://doi.org/10.3390/app11020721

Received: 2 December 2020
Accepted: 10 January 2021
Published: 13 January 2021

**Publisher's Note:** MDPI stays neutral with regard to jurisdictional claims in published maps and institutional affiliations.

**Copyright:** © 2021 by the authors. Licensee MDPI, Basel, Switzerland. This article is an open access article distributed under the terms and conditions of the Creative Commons Attribution (CC BY) license (https://creativecommons.org/licenses/by/4.0/).

## 1. Introduction

Speech enhancement is essential for various speech applications such as robust speech recognition, hearing aids, and mobile communications [1–4]. The main objective of speech enhancement is to improve the quality and intelligibility of the noisy speech by suppressing the background noise or interferences.

In the early studies on speech enhancement, the minimum mean-square error (MMSE)-based spectral amplitude estimator algorithms [5,6] were popular producing enhanced signal with low residual noise. However, the MMSE-based methods have been reported ineffective in non-stationary noise environments due to their stationarity assumption on speech and noise. An effective way to deal with the non-stationary noise is to utilize a priori information extracted from a speech or noise database (DB), called the template-based speech enhancement techniques. One of the most well-known template-based schemes is the non-negative matrix factorization (NMF)-based speech enhancement technique [7,8]. NMF is a latent factor analysis technique to discover the underlying part-based non-negative representations of the given data. Since there is no strict assumption on the speech and noise distributions, the NMF-based speech enhancement technique shows robustness to non-stationary noise environments. Since, however, the NMF-based algorithm assumes that all data is described as a linear combination of finite bases, it is known to suffer from speech distortion not covered by this representational form.

In the past few years, deep neural network (DNN)-based speech enhancement has received tremendous attention due to its ability to model complex mappings [9–12]. These methods map the noisy spectrogram to the clean spectrogram via the neural networks such as the convolutional neural network (CNN) [11] or recurrent neural network

(RNN) [12]. Although the DNN-based speech enhancement techniques have shown promising performance, most of the techniques typically focus on modifying the magnitude spectra. This could cause a phase mismatch between the clean and enhanced speech since the DNN-based speech enhancement methods usually reuse the noisy phase for waveform reconstruction. For this reason, there has been growing interest in phase-aware speech enhancement [13–15] that exploits the phase information during the training and reconstruction. To circumvent the difficulty of the phase estimation, end-to-end (E2E) speech enhancement technique which directly enhances noisy speech waveform in the time domain has been developed [16–18]. Since the E2E speech enhancement techniques are performed in a waveform-to-waveform manner without any consideration of the spectra, their performance is not dependant on the accuracy of the phase estimation.

The E2E approaches, however, rely on a distance-based loss functions between the time-domain waveforms. Since these distance-based costs do not take human perception into account, the E2E approaches are not guaranteed to achieve good human-perception-related metrics, e.g., the perceptual evaluation of speech quality (PESQ) [19], short-time objective intelligibility (STOI) [20], and etc. Recently, generative adversarial network (GAN) [21]-based speech enhancement techniques have been developed to overcome the limitation of the distance-based costs [22–26]. Adversarial losses of GAN provide an alternative objective function to reflect the human auditory property, which can make the distribution of the enhanced speech close to that of the clean speech. To our knowledge, SEGAN [22] was the first attempt to apply GAN to the speech enhancement task, which used the noisy speech as a conditional information for a conditional GAN (cGAN) [27]. In [26], an approach to replace a vanilla GAN with advanced GAN, such as Wasserstein GAN (WGAN) [28] or relativistic standard GAN (RSGAN) [29] was proposed based on the SEGAN framework.

Even though the GAN-based speech enhancement techniques have been found successful, there still remain two important issues: (1) training instability and (2) a lack in considering various speech characteristics. Since GAN aims at finding the Nash equilibrium to solve a mini-max problem, it has been known that the training is usually unstable. A number of efforts have been devoted to stabilize the GAN training in image processing, by modifying the loss function [28] or the generator and discriminator structures [30,31]. However, in speech processing, this problem has not been extensively studied yet. Moreover, since most of the GAN-based speech enhancement techniques directly employ the models used in image generation, it is necessary to modify them to suit the inherent nature of speech. For instance, the GAN-based speech enhancement techniques [22,24,26] commonly used U-Net generator originated from image processing tasks. Since the U-net generator consisted of multiple CNN layers, it was insufficient to reflect the temporal information of speech signal. In regression-based speech enhancement, the modified U-net structure adding RNN layers to capture the temporal information of speech showed prominent performances [32]. In [33] for the speech synthesis, an alternative loss function depended on multiple sizes of window length and fast Fourier transform (FFT) was proposed and generated a good quality of speech, which also considered speech characteristics in frequency domain.

In this paper, we propose novel generator and discriminator structures for the GAN-based speech enhancement which reflect the speech characteristics while ensuring stable training. The conventional generator is trained to find a mapping function from the noisy speech to the clean speech by using sequential convolution layers, which is considered an ineffective approach especially for speech data. In contrast, the proposed generator progressively estimates the wide frequency range of the clean speech via a novel up-sampling layer.

In the early stage of GAN training, it is too easy for the conventional discriminator to differentiate real samples from fake samples for high-dimensional data. This often lets GAN fail to reach the equilibrium point due to vanishing gradient [30]. To address this issue, we propose a multi-scale discriminator that is composed of multiple sub-discriminators processing speech samples at different sampling rates. Even if the training is in the early stage, the sub-discriminators at low-sampling rates can not easily differentiate the real samples from the fake, which contributes to stabilize the training. Empirical results showed that the proposed generator and discriminator were successful in stabilizing GAN training and outperformed the conventional GAN-based speech enhancement techniques. The main contributions of this paper are summarized as follows:

- We propose a progressive generator to reflect the multi-resolution characteristics of speech.
- We propose a multi-scale discriminator to stabilize the GAN training without additional complex training techniques.
- The experimental results showed that the multi-scale structure is an effective solution for both deterministic and GAN-based models, outperforming the conventional GAN-based speech enhancement techniques.

The rest of the paper is organized as follows: In Section 2, we introduce GAN-based speech enhancement. In Section 3, we present the progressive generator and multi-scale discriminator. Section 4 describes the experimental settings and performance measurements. In Section 5, we analyze the experimental results. We draw conclusions in Section 6.

## 2. GAN-Based Speech Enhancement

An adversarial network models the complex distribution of the real data via a two-player mini-max game between a generator and a discriminator. Specifically, the generator takes a randomly sampled noise vector $z$ as input and produces a fake sample $G(z)$ to fool the discriminator. On the other hand, the discriminator is a binary classifier that decides whether an input sample is real or fake. In order to generate a realistic sample, the generator is trained to deceive the discriminator, while the discriminator is trained to distinguish between the real sample and $G(z)$. In an adversarial training process, the generator and the discriminator are alternatively trained to minimize their respective loss functions. The loss functions for the standard GAN can be defined as follows:

$$L_G = \mathbb{E}_{z \sim \mathbb{P}_z(z)}[log(1 - D(G(z)))], \tag{1}$$

$$L_D = -\mathbb{E}_{x \sim \mathbb{P}_{clean}(x)}[log(D(x))] - \mathbb{E}_{z \sim \mathbb{P}_z(z)}[log(1 - D(G(z)))] \tag{2}$$

where $z$ is a randomly sampled vector from $\mathbb{P}_z(z)$ which is usually a normal distribution, and $\mathbb{P}_{clean}(x)$ is the distribution of the clean speech in the training dataset.

Since GAN was initially proposed for unconditional image generation that has no exact target, it is inadequate to directly apply GAN to speech enhancement which is a regression task to estimate the clean target corresponding to the noisy input. For this reason, GAN-based speech enhancement employs a conditional generation framework [27] where both the generator and discriminator are conditioned on the noisy waveform $c$ that has the clean waveform $x$ as the target. By concatenating the noisy waveform $c$ with the randomly sampled vector $z$, the generator $G$ can produce a sample that is closer to the clean waveform $x$. The training process of the cGAN-based speech enhancement is shown in Figure 1a, and the loss functions of the cGAN-based speech enhancement are

$$L_G = \mathbb{E}_{z \sim \mathbb{P}_z(z), c \sim \mathbb{P}_{noisy}(c)}[log(1 - D(G(z,c),c))], \tag{3}$$

$$L_D = -\mathbb{E}_{x \sim \mathbb{P}_{clean}(x), c \sim \mathbb{P}_{noisy}(c)}[log D(x,c)] - \mathbb{E}_{z \sim \mathbb{P}_z(z), c \sim \mathbb{P}_{noisy}(c)}[log(1 - D(G(z,c),c))] \tag{4}$$

where $\mathbb{P}_{clean}(x)$ and $\mathbb{P}_{noisy}(c)$ are respectively the distributions of the clean and noisy speech in the training dataset.

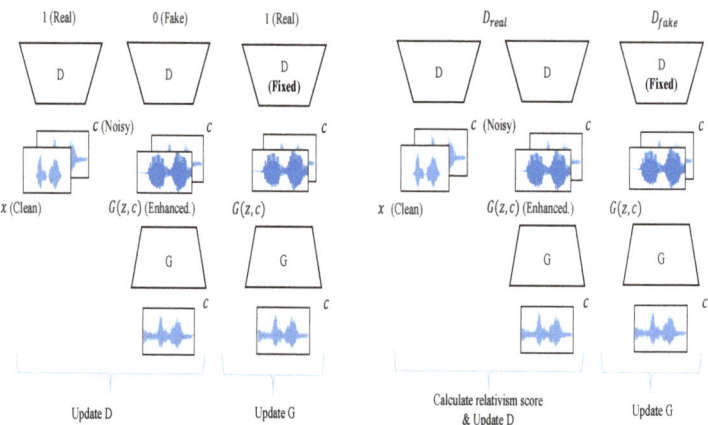

(a) cGAN-based speech enhancement   (b) RSGAN-based speech enhancement

**Figure 1.** Illustration of the conventional GAN-based speech enhancements. In the training of cGAN-based speech enhancement, the updates for generator and discriminator are alternated over several epochs. During the update of the discriminator, the target of discriminator is 0 for the clean speech and 1 for the enhanced speech. For the update of the generator, the target of discriminator is 1 with freezing discriminator parameters. In contrast, the RSGAN-based speech enhancement trains the discriminator to measure a relativism score of the real sample $D_{real}$ and generator to increase that of the fake sample $D_{fake}$ with fixed discriminator parameters.

In the conventional training of the cGAN, both the probabilities that a sample is from the real data $D(x,c)$ and generated data $D(G(z,c),c)$ should reach the ideal equilibrium point 0.5. However, unlike the expected ideal equilibrium point, they both have a tendency to become 1 because the generator can not influence the probability of the real sample $D(x,c)$. In order to alleviate this problem, RSGAN [29] proposed a discriminator to estimate the probability that the real sample is more realistic than the generated sample. The proposed discriminator makes the probability of the generated sample $D(G(z,c),c)$ increase when that of the real sample $D(x,c)$ decreases so that both the probabilities could stably reach the Nash equilibrium state. In [26], the experimental results showed that, compared to other conventional GAN-based speech enhancements, the RSGAN-based speech enhancement technique improved the stability of training and enhanced the speech quality. The training process of the RSGAN-based speech enhancement is given in Figure 1b, and the loss functions of RSGAN-based speech enhancement can be written as:

$$L_G = -\mathbb{E}_{(x_r,x_f)\sim(\mathbb{P}_r,\mathbb{P}_f)}[\log(\sigma(C(x_f) - C(x_r)))], \quad (5)$$

$$L_D = -\mathbb{E}_{(x_r,x_f)\sim(\mathbb{P}_r,\mathbb{P}_f)}[\log(\sigma(C(x_r) - C(x_f)))] \quad (6)$$

where the real and fake data-pairs are defined as $x_r \triangleq (x,c) \sim \mathbb{P}_r$ and $x_f \triangleq (G(z,c),c) \sim \mathbb{P}_f$, and $C(x)$ is the output of the last layer in discriminator before the sigmoid activation function $\sigma(\cdot)$, i.e., $D(x) = \sigma(C(x))$.

In order to stabilize GAN training, there are two penalties commonly used: A gradient penalty for discriminator [28] and $L_1$ loss penalty for generator [24]. First, the gradient penalty regularization for discriminator is used to prevent exploding or vanishing gradients. This regularization penalizes the model if the $L_2$ norm of the discriminator gradient moves away from 1 to satisfy the Lipschitz constraint. The modified discriminator loss functions with the gradient penalty are as follows:

$$L_{GP}(D) = \mathbb{E}_{\tilde{x},c \sim \tilde{\mathbb{P}}}\left[(\|\nabla_{\tilde{x},c} C(\tilde{x},c))\|_2 - 1)^2\right], \quad (7)$$

$$L_{D-GP}(D) = -\mathbb{E}_{(x_r,x_f) \sim (\mathbb{P}_r,\mathbb{P}_f)}[log(\sigma(C(x_r) - C(x_f)))] + \lambda_{GP} L_{GP}(D) \quad (8)$$

where $\tilde{\mathbb{P}}$ is the joint distribution of $c$ and $\tilde{x} = \epsilon x + (1 - \epsilon)\hat{x}$, $\epsilon$ is sampled from a uniform distribution in $[0, 1]$, and $\hat{x}$ is the sample from $G(z, c)$. $\lambda_{GP}$ is the hyper-parameter that controls the gradient penalty loss and the adversarial loss of the discriminator.

Second, several prior studies [22–24] found that it is effective to use an additional loss term that minimizes the $L_1$ loss between the clean speech $x$ and the generated speech $G(z, c)$ for the generator training. The modified generator loss with the $L_1$ loss is defined as

$$L_1(G) = \|G(z,c) - x\|_1, \quad (9)$$

$$L_{G-L_1}(G) = -\mathbb{E}_{(x_r,x_f) \sim (\mathbb{P}_r,\mathbb{P}_f)}[log(\sigma(C(x_f) - C(x_r)))] + \lambda_{L_1} L_1(G) \quad (10)$$

where $\|\cdot\|_1$ is $L_1$ norm, and $\lambda_{L_1}$ is a hyper-parameter for balancing the $L_1$ loss and the adversarial loss of the generator.

## 3. Multi-Resolution Approach for Speech Enhancement

In this section, we propose a novel GAN-based speech enhancement model which consists of a progressive generator and a multi-scale discriminator. The overall architecture of the proposed model is shown in Figure 2, and the details of the progressive generator and the multi-scale discriminator are given in Figure 3.

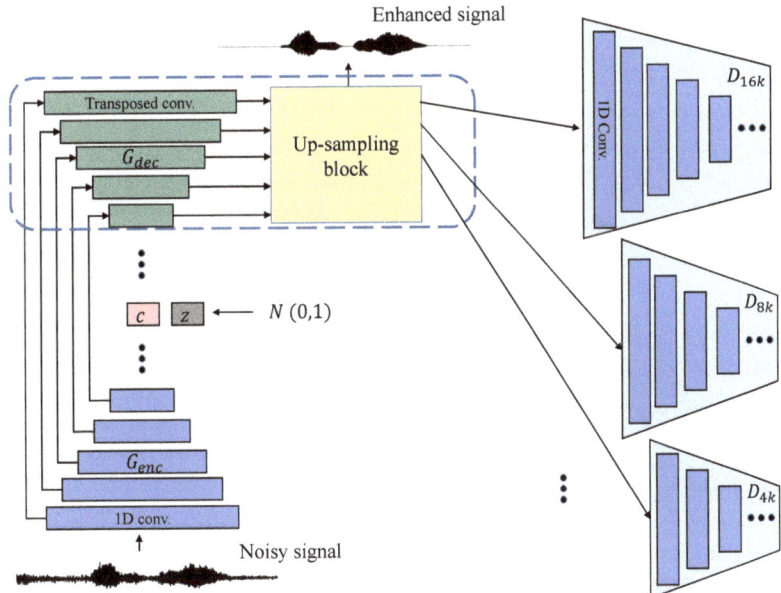

**Figure 2.** Overall architecture of the proposed GAN-based speech enhancement. The up-sampling block and the multiple discriminators $D_n$ are newly added, and the rest of the architecture is the same as that of [26]. The components within the dashed line will be addressed in Figure 3.

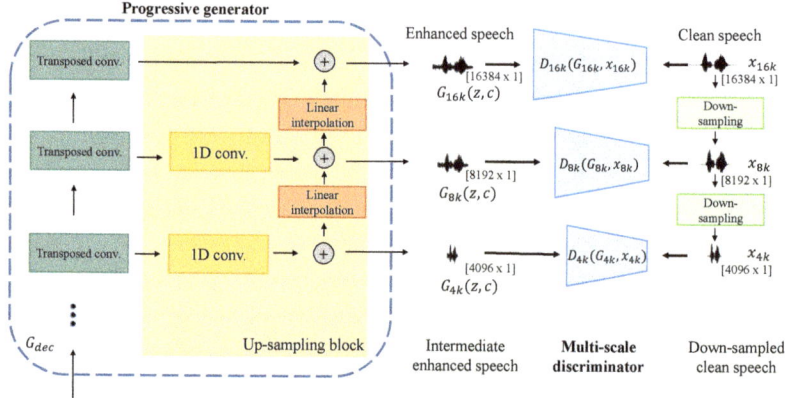

**Figure 3.** Illustration of the progressive generator and the multi-scale discriminator. Sub-discriminators calculate the relativism score $D_n(G_n, x_n) = \sigma(C_n(x_{r_n}) - C_n(x_{f_n}))$ at each layer. The figure is the case when $p, q = 4k$, but it can be extended for any $p$ and $q$. In our experiment, we covered that $p$ and $q$ were from $1k$ to $16k$.

*3.1. Progressive Generator*

Conventionally, GAN-based speech enhancement systems adopt U-Net generator [22] which is composed of two components: An encoder $G_{enc}$ and a decoder $G_{dec}$. The encoder $G_{enc}$ consists of repeated convolutional layers to produce compressed latent vectors from a noisy speech, and the decoder $G_{dec}$ contains multiple transposed convolutional layers to restore the clean speech from the compressed latent vectors. These transposed convolutional layers in $G_{dec}$ are known to be able to generate low-resolution data from the compressed latent vectors, however, the capability to generate a high-resolution data is severely limited [30]. Especially in the case of speech data, it is difficult for the transposed convolutional layers to generate the speech with a high-sampling rate because it should cover a wide frequency range.

Motivated from the progressive GAN, which starts with generating low-resolution images and then progressively increases the resolution [30,31], we propose a novel generator that can incrementally widen the frequency band of the speech by applying an up-sampling block to the decoder $G_{dec}$. As shown in Figure 3, the proposed up-sampling block consists of 1D-convolution layers, element-wise addition, and liner interpolation layers. The up-sampling block yields the intermediate enhanced speech $G_n(z, c)$ at each layer through the 1D convolution layer and element-wise addition so that the wide frequency band of the clean speech is progressively estimated. Since a sampling rate is increased through the linear interpolation layer, it is possible to generate the intermediate enhanced speech at the higher layer while maintaining estimated frequency components at the lower layer. This incremental process is repeated until the sampling rate reaches the target sampling rate which is $16kHz$ in our experiment. Finally, we exploit the down-sampled clean speech $x_n$ processed by low-pass filtering and decimation as the target for each layer to provide multi-resolution loss functions. We define the real and fake data-pairs at different sampling rates as $x_{r_n} \triangleq (x_n, c_n) \sim \mathbb{P}_{r_n}$ and $x_{f_n} \triangleq (G_n(z,c), c_n) \sim \mathbb{P}_{f_n}$, and the proposed multi-resolution loss functions with $L_1$ loss are given as follows:

$$L_G(p) = \sum_{\substack{n \geq p \\ n \in N_G}} L_{G_n} + \lambda_{L_1} L_1(G_n), N_G \in \{1k, 2k, 4k, 8k, 16k\},$$

$$= \sum_{\substack{n \geq p \\ n \in N_G}} -\mathbb{E}_{(x_{r_n}, x_{f_n}) \sim (\mathbb{P}_{r_n}, \mathbb{P}_{f_n})}[\log(\sigma(C_n(x_{f_n}) - C_n(x_{r_n})))] + \lambda_{L_1} \|G_n(z,c) - x_n\|_1$$

(11)

where $N_G$ is the possible set of $n$ for the proposed generator, and $p$ is the sampling rate at which the intermediate enhanced speech is firstly obtained.

### 3.2. Multi-Scale Discriminator

When generating high-resolution image and speech data in the early stage of training, it is hard for the generator to produce a realistic sample due to the insufficient model capacity. Therefore, the discriminator can easily differentiate the generated samples from the real samples, which means that the real and fake data distributions do not have substantial overlap. This problem often causes training instability and even mode collapses [30]. For the stabilization of the training, we propose a multi-scale discriminator that consists of multiple sub-discriminators treating speech samples at different sampling rates.

As presented in Figure 3, the intermediate enhanced speech $G_n(z,c)$ at each layer restores the down-sampled clean speech $x_n$. Based on this, we can utilize the intermediate enhanced speech and down-sampled clean speech as the input to each sub-discriminator $D_n$. Since each sub-discriminator can only access limited frequency information depending on the sampling rate, we can make each sub-discriminator solve different levels of discrimination tasks. For example, discriminating the real from the generated speech is more difficult at the lower sampling rate than at the higher rate. The sub-discriminator at a lower sampling rate plays an important role in stabilizing the early stage of the training. As the training progresses, the role shifts upwards to the sub-discriminators at higher sampling rates. Finally, the proposed multi-scale loss for discriminator with gradient penalty is given by

$$L_D(q) = \sum_{\substack{n \geq q \\ n \in N_D}} L_{D_n} + \lambda_{GP} L_{GP}(D_n), N_D \in \{1k, 2k, 4k, 8k, 16k\},$$

$$= \sum_{\substack{n \geq q \\ n \in N_D}} -\mathbb{E}_{(x_{r_n}, x_{f_n}) \sim (\mathbb{P}_{r_n}, \mathbb{P}_{f_n})}[\log(\sigma(C_n(x_{r_n}) - C(x_{f_n})))] + \lambda_{GP} \mathbb{E}_{\widetilde{x_n}, c_n \sim \widetilde{\mathbb{P}_n}}[(\|\nabla_{\widetilde{x_n}, c_n} C(\widetilde{x_n}, c_n)\|_2 - 1)^2]$$

(12)

where $\widetilde{\mathbb{P}_n}$ is the joint distribution of the down-sampled noisy speech $c_n$ and $\widetilde{x_n} = \epsilon x_n + (1-\epsilon)\hat{x}_n$, $\epsilon$ is sampled from a uniform distribution in $[0, 1]$, $x_n$ is the down-sampled clean speech, and $\hat{x}_n$ is the sample from $G_n(z,c)$. $N_D$ is the possible set of $n$ for the proposed discriminator, and $q$ is the minimum sampling rate at which the intermediate enhanced output was utilized as the input to a sub-discriminator for the first time. The adversarial losses $L_{D_n}$ are equally weighted.

## 4. Experimental Settings

### 4.1. Dataset

We used a publicly available dataset in [34] for evaluating the performance of the proposed speech enhancement technique. The dataset consists of 30 speakers from the Voice Bank corpus [35], and used 28 speakers (14 male and 14 female) for the training set (11572 utterances) and 2 speakers (one male and one female) for the test set (824 utterances). The training set simulated a total of 40 noisy conditions with 10 different noise sources (2 artificial and 8 from the DEMAND database [36]) at signal-to-noise ratios (SNRs) of 0, 5, 10, and 15 dB. The test set was created using 5 noise sources (living room, office, bus, cafeteria, and public square noise from the DEMAND database), which were different from the training noises, added at SNRs 2.5, 7.5, 12.5, and 17.5 dB. The training and test sets were down-sampled from 48 kHz to 16 kHz.

## 4.2. Network Structure

The configuration of the proposed generator is described in Table 1. We used the U-Net structure with 11 convolutional layers for the encoder $G_{enc}$ and the decoder $G_{dec}$ as in [22,26]. Output shapes at each layer were represented by the number of temporal dimensions and feature maps. Conv1D in the encoder denotes a one-dimensional convolutional layer, and TrConv in the decoder means a transposed convolutional layer. We used approximately 1 s of speech (16384 samples) as the input to the encoder. The last output of the encoder was concatenated with a noise which had the shape of 8 × 1024 randomly sampled from the standard normal distribution $N(0,1)$. In [27], it was reported that the generator usually learns to ignore the noise prior z in the CGAN, and we also observed a similar tendency in our experiments. For this reason, we removed the noise from the input, and the shape of the latent vector became 8 × 1024. The architecture of $G_{dec}$ was a mirroring of $G_{enc}$ with the same number and width of the filters per layer. However, skip connections from $G_{enc}$ made the number of feature maps in every layer to be doubled. The proposed up-sampling block $G_{up}$ consisted of 1D convolution layers, element-wise addition operations, and linear interpolation layers.

**Table 1.** Architecture of the proposed generator. Output shape represented temporal dimension and feature maps.

| Block | Operation | Output Shape | |
|---|---|---|---|
| | Input | 16,384 × 1 | |
| Encoder | Conv1D (filterlength = 31, stide = 2) | 8192 × 16 | |
| | | 4096 × 32 | |
| | | 2048 × 32 | |
| | | 1024 × 64 | |
| | | 512 × 64 | |
| | | 256 × 128 | |
| | | 128 × 128 | |
| | | 64 × 256 | |
| | | 32 × 256 | |
| | | 16 × 512 | |
| | Latent vector | 8 × 1024 | |
| Decoder | Trconv (filterlength = 31, stide = 2) | 16 × 1024 | |
| | | 32 × 512 | |
| | | 64 × 512 | |
| | | 128 × 256 | |
| | | 256 × 256 | |
| | | 512 × 128 | |
| | Trconv (filterlength = 31, stide = 2) | Conv1D (filterlength = 17, stide = 1) Element-wise addition Linear interpolation layer | 1024 × 128 | 1024 × 1 |
| | | | 2048 × 64 | 2048 × 1 |
| | | | 4096 × 64 | 4096 × 1 |
| | | | 8192 × 32 | 8192 × 1 |
| | | | 16,384 × 1 | |

In this experiment, the proposed discriminator had the same serial convolutional layers as $G_{enc}$. The input to the discriminator had two channels of 16,384 samples, which were the clean speech and enhanced speech. The rest of the temporal dimension and feature-maps were the same as those of $G_{enc}$. In addition, we used LeakyReLU activation function without a normalization technique. After the last convolutional layers, there were a $1 \times 1$ convolution, and its output was fed to a fully-connected layer. To construct the proposed multi-scale discriminator, we used 5 different sub-discriminators, which were $D_{16k}, D_{8k}, D_{4k}, D_{2k}, and D_{1k}$ trained according to in Equation (12). Each sub-discriminator had a different input dimension depending on the sampling rate.

The model was trained using the Adam optimizer [37] for 80 epochs with 0.0002 learning rate for both the generator and discriminator. The batch size was 50 with 1-s audio signals that were sliced using windows of length 16,384 with 8192 overlaps. We also applied a pre-emphasis filter with impulse response $[-0.95, 1]$ to all training samples. For inference, the enhanced signals were reconstructed through overlap-add. The hyper-parameters to balance the penalty terms were set as $\lambda_{L_1} = 200$ and $\lambda_{GP} = 10$ such that they could match the dynamic range of magnitude with respect to the generator and discriminator losses. Note that we gave the same weight to the adversarial losses, $L_{G_n}$ and $L_{D_n}$, for all $n \in \{1k, 2k, 4k, 8k, 16k\}$. We implemented all the networks using Keras with Tensorflow [38] back-end using the public code (The SERGAN framework is available at https://github.com/deepakbaby/se_relativisticgan). All training was performed on single Titan RTX 24 GB GPU, and it took around 2 days.

### 4.3. Evaluation Methods
#### 4.3.1. Objective Evaluation

The quality of the enhanced speech was evaluated using the following objective metrics:

- PESQ: Perceptual evaluation of speech quality defined in the ITU-T P.862 standard [19] (from $-0.5$ to $4.5$),
- STOI: Short-time objective intelligibility [20] (from 0 to 1),
- CSIG: Mean opinion score (MOS) prediction of the signal distortion attending only to the speech signal [39] (from 1 to 5),
- CBAK: MOS prediction of the intrusiveness of background noise [39] (from 1 to 5),
- COVL: MOS prediction of the overall effect [39] (from 1 to 5).

#### 4.3.2. Subjective Evaluation

To compare the subjective quality of the enhanced speech by baseline and proposed methods, we conducted two pairs of AB preference tests: AECNN versus the progressive generator and SERGAN versus the progressive generator with the multi-scale discriminator. Two speech in each pair were given in arbitrary order. For each listening test, 14 listeners participated, and 50 pairs of the speech were randomly selected. Listeners could listen to the speech pairs as many times as they wanted and were instructed to choose the speech with better perceptual quality. If the quality of the two samples was indistinguishable, listeners could select no preference.

## 5. Experiments and Results

In order to investigate the individual effect of the proposed generator and discriminator, we experimented on the progressive generator with and without the multi-scale discriminator. Furthermore, we plotted $L_1$ losses at each layer $L_1(G_n)$ to show that the proposed model makes training fast and stable. Finally, the performance of the proposed model is compared with that of the other GAN-based speech enhancement techniques.

### 5.1. Performance of Progressive Generator
#### 5.1.1. Objective Results

The purpose of these experiments is to show the effectiveness of the progressive generator. Table 2 presents the performance of the proposed generator when we minimized only the $L_1(G_n)$ in Equation (11). In order to better understand the influence of the progressive structure on the PESQ score, we conducted an ablation study with different $p$ in $\sum_{n \geq p} L_1(G_n)$. As illustrated in Table 2, compared to the auto-encoder CNN (AECNN) [26] that is the conventional U-net generator minimizing the $L_1$ loss only, the PESQ score of the progressive generator improved from 2.5873 to 2.6516. Furthermore, for the smaller $p$, we got a better PESQ score, and the best PESQ score was achieved when $p$ was the lowest, i.e., 1k. For enhancing high-resolution speech, we verified that it is very useful to progressively generate intermediate enhanced speech while maintaining the estimated information obtained at lower sampling rate. We used the best generator $p = 1k$ in Table 2 for the subsequent experiments.

**Table 2.** Comparison of results between different architectures of the progressive generator. The best model is shown in bold type.

| Model | $\sum_{n \geq p} L_1(G_n)$ | PESQ |
|---|---|---|
| AECNN [26] | $p = 16k$ | 2.5873 |
| Proposed | $p = 8k$ | 2.6257 |
|  | $p = 4k$ | 2.6335 |
|  | $p = 2k$ | 2.6407 |
|  | $p = 1k$ | **2.6516** |

5.1.2. Subjective Results

The preference score of AECNN and the progressive generator was shown in Figure 4a. The progressive generator was preferred to AECNN in 43.08% of the cases, while the opposite preference was 25.38% (no preference in 31.54% of the cases). From the results, we verified that the proposed generator could produce the speech with not only higher objective measurements but also better perceptual quality.

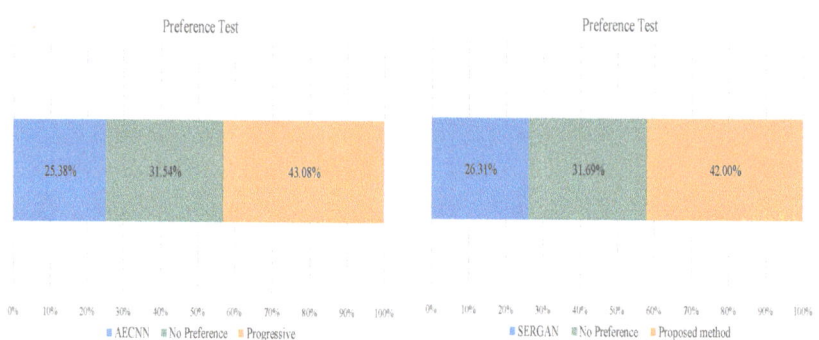

(a) AECNN versus Progressive generator  (b) SERGAN versus Proposed method

**Figure 4.** Results of AB preference test. A subset of test samples used in the evaluation is accessible on a webpage https://multi-resolution-SE-example.github.io.

5.2. Performance of Multi-Scale Discriminator

5.2.1. Objective Results

The goal of these experiments is to show the efficiency of the multi-scale discriminator compared to the conventional single discriminator. As shown in Table 3, we evaluated the performance of the multi-scale discriminator while varying $q$ of the multi-scale loss $L_D(q)$ in Equation (12), which means varying the number of sub-discriminators. Compared to the baseline proposed in [26], the progressive generator with the single discriminator showed an improved PESQ score from 2.5898 to 2.6514. The multi-scale discriminators outperformed the single discriminators, and the best PESQ score of 2.7077 was obtained when $q = 4k$. Interestingly, we could observe that the performance was degraded when the $q$ became below $4k$. One possible explanation for this phenomenon would be that since the progressive generator faithfully generated the speech below the 4 kHz sampling rate, it was difficult for the discriminator to differentiate the fake from the real speech. This let the additional sub-discriminators a little bit useless for performance improvement.

**Table 3.** Comparison of results between different architectures of the multi-scale discriminator. Except for the SERGAN, the generator of all architectures used the best model in Table 2. The best model is shown in bold type.

| Model | Generator | Discriminator | $L_D(q)$ | PESQ | RTF |
|---|---|---|---|---|---|
| SERGAN [26] | U-net | Single | $q = 16k$ | 2.5898 | 0.008 |
| Proposed | Progressive | Single | $q = 16k$ | 2.6514 | 0.010 |
| | Progressive | Multi-scale | $q = 8k$ | 2.6541 | |
| | | | $q = 4k$ | **2.7077** | |
| | | | $q = 2k$ | 2.6664 | |
| | | | $q = 1k$ | 2.6700 | |

5.2.2. Subjective Results

The preference scores of SERGAN and the progressive generator with multi-scale discriminator were shown in Figure 4b. The proposed method was preferred over SERGAN in 42.00% of the cases, while SERGAN was preferred in 26.31% of the cases (no preference in 31.69% of the cases). These results showed that the proposed method could enhance the speech with better objective metrics and subjective perceptual scores.

5.2.3. Real-Time Feasibility

SERGAN and the proposed method were evaluated in terms of the real-time factor(RTF) to verify the real-time feasibility, which is defined as the ratio of the time taken to enhance the speech to the duration of the speech (small factors indicate faster processing). CPU and graphic card used for the experiment were Intel Xeon Silver 4214 CPU 2.20 GHz and single Nvidia Titan RTX 24 GB. Since the generator of AECNN and SERGAN is the same, their RTF has the same value. Therefore, we only compared the RTF of SERGAN and the proposed method in Table 3. As the input window length was about 1 s of speech (16,384 samples), and the overlap was 0.5 s of speech (8192 samples), the total processing delay of all models can be computed by the sum of the 0.5 s and the actual processing time of the algorithm. In Table 3, we observed that the RTF of SERGAN and the proposed model was small enough for the semi-real-time applications. The similar value of the RTF for SEGAN and the proposed model also verified that adding the up-sampling network did not significantly increase the computational complexity.

5.3. Analysis and Comparison of Spectorgrams

An example of the spectrograms of clean speech, noisy speech, and the enhanced speech by different models are shown in Figure 5. First, we focused on the black box to verify the effectiveness of the progressive generator. Before 0.6 s, a non-speech period, we could observe that the noise containing wide-band frequencies was considerably reduced since the progressive generator incrementally estimated the wide frequency range of the clean speech. Second, when we compared spectrograms of the multi-scale discriminator and that of the single discriminator, the different pattern was presented in the red box. The multi-scale discriminator was able to suppress more noise than the single discriminator in the non-speech period. We could confirm that the multi-scale discriminator selectively reduced high-frequency noise in a speech period as the sub-discriminators in multi-scale discriminator differentiate the real and fake speech at the different sampling rates.

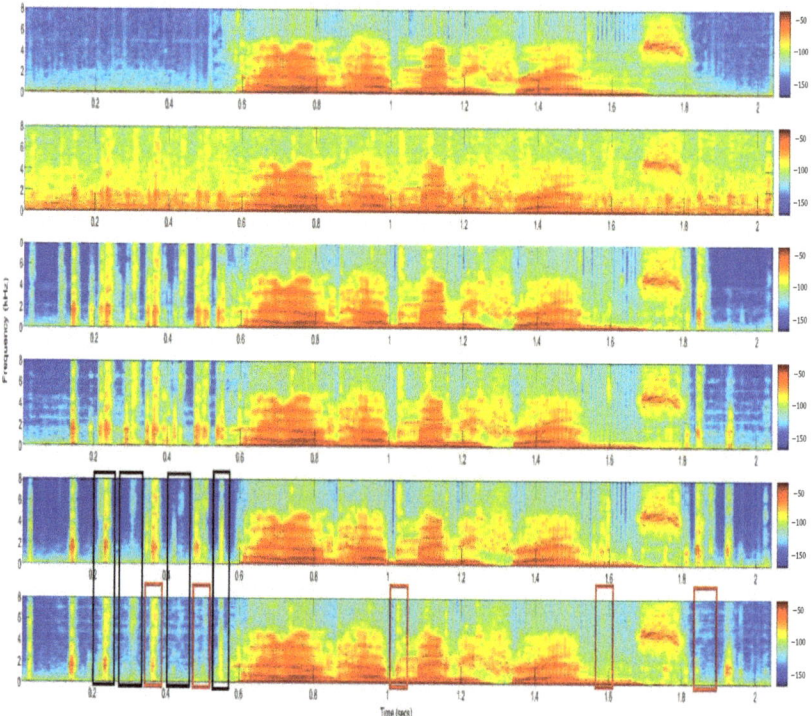

**Figure 5.** Spectrograms from the top to the bottom correspond to clean speech, noisy speech, enhanced speech by AECNN, SERGAN, progressive generator, progressive generator with multi-scale discriminator, respectively.

### 5.4. Fast and Stable Training of Proposed Model

To analyze the learning behavior of the proposed model in more depth, we plotted $L_1(G_n)$ in Equation (11) obtained from the best model in Table 3 and SERGAN [26] during the whole training periods. As the clean speech was progressively estimated by the intermediate enhanced speech, the stable convergence behavior of $L_1(G_n)$ was shown in Figure 6. With the help of $L_1(G_n)$ at low layers ($n = 1, 2, 4, 8$), $L_1(G_{16k})$ for the proposed model decreased faster and more stable than that of SERGAN. From the results, we can convince that the proposed model accelerates and stabilizes the GAN training.

### 5.5. Comparison with Conventional GAN-Based Speech Enhancement Techniques

Table 4 shows the comparison with other GAN-based speech enhancement methods that have the E2E structure. The GAN-based enhancement techniques which were evaluated in this experiment are as follows: **SEGAN** [22] has the U-net structure with conditional GAN. Similar to the structure of SEGAN, **AECNN** [26] is trained to only minimize $L_1$ loss, and **SERGAN** [26] is based on relativistic GAN. **CP-GAN** [40] has modified the generator and discriminator of SERGAN to utilize contextual information of the speech. The progressive generator without adversarial training even showed better results than CP-GAN on PESQ and CBAK. Finally, the progressive generator with the multi-scale discriminator outperformed the other GAN-based speech enhancement methods for three metrics.

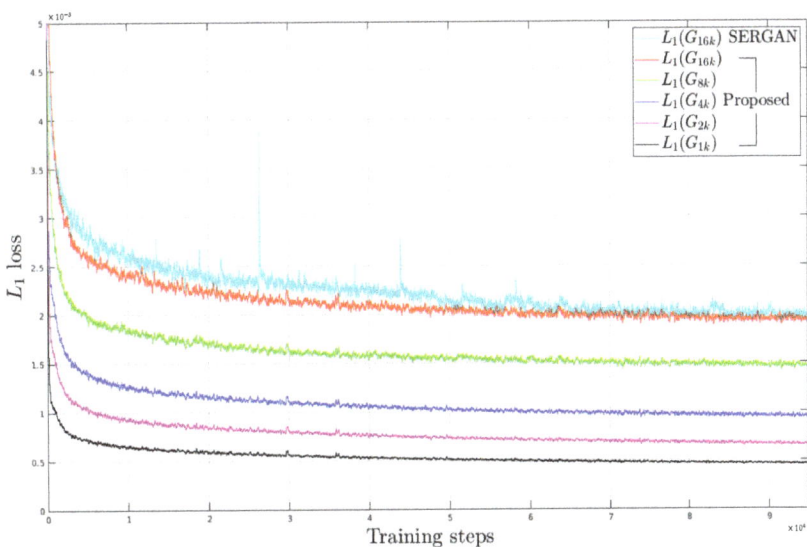

**Figure 6.** Illustration of $L_1(G_n)$ as a function of training steps.

**Table 4.** Comparison of results between different GAN-based speech enhancement Techniques. The best result is highlighted in bold type.

| Model | PESQ | CSIG | CBAK | COVL | STOI |
| --- | --- | --- | --- | --- | --- |
| Noisy | 1.97 | 3.35 | 2.44 | 2.63 | 0.91 |
| SEGAN [24] | 2.16 | 3.48 | 2.68 | 2.67 | 0.93 |
| AECNN [26] | 2.59 | 3.82 | **3.30** | 3.20 | 0.94 |
| SERGAN [26] | 2.59 | 3.82 | 3.28 | 3.20 | 0.94 |
| CP-GAN [38] | 2.64 | 3.93 | 3.29 | 3.28 | 0.94 |
| The progressive generator without adversarial training | 2.65 | 3.90 | **3.30** | 3.27 | 0.94 |
| The progressive generator with the multi-scale discriminator | **2.71** | **3.97** | 3.26 | **3.33** | 0.94 |

## 6. Conclusions

In this paper, we proposed a novel GAN-based speech enhancement technique utilizing the progressive generator and multi-scale discriminator. In order to reflect the speech characteristic, we introduced a progressive generator which can progressively estimate the wide frequency range of the speech by incorporating an up-sampling layer. Furthermore, for accelerating and stabilizing the training, we proposed a multi-scale discriminator which consists of a number of sub-discriminators operating at different sampling rates.

For performance evaluation of the proposed methods, we conducted a set of speech enhancement experiments using the VoiceBank-DEMAND dataset. From the results, it was shown that the proposed technique provides a more stable GAN training while showing consistent performance improvement on objective and subjective measures for speech enhancement. We also checked the semi-real-time feasibility by observing a small increment of RTF between the baseline generator and the progressive generator.

As the proposed network mainly focused on the multi-resolution attribute of speech in the time domain, one possible future study is to expand the proposed network to utilize the multi-scale attribute of speech in the frequency domain. Since the progressive generator and multi-scale discriminator can also be applied to the GAN-based speech reconstruction models such as neural vocoder for speech synthesis and codec, we will study the effects of the proposed methods.

**Author Contributions:** Conceptualization, H.Y.K. and N.S.K.; methodology, H.Y.K. and J.W.Y.; software, H.Y.K. and J.W.Y.; validation, H.Y.K. and N.S.K.; formal analysis, H.Y.K.; investigation, H.Y.K. and S.J.C.; resources, H.Y.K. and N.S.K.; data curation, H.Y.K. and W.H.K.; writing—original draft preparation, H.Y.K.; writing—review and editing, J.W.Y., W.H.K., S.J.C., and N.S.K.; visualization, H.Y.K.; supervision, N.S.K.; project administration, N.S.K.; funding acquisition, N.S.K. All authors have read and agreed to the published version of the manuscript.

**Funding:** This work was supported by Samsung Research Funding Center of Samsung Electronics under Project Number SRFCIT1701-04.

**Institutional Review Board Statement:** Not applicable.

**Informed Consent Statement:** Not applicable.

**Data Availability Statement:** Not applicable.

**Conflicts of Interest:** The authors declare no conflict of interest.

## References

1. Benesty, J.; Makino, S.; Chen, J.D. *Speech Enhancement*; Springer; New York, NY, USA, 2007.
2. Boll, S.F. Suppression of acoustic noise in speech using spectral subtraction. *IEEE Trans. Acoust. Speech Signal Process.* **1979**, *27*, 113–120. [CrossRef]
3. Lim, J.S.; Oppenheim, A.V. Enhancement and bandwidth compression of noisy speech. *Proc. IEEE* **1979**, *67*, 1586–1604. [CrossRef]
4. Scalart, P. Speech enhancement based on a priori signal to noise estimation. In Proceedings of the IEEE International Conference on Acoustics, Speech and Signal Processing (ICASSP), Atlanta, GA, USA, 7–10 May 1996; pp. 629–632.
5. Ephraim, Y.; Malah, D. Speech enhancement using a minimum-mean square error short-time spectral amplitude estimator. *IEEE Trans. Acoust. Speech Signal Process.* **1984**, *32*, 1109–1121. [CrossRef]
6. Kim, N.S.; Chang, J.H. Spectral enhancement based on global soft decision. *IEEE Signal Process. Lett.* **2000**, *7*, 108–110.
7. Kwon, K.; Shin, J.W.; Kim, N.S. NMF-based speech enhancement using bases update. *IEEE Signal Process. Lett.* **2015**, *22*, 450–454. [CrossRef]
8. Wilson, K.; Raj, B.; Smaragdis, P.; Divakaran, A. Speech denoising using nonnegative matrix factorization with priors. In Proceedings of the IEEE International Conference on Acoustics, Speech and Signal Processing (ICASSP), Las Vegas, NV, USA, 30 March–4 April 2008; pp. 4029–4032.
9. Xu, Y.; Du, J.; Dai, L.R.; Lee, C.H. A regression approach to speech enhancement based on deep neural networks. *IEEE Trans. Audio Speech Lang. Process.* **2015**, *23*, 7–19. [CrossRef]
10. Grais, E.M.; Sen, M.U.; Erdogan, H. Deep neural networks for single channel source separation. In Proceedings of the IEEE International Conference on Acoustics, Speech and Signal Processing (ICASSP), Florence, Italy, 4–9 May 2014; pp. 3734–3738.
11. Zhao, H.; Zarar, S.; Tashev, I.; Lee, C.H. Convolutional-recurrent neural networks for speech enhancement. In Proceedings of the IEEE International Conference on Acoustics, Speech and Signal Processing (ICASSP), Calgary, AB, Canada, 15–20 April 2018; pp. 2401–2405.
12. Huang, P.S.; Kim, M.; Hasegawa. J.M.; Smaragdis, P. Joint optimization of masks and deep recurrent neural networks for monaural source separation. *IEEE Trans. Audio Speech Lang. Process.* **2015**, *23*, 2136–2147. [CrossRef]
13. Roux, J.L.; Wichern, G.; Watanabe, S.; Sarroff, A.; Hershey, J. The Phasebook: Building complex masks via discrete representations for source separation. In Proceedings of the IEEE International Conference on Acoustics, Speech and Signal Processing (ICASSP), Brighton, UK, 12–17 May 2019; pp. 66–70.
14. Wang, Z.; Tan, K.; Wang, D. Deep Learning based phase reconstruction for speaker separation: A trigonometric perspective. In Proceedings of the IEEE International Conference on Acoustics, Speech and Signal Processing (ICASSP), Brighton, UK, 12–17 May 2019; pp. 71–75.
15. Wang, Z.; Roux, J.L.; Wang, D.; Hershey, J. End-to-End Speech Separation with Unfolded Iterative Phase Reconstruction. *arXiv* **2018**, arXiv:1804.10204.
16. Pandey, A.; Wang, D. A new framework for supervised speech enhancement in the time domain. In Proceedings of the INTERSPEECH, Hyderabad, India, 2–6 September 2018; pp. 1136–1140.
17. Stoller, D.; Ewert, S.; Dixon, S. Wave-U-net: A multi-scale neural network for end-to-end audio source separation. In Proceedings of the International Society for Music Information Retrieval Conference (ISMIR), Paris, France, 23–27 September 2018; pp. 334–340.
18. Rethage, D.; Pons, J.; Xavier, S. A wavenet for speech denoising. In Proceedings of the IEEE International Conference on Acoustics, Speech and Signal Processing (ICASSP), Calgary, AB, Canada, 15–20 April 2018; pp. 5069–5073.

29. ITU-T. Perceptual Evaluation of Speech Quality (PESQ): An Objective Method for End-to-End Speech Quality Assessment of Narrow-Band Telephone Networks and Speech Codecs. Rec. ITU-T P. 862; 2000. Available online: https://www.itu.int/rec/T-REC-P.862 (accessed on 18 February 2019).
30. Jensen, J.; Taal, C.H. An algorithm for predicting the intelligibility of speech masked by modulated noise maskers. *IEEE Trans. Audio Speech Lang. Process.* **2016**, *24*, 2009–2022. [CrossRef]
31. Goodfellow, I.; Pouget-Abadie, J.; Mirza, M.; Xu, B.; Warde-Farley, D.; Ozair, S.; Courville, A.; Bengio, Y. Generative adversarial nets. In Proceedings of the Advances in Neural Information Processing Systems (NeurIPS), Montreal, QC, Canada, 8–13 December 2014; pp. 2672–2680.
32. Pascual, S.; Bonafonte, A.; Serrà, J. SEGAN: Speech enhancement generative adversarial network. In Proceedings of the INTERSPEECH, Stockholm, Sweden, 20–24 August 2017; pp. 3642–3646.
33. Soni, M.H.; Shah, N.; Patil, H.A. Time-frequency masking-based speech enhancement using generative adversarial network. In Proceedings of the IEEE International Conference on Acoustics, Speech and Signal Processing (ICASSP), Calgary, AB, Canada, 15–20 April 2018; pp. 5039–5043.
34. Pandey, A.; Wang, D. On adversarial training and loss functions for speech enhancement. In Proceedings of the IEEE International Conference on Acoustics, Speech and Signal Processing (ICASSP), Calgary, AB, Canada, 15–20 April 2018; pp. 5414–5418
35. Fu, S.-W.; Liao, C.-F.; Yu, T.; Lin, S.-D. MetricGAN: Generative adversarial networks based black-box metric scores optimization for speech enhancement. In Proceedings of the International Conference on Machine Learning (ICML), Long Beach, CA, USA, 9–15 September 2019.
36. Baby, D.; Verhulst, S. Sergan: speech enhancement using relativistic generative adversarial networks with gradient penalty. In Proceedings of the IEEE International Conference on Acoustics, Speech and Signal Processing (ICASSP), Brighton, UK, 12–17 May 2019; pp. 106–110.
37. Isola, P.; Zhu, J.-Y.; Zhou, T.; Efros, A.A. Image-to-image translation with conditional adversarial networks. In Proceedings of the IEEE/CVF Conference on Computer Vision and Pattern Recognition (CVPR), Las Vegas, NV, USA, 26 June–1 July 2016; pp. 1125–1134.
38. Gulrajani, I.; Ahmed, F.; Arjovsky, M.; Dumoulin, V.; Courville, A.C. Improved training of wasserstein gans, In Proceedings of the Advances in Neural Information Processing Systems (NeurIPS), Long Beach, CA, USA, 4–9 December 2017; pp. 5769–5779.
39. Jolicoeur-Martineau, A. The Relativistic Discriminator: A Key Element Missing from Standard GAN. *arXiv* **2018**, arxiv:1807.00734.
40. Karras, T.; Aila, T.; Laine, S.; Lehtinen, J. Progressive Growing of GANs for Improved Quality, Stability, and Variation. *arXiv* **2018**, arxiv:1710.10196.
41. Karras, T.; Laine, S.; Aittala, M.; Hellsten, J.; Lehtinen, J.; Aila, T. Analyzing and improving the image quality of styleGAN. In Proceedings of the IEEE/CVF Conference on Computer Vision and Pattern Recognition (CVPR), Seattle, WA, USA, 13–19 June 2020; pp. 8107–8116.
42. Alexandre, D.; Gabriel, S.; Yossi, A. Real time speech enhancement in the Waveform domain. In Proceedings of the INTERSPEECH, Shanghai, China, 25–29 October 2020; pp. 3291–3295.
43. Yamamoto, R.; Song, E.; Kim, J.-M. Parallel wavegan: A fast waveform generation model based on generative adversarial networks with multi-resolution spectrogram. In Proceedings of the IEEE International Conference on Acoustics, Speech and Signal Processing (ICASSP), Barcelona, Spain, 4–8 May 2020; pp. 6199–6203.
44. Valentini-Botinhao, C.; Wang, X.; Takaki, S.; Yamagishi, J. Investigating rnn-based speech enhancement methods for noise robust text-to-speech. In Proceedings of the International Symposium on Computer Architecture, Seoul, Korea, 18–22 June 2016; pp. 146–152.
45. Veaux, C.; Yamagishi, J.; King, S. The voice bank corpus: Design, collection and data analysis of a large regional accent speech database. In Proceedings of the International Conference Oriental COCOSDA held jointly with 2013 Conference on Asian Spoken Language Research and Evaluation (O-COCOSDA/CASLRE), Gurgaon, India, 25–27 November 2013; pp. 1–4.
46. Thiemann, J.; Ito, N.; Vincent, E. The diverse environments multi-channel acoustic noise database (DEMAND): A database of multichannel environmental noise recordings. In Proceedings of the Meetings on Acoustics (ICA2013), Montreal, QC, Canada, 2–7 June 2013; Volume 19, p. 035081.
47. Kingma, D.; Ba, J. Adam: A method for stochastic optimization. In Proceedings of the International Conference on Learning Representations (ICLR), San Diego, CA, USA, 7–9 May 2015.
48. Abadi, M.; Agarwal, A.; Barham, P.; Brevdo, E.; Chen, Z.; Citro, C.; Corrado, G.S.; Davis, A.; Dean, J.; Devin, M.; et al. Tensorflow: Large-Scale Machine Learning on Heterogeneous Distributed Systems. *arXiv* **2016**, arXiv:1603.04467.
49. Hu, Y.; Loizou, P.C. Evaluation of objective quality measures for speech enhancement. *IEEE Trans. Audio Speech Lang. Process.* **2008**, *16*, 229–238. [CrossRef]
50. Liu, G.; Gong, K.; Liang, X.; Chen, Z. CP-GAN: Context pyramid generative adversarial network for speech enhancement. In Proceedings of the IEEE International Conference on Acoustics, Speech and Signal Processing (ICASSP), Barcelona, Spain, 4–8 May 2020; pp. 6624–6628.

*Article*

# Multimodal Unsupervised Speech Translation for Recognizing and Evaluating Second Language Speech

**Yun Kyung Lee * and Jeon Gue Park**

Artificial Intelligence Research Laboratory, Electronics and Telecommunications Research Institute (ETRI), Daejeon 34129, Korea; jgp@etri.re.kr
\* Correspondence: yunklee@etri.re.kr

**Citation:** Lee, Y.K.; Park, J.G. Multimodal Unsupervised Speech Translation for Recognizing and Evaluating Second Language Speech. *Appl. Sci.* **2021**, *11*, 2642. https://doi.org/10.3390/app11062642

Academic Editor: Byung-Gyu Kim

Received: 16 February 2021
Accepted: 12 March 2021
Published: 16 March 2021

**Publisher's Note:** MDPI stays neutral with regard to jurisdictional claims in published maps and institutional affiliations.

**Copyright:** © 2021 by the authors. Licensee MDPI, Basel, Switzerland. This article is an open access article distributed under the terms and conditions of the Creative Commons Attribution (CC BY) license (https://creativecommons.org/licenses/by/4.0/).

**Abstract:** This paper addresses an automatic proficiency evaluation and speech recognition for second language (L2) speech. The proposed method recognizes the speech uttered by the L2 speaker, measures a variety of fluency scores, and evaluates the proficiency of the speaker's spoken English. Stress and rhythm scores are one of the important factors used to evaluate fluency in spoken English and are computed by comparing the stress patterns and the rhythm distributions to those of native speakers. In order to compute the stress and rhythm scores even when the phonemic sequence of the L2 speaker's English sentence is different from the native speaker's one, we align the phonemic sequences based on a dynamic time-warping approach. We also improve the performance of the speech recognition system for non-native speakers and compute fluency features more accurately by augmenting the non-native training dataset and training an acoustic model with the augmented dataset. In this work, we augment the non-native speech by converting some speech signal characteristics (style) while preserving its linguistic information. The proposed variational autoencoder (VAE)-based speech conversion network trains the conversion model by decomposing the spectral features of the speech into a speaker-invariant content factor and a speaker-specific style factor to estimate diverse and robust speech styles. Experimental results show that the proposed method effectively measures the fluency scores and generates diverse output signals. Also, in the proficiency evaluation and speech recognition tests, the proposed method improves the proficiency score performance and speech recognition accuracy for all proficiency areas compared to a method employing conventional acoustic models.

**Keywords:** fluency evaluation; speech recognition; data augmentation; variational autoencoder; speech conversion

## 1. Introduction

As the demand for untact technology in various fields increases and machine learning technologies advance, the need for computer-assisted second language (L2) learning contents has increased [1–4]. The widely used method for learning a second language is to practice listening, repeating, and speaking language. A GenieTutor, one of the second language (English at present) systems, plays the role of a language tutor by asking questions to the learners, recognizing their speech, which is answered in second language, checking grammatical errors, evaluating the learners' spoken English proficiency, and providing feedbacks to help L2 learners practice their English proficiency. The system comprises several topics, and the learners can select a topic to have communication with the system based on the role-play scenarios. After the learner finishes the speaking of each sentence, the system measures various fluency factors such as pronunciation score, word score, grammar error, stress pattern, and intonation curve, and provides feedback to learners about them compared with the fluency factors of the native speakers [5–7].

The stress and rhythm scores are one of the important factors for fluency evaluation in English speaking, and they are computed by comparing the stress pattern and the rhythm distribution of the L2 speaker with those of native speakers. However, in some cases, the phonemic sequences of speeches uttered by the L2 speaker and the native speaker are recognized differently according to the pronunciation of the learner. Learners may mispronounce or pronounce different words from the referred one. In such cases, the stress and rhythm scores cannot be computed using the previous pattern comparison methods [8–14].

In order to solve this problem, we proposed a dynamic time-warping (DTW)-based stress and rhythm scores measurement method. We aligned the learner's phonemic sequence with the native speaker's phonemic sequence by using the DTW approach, and then computed the stress patterns and rhythm distributions from the aligned phonemic sequences [15–19]. By using the aligned phonemic sequences, we detected the learner's pronunciation error phonemes and computed an error-tagged stress pattern and scores which are deducted by the presence of error phonemes if there is an erroneous phoneme. We computed two stress scores: a word stress score and a sentence stress score. The word stress score is measured by comparing the stress patterns of the content words, and the sentence stress score was computed for the entire sentence. The rhythm scores are measured by computing the mean and standard deviation of the time distances between stressed phonemes. Two stress scores and rhythm scores are used to evaluate the English-speaking proficiency of the learner with other fluency features.

The proposed method uses an automatic speech recognition (ASR) system to recognize the speech uttered by the L2 learner and perform a forced-alignment to obtain the time-aligned phonemic sequences. Deep learning has been applied successfully to ASR systems by relying on hierarchical representations that are commonly learned with a large amount of training data. However, non-native speakers' speech significantly degrades the performance of the ASR due to the pronunciation variability in non-native speech, and it is difficult to collect enough non-native data to train. For better performance of the ASR for non-native speakers, we augment the non-native training speech dataset by using a variational autoencoder-based speech conversion model and train an acoustic model (AM) with the augmented training dataset. Data augmentation has been proposed as a method to generate additional training data, increase the quantity of training data, and reduce overfitting for ASR systems [20–25].

The speech conversion (SC) technique is to convert the speech signal from a source domain to that of a target domain, while preserving its linguistic content information. Variational autoencoder (VAE) is a widely used method for speech modeling and conversion. In the VAE framework, the spectral features of the speech are encoded to a speaker-independent latent variable space. After sampling from the latent variable space, sampled features are decoded back to the speaker-dependent spectral features. A conditional VAE, one of the VAE-based speech conversion methods, employs speaker identity information to feed the decoder with the sampled features. With this conditional decoder, the VAE framework can reconstruct or convert input speech by choosing speaker identity information [26–33]. Recently, generative adversarial networks (GANs)-based SC and some frameworks that jointly train a VAE and GAN were proposed [34–44]. However, most conversion frameworks usually assume and learn a deterministic or unimodal mapping, so their significant limitation is the lack of diversity in the converted outputs.

We build up the proposed conversion model based on the VAE, due to its potential in employing latent space to represent hidden aspects of speech signal. In order to improve speech conversion without conditional information and learn more meaningful speaker characteristic information, we proposed a VAE-based multimodal unsupervised speech conversion method. In the proposed method, we assume that the spectral features of speech are divided into a speaker-invariant content factor (phonetic information in speech) and a speaker-specific style factor [45–47]. We employ a single shared content encoder network and an individual style encoder network for each speaker to train the encoder models robustly. The encoded content factor is fed into a decoder with a target style factor to generate converted spectral features. By sampling different style factors, the proposed model is able to generate diverse and multimodal outputs. In addition, we train our speech conversion model from nonparallel data because parallel data of the source and target speakers are not available in most practical applications and it is difficult to collect such data. By transferring some speech characteristics and converting the speech, we generate additional training data with nonparallel data and train the AM with the augmented training dataset.

We evaluated the proposed method on the corpus of English read speech for the spoken English proficiency assessment [48]. In our experiments, we evaluated the fluency scoring ability of the proposed method by measuring fluency scores and comparing them with the fluency scores of native speakers, and the results demonstrate that the proposed DTW-based fluency scoring method can compute stress patterns and measure stress and rhythm scores effectively even if there are pronunciation errors in the learner's utterances. The spectral feature-related outputs demonstrate that the proposed conversion model can efficiently generate diverse signals while keeping the linguistic information of the original signal. Proficiency evaluation test and speech recognition results with and without an augmented speech dataset also show that the data augmentation with the proposed speech conversion model contributed to improving speech recognition accuracy and proficiency evaluation performance compared to a method employing conventional AMs.

The remainder of this paper is organized as follows. Section 2 briefly describes the second language learning system used in this work. Section 3 describes a description of the proposed DTW-based fluency scoring and VAE-based nonparallel speech conversion method. In Section 4, experimental results are reported, and finally, we conclude and discuss this paper in Section 5.

## 2. Previous Work

GenieTutor is a computer-assisted second language (English at present) learning system. In order to help learners practice their English proficiency, the system recognizes the learners' spoken English responses for given questions, checks content properness, automatically checks and corrects grammatical errors, evaluates spoken English proficiency, and provides educational feedback to learners. Figure 1 shows the schematic diagram of the system [7].

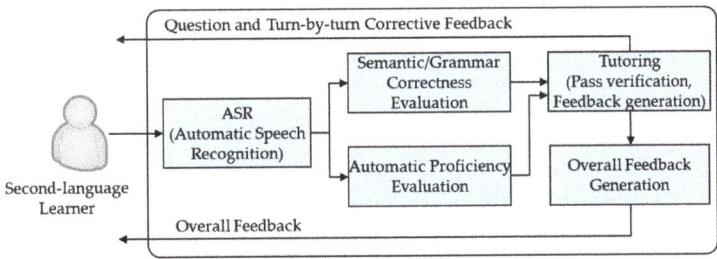

**Figure 1.** Schematic diagram of the GenieTutor system.

The system comprises two learning stages: Think&Talk and Look&Talk. The Think&Talk stage has various subjects, and each subject comprises several fixed role-play dialogues. In this stage, an English learner can select a study topic and a preferred scenario, and then talk with the system based on the selected role-play scenario. After the learner's spoken English response for each given question is completed, the system computes an intonation curve, a sentence stress pattern, and word pronunciation scores. The learner's and a native speaker's intonation curve patterns are plotted as a graph, and the stress patterns of the learner and native speaker are plotted by circles with different sizes below the corresponding word to represent the intensity of each word at a sentence stress level. Once the learner has finished all conversations on the selected subject or all descriptions of the selected picture, the system semantically and grammatically evaluates the responses, and provides overall feedback.

Figure 2 shows an example of a role-play scenario and educational overall feedback with the system. In the Look&Talk stage, the English learner can select a picture and then describe the selected picture to the system.

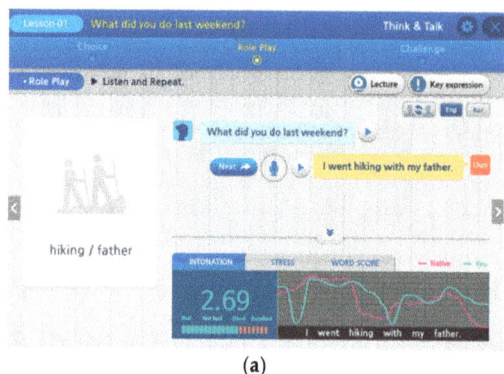
(a)

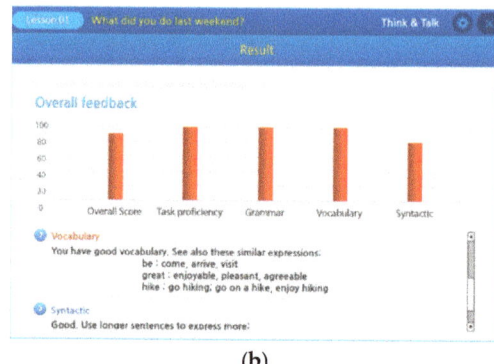
(b)

**Figure 2.** Example of a role-play scenario and the overall feedback. (**a**) Example of a role-play dialogue exercise and intonation feedback of the learner with the native speaker. (**b**) Overall feedback of the role-play dialogue exercise.

## 3. Proposed Fluency Scoring and Automatic Proficiency Evaluation Method

Proficiency evaluation with the proposed method consists of fluency features extraction for scoring each proficiency area, proficiency evaluation model training with fluency features, and automatic evaluation of pronunciation proficiency. The proposed method computes various acoustic features, such as speech rate, intonation, and segmental features, from spoken English uttered by non-native speakers according to a rubric designed to evaluate pronunciation proficiency. In order to compute the fluency features, speech signals are recognized using the automatic speech recognition system and time-aligned sequences of words and phonemes are computed using a forced-alignment algorithm. Each time-aligned sequence contains start and end times for each word and phoneme and acoustic scores. Using the time-aligned sequences, the fluency features are extracted in various aspects of each word and sentence. Proficiency evaluation models are trained using the extracted fluency features and scores from human expert raters, and proficiency scores are computed using the fluency features and scoring models. Figure 3 shows a block diagram of the proficiency evaluation model training and evaluating system for automatic proficiency evaluation.

### 3.1. DTW-Based Feature Extraction for Fluency Scoring

Most language learning systems evaluate and score the learners' spoken English compared to the native speaker's one. However, in realistic speaking situations, a learner's English pronunciation often differs from that of native speakers. For example, learners may pronounce given words incorrectly or pronounce different words from the referred one. In such cases, some fluency features, especially stress and rhythm scores, cannot be measured using previous pattern comparison methods. To solve this problem and measure more meaningful scores, the proposed method aligns the phonemic sequence of the sentence uttered by the learner with the native speaker's phonemic sequence through dynamic time-warping (DTW) alignment and computes the stress patterns, stress scores, and rhythm scores from the aligned phonemic sequences. Figure 4 shows a block diagram of the proposed DTW-based stress and rhythm scoring method.

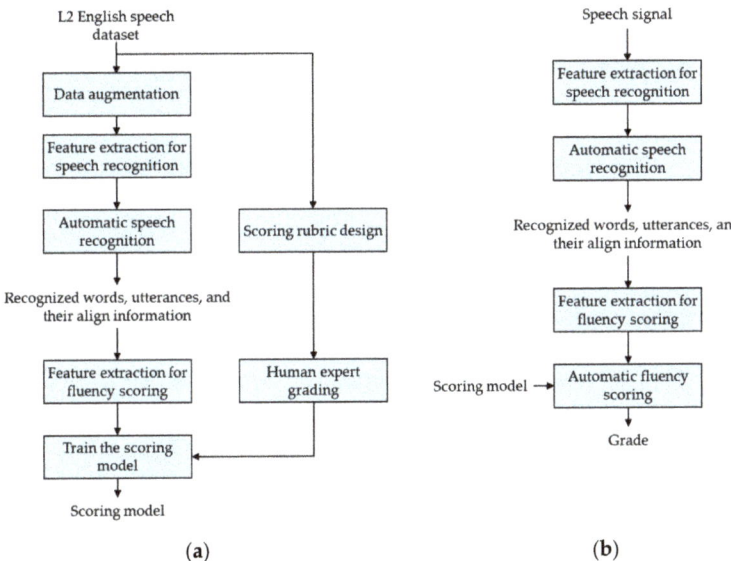

**Figure 3.** Block diagram of the proficiency evaluation model training and evaluating system for automatic proficiency evaluation. (**a**) Flow of the proficiency evaluation model training. (**b**) Flow of the automatic proficiency evaluation.

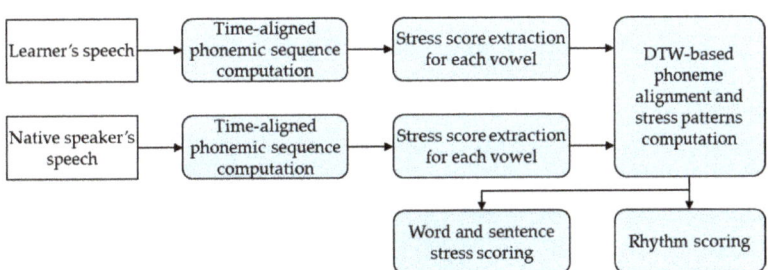

**Figure 4.** Block diagram of the proposed dynamic time-warping (DTW)-based stress and rhythm scoring method.

### 3.1.1. DTW-Based Phoneme Alignment

Dynamic time-warping is a well-known technique for finding an optimal alignment between two time-dependent sequences by comparing them [15]. To compare and align two phonemic sequences uttered by learner and native speaker (reference), we compute a local cost matrix of two phonemic sequences defined by the Euclidean distance. Typically, if a learner and a native speaker are similar to each other, the local cost matrix is small, and otherwise, the local cost matrix is large. The total cost of an alignment path between the learner's and native speaker's phonemic sequences is obtained by summing the local cost measurement values for each pair of elements in two sequences. An optimal alignment path is the alignment path having minimal total cost among all possible alignment paths, and the goal is to find the optimal alignment path and align between two phonemic sequences with the minimal overall cost.

Figure 5 shows an example of DTW-based phonemic sequence alignment and stress patterns' computation results. Error phonemes caused by phoneme mismatch are marked in red, and the stress value of error phonemes was set to 3, which is not a standard stress value indicated by 0 (no stress), 1 (secondary stress), or 2 (primary stress) in order to compute the deducted stress score according to the presence of the phonemic errors. By tagging error phonemes, the proposed method evaluates the learner's utterance more accurately and helps L2 learners practice their English pronunciation.

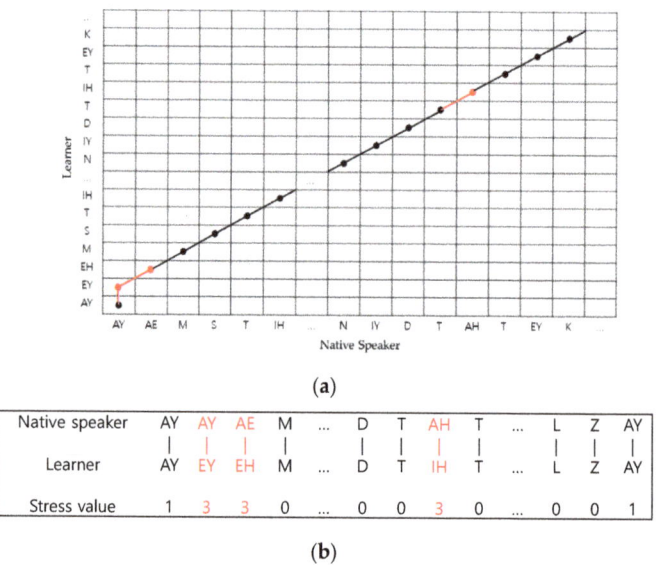

**Figure 5.** Example of DTW-based phonemic sequence alignment in sentence "I am still very sick. I need to take some pills." (**a**) Aligned alignment path of the phonemic sequences. (**b**) Aligned phonemic sequence and stress values.

3.1.2. Stress and Rhythm Scoring

Given the aligned phonemic sequences, the proposed method computes the word stress score, sentence stress score, and rhythm scores. In order to measure the word and sentence stress scores, word stress patterns are computed for each content word in the given sentence, and sentence stress patterns are computed for the entire sentence. Then, the word and sentence stress scores are measured by computing the similarity between the learner's stress patterns and the native speaker's stress patterns.

The rhythm scores are measured by computing the mean and standard deviation of the time intervals between the stressed phonemes. An example of computing the rhythm score in the sentence "I am still very sick. I need to take some pills." is as follows:

- Compute the stress patterns from the aligned phonemic sequences. Table 1 shows an example of the sentence stress pattern. The start times of the stressed phonemes, including the start and end times of the sentence, are highlighted (bold in Table 1) to compute the rhythm features.
- Select the stressed phonemes (highlighted point in Table 1), and compute the mean and standard deviation of the time intervals between them:

    (1) Mean time interval: $(0.2 + 0.2 + 0.6 + 0.5 + 1.5)/5 = 0.6$
    (2) Standard deviation of the time interval: $(0.16 + 0.16 + 0.0 + 0.01 + 0.81)/5 = 0.23$

**Table 1.** Sentence stress pattern in the sentence "I am still very sick. I need to take some pills."

| Word | Start Time | End Time | Stress Value |
| --- | --- | --- | --- |
| I | 0.0 | 0.2 | 1 |
| am | 0.2 | 0.4 | 1 |
| still | 0.4 | 0.8 | 2 |
| very | 0.8 | 1.0 | 0 |
| sick | 1.0 | 1.3 | 1 |
| I | 1.4 | 1.5 | 0 |
| need | 1.5 | 1.8 | 1 |
| to | 1.8 | 2.0 | 0 |
| take | 2.0 | 2.2 | 0 |
| some | 2.2 | 2.5 | 0 |
| pills | 2.5 | 3.0 | 0 |

Table 2 shows an example of the mean values of the mean time interval and the standard deviation of the time interval for each pronunciation proficiency level evaluated by human raters. Proficiency scores 1, 2, 3, 4, and 5 indicate very poor, poor, acceptable, good, and perfect, respectively. As shown in Table 2, the lower the proficiency level, the greater the mean and standard deviation values of the time intervals between the stressed phonemes. Two stress scores and rhythm scores are used for spoken English proficiency evaluation with other features.

**Table 2.** Example of the mean rhythm scores.

| Proficiency Score | Mean Time Interval | Standard Deviation of the Time Interval |
| --- | --- | --- |
| 1 | 1.12 | 0.62 |
| 2 | 0.82 | 0.42 |
| 3 | 0.68 | 0.34 |
| 4 | 058 | 0.32 |
| 5 | 0.56 | 0.31 |

### 3.2. Automatic Proficiency Evaluation with Data Augmentation

The speech recognition system is optimized for non-native speakers as well as natives for educational purposes and smooth interaction. Speech features for computing fluency scores are extracted and decoded into time-aligned sequences by forced-alignment using the non-native acoustic model (AM). In addition, multiple AM scores are used to evaluate proficiency. In order to improve speech recognition accuracy and time-alignment performance, and to compute AM scores more accurately and meaningfully, we augment the training speech dataset and train non-native AM using the augmented training dataset.

In this work, we convert some speech characteristics (style) to generate speech data for augmentation. In the proposed speech conversion model, we assume that each spectral feature of the speech signal is decomposed into a speaker-independent content factor desired to be maintained and each speaker-specific style factor we want to change in latent space. After extracting the content factor from the source speech signal, the proposed conversion model converts the source speech to the desired speech style by extracting the style factor of target speech and recombining it with the extracted content factor. By simply choosing the style factor for this recombination as a source style factor or target style factor, the conversion model can reconstruct or convert speech:

$$\hat{x}_{s \to s} = D(E_c(x_s), E_s^s(x_s)), \qquad (1)$$

$$\hat{x}_{s \to t} = D(E_c(x_s), E_s^t(x_t)), \qquad (2)$$

where $\hat{x}_{s \to s}$ and $\hat{x}_{s \to t}$ are the reconstructed and converted spectra, $x_s$ and $x_t$ are the source and target speech spectra, $D$ is the decoder, and $E_c$, $E_s^s$, and $E_s^t$ denote the content encoder, source style encoder, and target style encoder, respectively. The content encoder network is shared across both speakers, and the style encoder networks are domain-specific networks for individual speakers. Figure 6 shows a block diagram of the proposed speech conversion method.

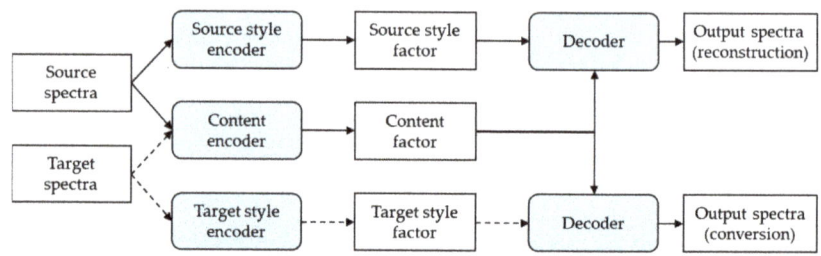

**Figure 6.** Flow of the proposed variational autoencoder (VAE)-based nonparallel speech conversion method. Each of the decoder networks are shared. The arrows indicate flow of the proposed method related to the source spectra or common elements (e.g., source style factor, content factor), and the dashed arrows indicate the flow of the method belonging to the target spectra (e.g., target spectra, target style factor).

As shown in Figure 6, the content encoder network extracts the content factor and is shared across all domains. All convolutional layers of the content encoder were followed by instance normalization (IN) to remove the speech style information and learn domain-independent content information (phoneme in speech). The style encoder network computes the domain-specific style factor for each domain and is composed of multiple separate style encoders (source style encoder and target style encoder in Figure 6) for individual domains. In the style encoders, IN was not used, because it removes the speech style information.

We jointly train the encoders and decoder with multiple losses. To keep encoder and decoder as inverse operations and ensure the proposed system should be able to reconstruct the input spectral features after encoding and decoding, we consider reconstruction loss as follows:

$$L_{recon_s} = E_{s \sim p(s)}[\|D(E_c(x_s), E_s^s(x_s)) - x_s\|_1]. \tag{3}$$

For the content factor and style factors, we apply a semi-cycle loss in latent variable → speech spectra → latent variable coding direction as the latent space is partially shared. Here, a content reconstruction loss encourages the translated content latent factor to preserve the semantic content information of input spectral features, and a style reconstruction loss encourages style latent factors to extract and change speaker-specific speaking style information. Two semi-cycle losses for source speech are computed as follows:

$$L_{recon_s}^c = E_{c \sim p(c), s_t \sim q(s_t)}[\|E_c(D(c, s_t)) - c\|], \tag{4}$$

$$L_{recon_s}^s = E_{c \sim p(c), s_t \sim q(s_t)}[\|E_s^t(D(c, s_t)) - s_t\|], \tag{5}$$

where $c$ denotes the content factor, and $s_t$ and $s_s$ denote the target style factor and source style factor, respectively. The losses for target speech are similarly computed. The full loss of the proposed speech conversion method is the weighted sum of all losses, which is defined as follows:

$$L_{VAE} = \lambda_1(L_{recon_s} + L_{recon_t}) + \lambda_2(L_{recon_s}^c + L_{recon_t}^c) + \lambda_3(L_{recon_s}^s + L_{recon_t}^s), \tag{6}$$

where $\lambda_1$, $\lambda_2$, and $\lambda_3$ control the weights of the components.

## 4. Experimental Results

### 4.1. Dialogue Exercise Result

To validate the effectiveness of the proposed method, we performed computer-assisted fluency scoring experiments with spoken English sentences collected in dialogue scenarios of the GenieTutor system. Figure 7 shows an example of a role-play scenario and fluency scores feedback with the proposed method. Once the learner completes a sentence utterance, the system computes several aspects of pronunciation evaluation and displays them in diagram forms. Learners can check their fluency scores by selecting the sentences they want to check. Learners are provided with overall feedback after finishing all conversations. As shown in Figure 7, the proposed method can efficiently compute the intonation curves and stress patterns of the sentences uttered by the learner even when pronunciation errors occur. In addition, the error words are marked in red, so the learner can see the error parts.

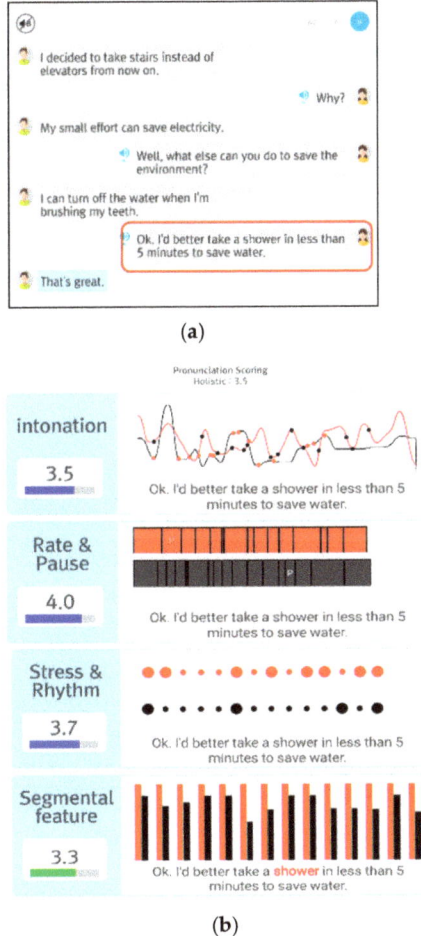

**Figure 7.** Example of dialogue exercise and fluency scores of the learner and the native speaker with the proposed method: (**a**) example of a role-play dialogue exercise, (**b**) fluency scores feedback.

## 4.2. Proficiency Evaluation Test

### 4.2.1. Speech Database

We also performed the proficiency evaluation test using the rhythm and stress scores with other fluency features. A speech dataset was selected from the English read speech dataset read by non-native and native speakers for the spoken proficiency assessment. The dataset is a corpus of English speech sounds spoken by Koreans and 7 American English native speakers (references) for experimental phonetics, phonology, and English education, and is designed to see Korean speakers' intonation and rhythmic patterns in English connected speech and the errors which Korean speakers are apt to make in pronunciation of segments. Each utterance was scored by human expert raters on a scale of 1 to 5. In this study, the gender and spoken language proficiency levels were evenly distributed among the speakers. Table 3 shows scripts samples. The speech dataset comprised 100 non-native speakers, and for each speaker, 80 sentences were used for training and another 20 sentences, not included in the training dataset, were used for testing.

**Table 3.** Samples of the scripts.

| No. | Sentence |
| --- | --- |
| 1 | My pet bird sleeps in the cage. |
| 2 | I eat fruits when I am hungry. |
| 3 | Miss Henry drank a cup of coffee. |
| ... | |
| 100 | They suspect that the suspect killed Ted. |

For speech conversion and augmentation, an additional 7 American English native speakers (3 males and 4 females), and for each speaker, 100 sentences, were used, and frame alignment of the dataset was not performed. We used the WORLD package [49] to perform speech analysis. The sampling rate of all speech signals reported in this paper was 16 kHz. The frame shift length was 5 ms and the number of fast Fourier transform (FFT) points was 1024. For each extracted spectral sequence, 80 Mel-cepstral coefficients (MCEPs) were derived.

### 4.2.2. Human Expert Rater

Each spoken English sentence uttered by non-native learners was annotated by four human expert raters who have English teaching experience or are currently English teachers. Each non-native utterance was rated for five proficiency area scores: holistic impression of proficiency, intonation, stress and rhythm, speech rate and pause, and segmental accuracy. In addition, each proficiency score was measured on a fluency level scale of 1–5. A holistic score for each utterance is calculated as an average of all proficiency scores and used for proficiency evaluation in this paper. Table 4 shows a mean of the correlation between human expert raters' holistic scores.

**Table 4.** Inter-rater correlation.

| Rater | 2 | 3 | 4 |
| --- | --- | --- | --- |
| 1 | 0.79 | 0.75 | 0.80 |
| 2 | - | 0.81 | 0.83 |
| 3 | - | - | 0.80 |

### 4.2.3. Data Augmentation

The proposed VAE-based speech conversion model consisted of a content encoder, style encoders, and a joint decoder. The content encoder comprised two dilated convolutional layers and a gated recurrent unit (GRU) based on a recurrent neural network. In order to remove the speech style information, all convolutional layers were followed by instance normalization (IN) [50]. The style encoder comprised a global average pooling layer,

3-layer multi-layer perceptron (MLP), and a fully connected layer. In the style encoder, IN was not used because it removes the original feature mean and variance that represent speech style information. Then, content and style factors were fed into the decoder to reconstruct or convert the speech. The decoder comprised two dilated convolutional layers and the recurrent neural network-based GRU. All convolutional layers were used with an Adaptive Instance Normalization layer generated by the MLP from the style factor [50].

$$AdaIN(z,s) = \sigma(s)\left(\frac{z - \mu(z)}{\sigma(z)}\right) + \mu(s), \qquad (7)$$

where $z$ is the activation of the previous convolutional layer, and $\mu(.)$ and $\sigma(.)$ denote the mean and variance, respectively.

Figure 8 shows an example of Mel-spectrograms obtained by the proposed method. Comparing the decoding results, we confirmed that the proposed method reconstructs and converts the spectral features efficiently.

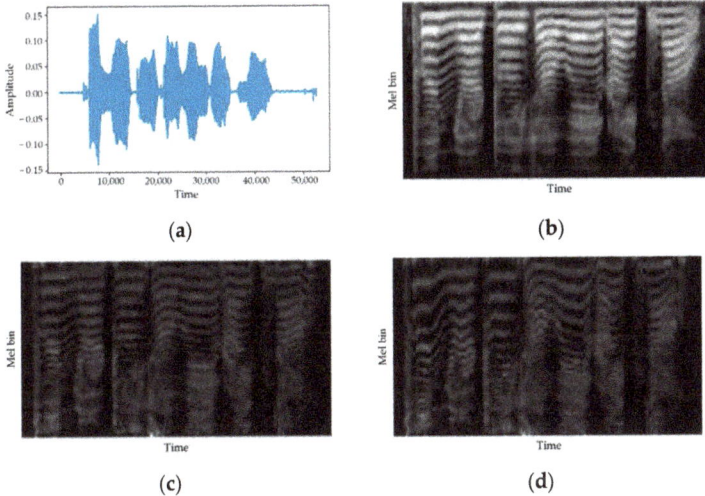

**Figure 8.** Waveform and Mel-spectrograms. (**a**) Waveform of the input signal, (**b**) Mel-spectrogram of the input signal, (**c**) reconstructed Mel-spectrogram, and (**d**) converted Mel-spectrogram.

We performed the perception test to compare the sound quality and speaker similarity of converted speech between the proposed VAE-based speech conversion method and the conventional conditional VAE-based speech conversion (CVAE-SC) method [29], which is one of the most common speech conversion methods. We conducted an AB test and an ABX test. "A" and "B" were outputs from the proposed method and the CVAE, and "X" was a real speech sample. To eliminate bias in the order, "A" and "B" were presented in random orders. In the AB test, each listener was presented with "A" and "B" audios at a time, and was asked to select "A", "B", or "fair" by considering both speech naturalness and intelligibility. In the ABX test, each listener was presented with two audios and a reference audio "X", and then, was asked to select a preferred audio or "fair" by considering the one closer to the reference. We used 24 utterance pairs for the AB test and another 24 utterance pairs, not included in the AB test, for the ABX test. The number of listeners was 20. Figure 9 shows the results, and we confirmed that the proposed method outperforms the baseline in both sound quality and speaker similarity terms.

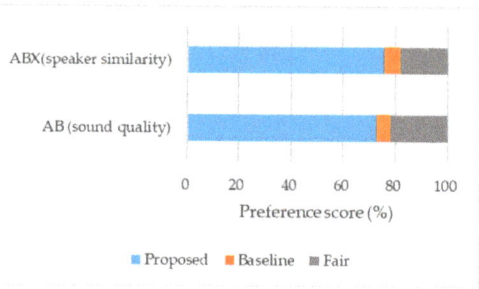

**Figure 9.** Results of the AB test and the ABX test.

We also performed the speech recognition test to validate that the spectral features were converted meaningfully using the English read speech dataset. We used the ESPnet [51] for an end-to-end ASR system. We trained the AM using only the training dataset ("Train database only" in Table 5) and evaluated the test dataset, and we compared the recognition results to those obtained by evaluating the same test dataset using the AM trained with the augmented dataset ("Augmentation" in Table 5). Table 5 shows the word error rate (WER) results. For comparison, SpecAugment [21], speed perturbation method [20], and CVAE-SC were used as a reference. As shown in Table 5, we confirmed that the data augmentation with the proposed method improves the speech recognition accuracy for all proficiency score levels compared to a method employing conventional AM and the other augmentation methods. By sampling different style factors, the proposed speech conversion method is able to generate diverse outputs, but the computational complexity is higher than that of other methods.

**Table 5.** Speech augmentation and word error rate (%) results.

| Proficiency Level | Applied Method | 1 | 2 | 3 | 4 | 5 |
| --- | --- | --- | --- | --- | --- | --- |
| Train database only | - | 57.3 | 53.4 | 30.1 | 23.4 | 22.7 |
| Augmentation | SpecAugment | 52.1 | 45.4 | 27.2 | 21.9 | 20.6 |
|  | Speed perturbation | 51.3 | 44.9 | 27.0 | 21.8 | 20.7 |
|  | CVAE-SC | 49.3 | 43.1 | 26.0 | 21.5 | 20.4 |
|  | Proposed method | 40.9 | 37.7 | 24.5 | 20.5 | 19.1 |

#### 4.2.4. Features for Proficiency Scoring

All features for proficiency scoring are computed based on the time-aligned phone sequence and its time information [11,12,14]. Table 6 shows the proficiency scoring feature list used to train the automatic proficiency scoring models in this work.

**Table 6.** Features for the proficiency scoring modeling.

| Feature Name | Description |
| --- | --- |
| Genie_pron | Pronunciation score |
| SLLR | Sentence-level log-likelihood ratio |
| amscore0/amscore1 | Acoustic model (AM) score/anti-model-based AM score |
| Uttsegdur | Duration of entire transcribed segment but without inter-utterance pauses |
| Globsegdur | Duration of entire transcribed segment, including all pauses |
| wpsec/wpsecutt | Speech articulation rate/words per second in utterance |
| Silpwd/Silpsec | Number of silences per word/second |
| Numsil | Number of silences |
| Silmean/Silmeandev/Silstddev | Mean/mean deviation/standard deviation of silence duration in second |
| Longpfreq/Longpwd | Frequency/number of long pauses per word (0.495 s $\leq$ duration) |
| Longpmn/Longpmeandev/Longpstddev | Mean/mean deviation/standard deviation of long pauses |
| Wdpchk/Secpchk | Average speech chunk length in words/seconds |
| Wdpchkmeandev/Secpchkmeandev | Mean deviation of chunks in words/seconds |
| Repfeq | Number of repetitions divided by number of words |

Table 6. Cont.

| Feature Name | Description |
|---|---|
| Tpsec | Types (unique words) per second |
| Tpsecutt | Types normalized by speech duration |
| Wdstress/Uttstress | Word/sentence stress score |
| Rhymean/Rhystddev | Mean/standard deviation of rhythm score |
| FlucMean/Flucstddev | Mean/standard deviation of the range of fluctuation |
| propV | Vocalic segment duration ratio in sentence |
| deltaV/deltaC | Standard deviation of vocalic/consonantal segment duration |
| varcoV/varcoC | Standard deviation of vocalic/consonantal segment duration normalized by speech rate |
| Genie_amscoreK0/Genie_amscoreK1 | AM score/anti-model AM score reflecting Korean pronunciation characteristics of English |
| Numdff | Number of disfluencies |
| Dpsec | Disfluencies per second |

### 4.2.5. Proficiency Scoring Model

We used two modeling methods: (1) multiple linear regression (MLR) and (2) deep neural network, to train scoring models with high agreement with human expert raters. MLR is simple and has been used for a long time for automatic proficiency scoring purposes. Based on the MLR scoring model, the proficiency score is computed as follows:

$$Score = \sum_i \alpha_i \cdot f_i + \beta, \quad (8)$$

where $i$ is the index of each feature, $\alpha_i$ is the weight associated with each scoring feature $f_i$, and $\beta$ is a constant intercept.

We also used a neural network to train the proficiency scoring model nonlinearly and more accurately. The neural network comprised a convolutional layer with 1 hidden layer and 3 hidden units and a fully connected layer. Given 41 features, the neural network trains the proficiency scoring model.

### 4.2.6. Proficiency Evaluation Results

In order to validate that the proposed automatic proficiency evaluation system measured the proficiency scores effectively and meaningfully, we computed and compared a Pearson's correlation coefficient between the proficiency scores of the proposed system and those of human raters. The Pearson's correlation coefficient is a commonly used metric for evaluating the performance of proficiency assessment methods [52–54]. Tables 7 and 8 show the proficiency evaluation results obtained by the proposed method with and without data augmentation. For comparison, the range of correlation coefficients of the inter-rater scores ("Human" in Tables 7 and 8) were used as a reference. As shown in Tables 7 and 8, we confirmed that the proposed automatic proficiency evaluation method measures proficiency scores efficiently for all proficiency area scores. In addition, we confirmed that data augmentation for AM training with the proposed speech conversion method improves the averaged correlation performance for all proficiency area scores compared to the method employing conventional AM trained without data augmentation. By automatically evaluating the proficiency of the L2 speaker's utterance, the proposed proficiency scoring system is able to perform fast and consistent evaluation in various environments.

Table 7. Correlation between human rater and proposed proficiency scoring system without data augmentation.

| Rater | Human | MLR | Neural Network |
|---|---|---|---|
| Holistic | 0.68~0.79 | 0.78 | 0.82 |
| Intonation | 0.64~0.72 | 0.73 | 0.77 |
| Stress and rhythm | 0.71~0.74 | 0.75 | 0.78 |
| Speech rate and pause | 0.71~0.75 | 0.75 | 0.77 |
| Segmental features | 0.59~0.67 | 0.69 | 0.73 |

Table 8. Correlation between human rater and proposed proficiency scoring system with data augmentation.

| Rater | Human | MLR | Neural Network |
|---|---|---|---|
| Holistic | 0.68~0.79 | 0.83 | 0.84 |
| Intonation | 0.64~0.72 | 0.78 | 0.80 |
| Stress and rhythm | 0.71~0.74 | 0.78 | 0.79 |
| Speech rate and pause | 0.71~0.75 | 0.81 | 0.82 |
| Segmental features | 0.59~0.67 | 0.73 | 0.76 |

## 5. Conclusions and Future Work

We proposed an automatic proficiency evaluation method for L2 learners in spoken English. In the proposed method, we augmented the training dataset using the VAE-based speech conversion model and trained the acoustic model (AM) with an augmented training dataset to improve the speech recognition accuracy and time-alignment performance for non-native speakers. After recognizing the speech uttered by the learner, the proposed method measured various fluency features and evaluated the proficiency. In order to compute the stress and rhythm scores even when the phonemic sequence errors occur in the learner's speech, the proposed method aligned the phonemic sequences of the spoken English sentences by using the DTW, and then computed the error-tagged stress patterns and the stress and rhythm scores. In computer experiments with the English read speech dataset, we showed that the proposed method effectively computed the error-tagged stress patterns, stress scores, and rhythm scores. Moreover, we showed that the proposed method efficiently measured proficiency scores and improved the averaged correlation between human expert raters and the proposed method for all proficiency areas compared to the method employing conventional AM trained without data augmentation.

The proposed method can also be used for most signal processing and generation problems, such as sound conversion between instruments or generation of various images. However, the current style conversion framework has a limitation that the conversion model learns the domain-level style factors and generates the converted speech signal rather than diverse pronunciation styles of multiple speakers included in each domain. In order to learn more meaningful and diverse style factors and perform many-to-many speech conversion, we plan to address the issues of automatic speaker label estimation and expansion to each speaker-specific style encoder in the future work.

**Author Contributions:** Conceptualization, methodology, validation, formal analysis, writing—original draft preparation, and writing—review and editing, Y.K.L.; supervision and project administration, J.G.P. All authors have read and agreed to the published version of the manuscript.

**Funding:** This work was supported by an Electronics and Telecommunications Research Institute (ETRI) grant funded by the Korean government (21ZS1100, Core Technology Research for Self-Improving Integrated Artificial Intelligence System), and by the Institute of Information & Communications Technology Planning & Evaluation (IITP) grant funded by the Korean Government (MSIT) (2019-0-2019-0-00004, Development of semi-supervised learning language intelligence technology and Korean tutoring service for foreigners).

**Institutional Review Board Statement:** Not applicable.

**Informed Consent Statement:** Not applicable.

**Data Availability Statement:** Not applicable.

**Conflicts of Interest:** The authors declare no conflict of interest.

## References

1. Eskenazi, M. An overview of spoken language technology for education. *Speech Commun.* **2009**, *51*, 832–844. [CrossRef]
2. Kannan, J.; Munday, P. New Trends in Second Language Learning and Teaching through the Lens of ICT, Networked Learning, and Artificial Intelligence. *Círculo de Lingüística Aplicada a la Comunicación* **2018**, *76*, 13–30. [CrossRef]

3. Gabriel, S.; Ricardo, T.; Jose, A.G. Automatic code generation for language-learning applications. *IEEE Lat. Am. Trans.* **2020**, *18*, 1433–1440.
4. Chen, M.; Chen, W.; Ku, L. Application of sentiment analysis to language learning. *IEEE Access* **2018**, *6*, 24433–24442. [CrossRef]
5. Song, H.J.; Lee, Y.K.; Kim, H.S. Probabilistic Bilinear Transformation Space-based Joint Maximum a Posteriori Adaptation. *ETRI J.* **2010**, *34*, 783–786. [CrossRef]
6. Lee, S.J.; Kang, B.O.; Chung, H.; Lee, Y. Intra-and Inter-Frame Features for Automatic Speech Recognition. *ETRI J.* **2014**, *36*, 514–517. [CrossRef]
7. Kwon, O.W.; Lee, K.; Kim, Y.-K.; Lee, Y. GenieTutor: A computer assisted second-language learning system based on semantic and grammar correctness evaluations. In Proceedings of the 2015 EUROCALL Conference, Padova, Italy, 26–29 August 2015; pp. 330–335.
8. Deshmukh, O.; Kandhway, K.; Verma, A.; Audhkhasi, K. Automatic evaluation of spoken English fluency. In Proceedings of the 2009 IEEE International Conference on Acoustics, Speech and Signal Processing, Taipei, Taiwan, 19–24 April 2009; pp. 4829–4832.
9. Müller, M. *Information Retrieval for Music and Motion*; Springer: New York, NY, USA, 2007; pp. 69–84.
10. Rahman, M.M.; Saha, S.K.; Hossain, M.Z.; Islam, M.B. Performance Evaluation of CMN for Mel-LPC based Speech Recognition in Different Noisy Environments. *Int. J. Comput. Appl.* **2012**, *58*, 6–10.
11. Hermansky, H.; Morgan, N. RASTA Processing of Speech. *IEEE Trans. Speech Audio Process.* **1994**, *2*, 578–589. [CrossRef]
12. You, H.; Alwan, A. Tenporal Modulation Processing of Speech Signals for Noise Robust ASR. In Proceedings of the Tenth Annual Conference of the International Speech Communication Association, Brighton, UK, 6–10 September 2009; pp. 36–39.
13. Cadzow, J.A. Blind Deconvolution via Cumulant Extrema. *IEEE Signal Process. Mag.* **1993**, *13*, 24–42. [CrossRef]
14. Chen, L.; Zechner, K.; Yoon, S.Y.; Evanini, K.; Wang, X.; Loukina, A.; Tao, J.; Davis, L.; Lee, C.M.; Ma, M.; et al. Automated scoring of nonnative speech using the SpeechRatorSM v. 5.0 Engine. *ETS Res. Rep. Ser.* **2018**, *2018*, 1–31. [CrossRef]
15. Bell, A.J.; Sejnowski, T.J. An Information-Maximization Approach to Blind Separation and Blind Deconvolution. *Neural Comput.* **1995**, *7*, 1129–1159. [CrossRef] [PubMed]
16. Yang, H.H.; Amari, S. Adaptive on-Line Learning Algorithms for Blind Separation—Maximum Entropy and Minimum Mutual Information. *Neural Comput.* **1997**, *9*, 1457–1482. [CrossRef]
17. Loizou, P.C. *Speech Enhancement*; CRC Press: Boca Raton, FL, USA, 2007; pp. 97–289.
18. Papoulis, A. *Probability, Random Variables, and Stochastic Processes*; McGraw-Hill: New York, NY, USA, 1991.
19. Oppenheim, A.V.; Schaefer, R.W. *Digital Signal Processing*; Prentice-Hall: Upper Saddle River, NJ, USA, 1989.
20. Ko, T.; Peddinti, V.; Povey, D.; Khudanpur, S. Audio Augmentation for Speech Recognition. In Proceedings of the Sixteenth Annual Conference of the International Speech Communication Association, Dresden, Germany, 6–10 September 2015; pp. 3586–3589.
21. Park, D.S.; Chan, W.; Zhang, Y.; Chiu, C.-C.; Zoph, B.; Cubuk, E.D.; Le, Q.V. Specaugment: A simple data augmentation method for automatic speech recognition. In Proceedings of the Interspeech, Graz, Austria, 15–19 September 2019.
22. Celin, T.; Nagarajan, T.; Vijayalakshmi, P. Data Augmentation Using Virtual Microphone Array Synthesis and Multi-Resolution Feature Extraction for Isolated Word Dysarthric Speech Recognition. *IEEE J. Sel. Top. Signal Process.* **2020**, *14*, 346–354.
23. Oh, Y.R.; Park, K.; Jeon, H.-B.; Park, J.G. Automatic proficiency assessment of Korean speech read aloud by non-natives using bidirectional LSTM-based speech recognition. *ETRI J.* **2020**, *42*, 761–772. [CrossRef]
24. Sun, X.; Yang, Q.; Liu, S.; Yuan, X. Improving low-resource speech recognition based on improved NN-HMM structures. *IEEE Access* **2020**, *8*, 73005–73014. [CrossRef]
25. Yang, S.; Wang, Y.; Xie, L. Adversarial feature learning and unsupervised clustering based speech synthesis for found data with acoustic and textual noise. *IEEE Signal Process. Lett.* **2020**, *27*, 1730–1734. [CrossRef]
26. Hsu, C.; Hwang, H.; Wu, Y.; Tsao, Y.; Wang, H. Voice conversion from non-parallel corpora using variational autoencoder. In Proceedings of the Asia-Pacific Signal and Information Processing Association Annual Summit and Conference, Jeju, Korea, 13–15 December 2016; pp. 1–6.
27. Hsu, W.-N.; Zhang, Y.; Glass, J. Unsupervised learning of disentangled and interpretable representations from sequential data. In Proceedings of the Advances in Neural Information Processing Systems, Long Beach, CA, USA, 4–9 December 2017; pp. 1878–1889.
28. Saito, Y.; Ijima, Y.; Nishida, K.; Takamichi, S. Non-parallel voice conversion using variational autoencoders conditioned by phonetic posteriorgrams and d-vectors. In Proceedings of the ICASSP, Calgary, AB, Canada, 15–20 April 2018; pp. 5274–5278.
29. Tobing, P.L.; Wu, Y.-C.; Hayashi, T.; Kobayashi, K.; Toda, T. Non-parallel voice conversion with cyclic variational autoencoder. In Proceedings of the Interspeech, Graz, Austria, 15–19 September 2019; pp. 674–678.
30. Kang, B.O.; Jeon, H.B.; Park, J.G. Speech recognition for task domains with sparse matched training data. *Appl. Sci.* **2020**, *10*, 6155. [CrossRef]
31. Wang, Y.; Dai, B.; Hua, G.; Aston, J.; Wipf, D. Recurrent variational autoencoders for learning nonlinear generative models in the presence of outliers. *IEEE J. Sel. Top. Signal Process.* **2018**, *12*, 1615–1627. [CrossRef]
32. Cristovao, P.; Nakada, H.; Tanimura, Y.; Asoh, H. Generating in-between images through learned latent space representation using variational autoencoders. *IEEE Access* **2020**, *8*, 149456–149467. [CrossRef]
33. Kameoka, H.; Kaneko, T.; Tanaka, K.; Hojo, N. ACVAE-VC: Non-parallel voice conversion with auxiliary classifier variational autoencoder. *IEEE Acm Trans. Audio Speech Lang. Process.* **2019**, *27*, 1432–1443. [CrossRef]

34. Kameoka, H.; Kaneko, T.; Tanaka, K.; Hojo, N. Stargan-vc: Non-parallel many-to-many voice conversion using star generative adversarial networks. In Proceedings of the 2018 IEEE Spoken Language Technology Workshop (SLT), Athens, Greece, 18–21 December 2018; pp. 266–273.
35. Zhu, J.-Y.; Park, T.; Isola, P.; Efros, A.A. Unpaired image-to-image translation using cycle-consistent adversarial networks. In Proceedings of the IEEE International Conference on Computer Vision (ICCV), Venice, Italy, 22–29 October 2017.
36. Goodfellow, I.; Pouget-Abadie, J.; Mirza, M.; Xu, B.; Warde-Farley, D.; Ozair, S.; Courville, A.; Bengio, Y. Generative adversarial nets. In Proceedings of the Advances in Neural Information Processing Systems, Montreal, QC, Canada, 8–13 December 2014.
37. Saito, Y.; Takamichi, S.; Saruwatari, H. Statistical parametric speech synthesis incorporating generative adversarial networks. *IEEE Acm Trans. Audio Speech Lang. Process.* **2018**, *26*, 84–96. [CrossRef]
38. Oyamada, K.; Kameoka, H.; Kaneko, T.; Tanaka, K.; Hojo, N.; Ando, H. Generative adversarial network-based approach to signal reconstruction from magnitude spectrograms. In Proceedings of the EUSIPCO, Rome, Italy, 3–7 September 2018; pp. 2514–2518.
39. Gu, J.; Shen, Y.; Zhou, B. Image processing using multi-code gan prior. In Proceedings of the IEEE/CVF Conference on Computer Vision and Pattern Recognition, Seattle, WA, USA, 16–18 June 2020.
40. Isola, P.; Zhu, J.-Y.; Zhou, T.; Efros, A.A. Image-to-image translation with conditional adversarial networks. In Proceedings of the Computer Vision and Pattern Recognition, Honolulu, HI, USA, 21–26 June 2017.
41. Hsu, C.-C.; Hwang, H.-T.; Wu, Y.-C.; Tsao, Y.; Wang, H.-M. Voice conversion from unaligned corpora using variational autoencoding Wasserstein generative adversarial networks. In Proceedings of the Interspeech, Stockholm, Sweden, 20–24 August 2017; pp. 3164–3168.
42. Liu, X.; Gherbi, A.; Wei, Z.; Li, W.; Cheriet, M. Multispectral image reconstruction from color images using enhanced variational autoencoder and generative adversarial network. *IEEE Access* **2021**, *9*, 1666–1679. [CrossRef]
43. Wang, X.; Tan, K.; Du, Q.; Chen, Y.; Du, P. CVA2E: A conditional variational autoencoder with an adversarial training process for hyperspectral imagery classification. *IEEE Trans. Geosci. Remote Sens.* **2020**, *58*, 5676–5692. [CrossRef]
44. Weng, Z.; Zhang, W.; Dou, W. Adversarial attention-based variational graph autoencoder. *IEEE Access* **2020**, *8*, 152637–152645. [CrossRef]
45. Gao, J.; Chakraborty, D.; Tembine, H.; Olaleye, O. Nonparallel emotional speech conversion. In Proceedings of the Interspeech 2019, 20th Annual Conference of the International Speech Communication Association, Graz, Austria, 15–19 September 2019; pp. 2858–2862.
46. Huang, X.; Liu, M.; Belongie, S.J.; Kautz, J. Multimodal unsupervised image-to-image translation. In Proceedings of the ECCV, Munich, Germany, 8–14 September 2018; pp. 172–189.
47. Lee, Y.K.; Kim, H.W.; Park, J.G. Many-to-many unsupervised speech conversion from nonparallel corpora. *IEEE Access* **2021**, *9*, 27278–27286. [CrossRef]
48. Chung, H.; Lee, Y.K.; Lee, S.J.; Park, J.G. Spoken english fluency scoring using convolutional neural networks. In Proceedings of the Oriental Chapter of the International Coordinating Committee on Speech Databases and Speech I/O Systems and Assessment (O-COCOSDA), Seoul, Korea, 1–3 November 2017; pp. 1–6.
49. Huang, C.; Akagi, M. A three-layered model for expressive speech perception. *Speech Commun.* **2008**, *50*, 810–828. [CrossRef]
50. Huang, X.; Belongie, S.J. Arbitrary style transfer in real-time with adaptive instance normalization. In Proceedings of the ICCV, Venice, Italy, 22–29 October 2017; pp. 1501–1510.
51. Watanabe, S.; Hori, T.; Karita, S.; Hayashi, T.; Nishitoba, J.; Unno, Y.; Enrique Yalta Soplin, N.; Heymann, J.; Wiesner, M.; Chen, N.; et al. ESPnet: End-toend speech processing toolkit. In Proceedings of the Interspeech, Hyderabad, India, 2–6 September 2018; pp. 2207–2211.
52. Sisman, B.; Zhang, M.; Li, H. Group sparse representation with WaveNet Vocoder adaptation for spectrum and prosody conversion. *IEEE ACM Trans. Audio Speech Lang. Process.* **2019**, *27*, 1085–1097. [CrossRef]
53. Benesty, J.; Chen, J.; Huang, Y.; Cohen, I. Pearson correlation coefficient. In *Noise Reduction Speech Processing*; Springer: Berlin, Germany, 2009; pp. 1–4.
54. Zhou, K.; Sisman, B.; Li, H. Transforming Spectrum and Prosody for Emotional Voice Conversion with Non-Parallel Training Data. In Proceedings of the Odyssey 2020 the Speaker and Language Recognition Workshop, Tokyo, Japan, 2–5 November 2020; pp. 230–237.

Article

# 3D Avatar Approach for Continuous Sign Movement Using Speech/Text

Debashis Das Chakladar [1], Pradeep Kumar [1], Shubham Mandal [1], Partha Pratim Roy [1], Masakazu Iwamura [2] and Byung-Gyu Kim [3,*]

[1] Department of Computer Science and Engineering, Indian Institute of Technology Roorkee, Uttarakhand 247667, India; dchakladar@cs.iitr.ac.in (D.D.C.); pradeep.iitr7@gmail.com (P.K.); shubhammandal96@gmail.com (S.M.); proy.fcs@iitr.ac.in (P.P.R.)

[2] Department of Computer Science and Intelligent Systems, Osaka Prefecture University, Osaka 599-8531, Japan; masa@cs.osakafu-u.ac.jp

[3] Department of IT Engineering, Sookmyung Women's University, Seoul 140-742, Korea

\* Correspondence: bg.kim@sookmyung.ac.kr

**Citation:** Das Chakladar, D.; Kumar, P.; Mandal, S.; Roy, P.P.; Iwamura, M.; Kim, B.-G. 3D Avatar Approach for Continuous Sign Movement Using Speech/Text. *Appl. Sci.* **2021**, *11*, 3439. https://doi.org/10.3390/app11083439

Academic Editor: Andrea Prati

Received: 15 February 2021
Accepted: 2 April 2021
Published: 12 April 2021

**Publisher's Note:** MDPI stays neutral with regard to jurisdictional claims in published maps and institutional affiliations.

**Copyright:** © 2021 by the authors. Licensee MDPI, Basel, Switzerland. This article is an open access article distributed under the terms and conditions of the Creative Commons Attribution (CC BY) license (https:// creativecommons.org/licenses/by/ 4.0/).

**Abstract:** Sign language is a visual language for communication used by hearing-impaired people with the help of hand and finger movements. Indian Sign Language (ISL) is a well-developed and standard way of communication for hearing-impaired people living in India. However, other people who use spoken language always face difficulty while communicating with a hearing-impaired person due to lack of sign language knowledge. In this study, we have developed a 3D avatar-based sign language learning system that converts the input speech/text into corresponding sign movements for ISL. The system consists of three modules. Initially, the input speech is converted into an English sentence. Then, that English sentence is converted into the corresponding ISL sentence using the Natural Language Processing (NLP) technique. Finally, the motion of the 3D avatar is defined based on the ISL sentence. The translation module achieves a 10.50 SER (Sign Error Rate) score.

**Keywords:** Indian Sign Language (ISL); natural language processing; avatar; sign movement; context-free grammar

## 1. Introduction

A sign is a sequential or parallel construction of its manual and non-manual components. A manual component can be defined by hand shape, orientation, position, and movements, whereas non-manual components are defined by facial expressions, eye gaze, and head/body posture [1–5]. Hearing-impaired people use sign language for their communication. Every country has its sign language based on its vocabulary and syntax. Therefore, sign translation from speech/text is specific to the particular targeted country. Indian Sign Language (ISL) is one of the sign languages that can be efficiently translated from English. Moreover, ISL is recognized as a widely accepted natural language for its well-defined grammar, syntax, phonetics, and morphology structure over others [6]. ISL is a visual–spatial language that provides linguistic information using the hands, arms, face, and head/body postures. The ISL open lexicon can be categorized into three parts: (i) Signs whose place of articulation is fixed, (ii) signs whose place of articulation can change, and (iii) directional signs, where there is a movement between two points in space [7]. However, people who use English as a spoken language do not understand the ISL. Therefore, an English to ISL sign movement translation system is required for assistance and learning purposes.

In India, more than 1.5 million people are hearing-impaired who use ISL as their primary means of communication [8]. Some studies [6,8,9] implemented ISL videos for sign representation from English text. To generate a robust sign language learning system from

English to ISL, output sign representation should be efficient, such as being able to generate proper signs without delay for complete sentences. However, sign language translation from ISL video recordings requires notable processing time [6]. By contrast, the sign representations using a 3D avatar require minimum computational time, and the avatar can be easily reproduced as per the translation system [10]. Moreover, most of the existing studies [11,12] have not considered complete sentences for sign language conversion. To overcome these shortcomings, in this paper, we propose a 3D avatar-based ISL learning model that can perform sign movements not only for isolated words but also for complete sentences through input text or speech. The flow diagram of such an assisting system is depicted in Figure 1, where the input to the system is either English speech or text, which is then processed using a text processing technique to obtain ISL representation. Next, a gesture model is used to perform the sign movement corresponding to ISL with the help of an avatar. The main contributions of the work are defined as follows.

- Our first contribution is the development of a 3D avatar model for Indian Sign Language (ISL).
- The proposed 3D avatar model can generate sign movements from three different inputs, namely speech, text, and complete sentences. The complete sentence obtained is made up of continuous signs corresponding to a sentence of spoken language.

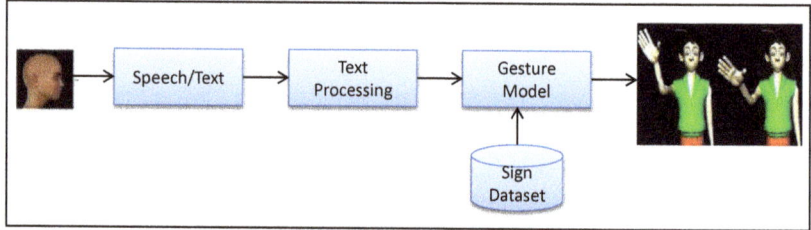

**Figure 1.** An assistive model for generating sign movements using the 3D avatar from English speech/text.

The rest of the paper is organized as follows. The related work of sign language translation is discussed in Section 2. In Section 3, we describe the proposed model of speech to sign movement for ISL sentences. In Section 4, we analyze each module of our proposed model. Finally, Section 5 presents the conclusion of this paper.

## 2. Related Work

This section consists of two subsections: sign language translation systems and performance analysis of the sign language translation systems. A detailed description of each module is given below.

### 2.1. Sign Language Translation Systems

Sign movement can be effectively generated from input speech. In [13], the authors have designed a speech–sign translation system for Spanish Sign Language (SSL) using a speech recognizer, a natural language translator, and a 3D avatar animation module. In [11,14], the authors have implemented the conversion of Arabic Sign Language (ArSL) from Arabic text using an Arabic text-to-sign translation system. The translation system uses the set of translation rules and linguistic language models for detecting different signs from the text. An American Sign Language (ASL)-based animation sequence has been proposed in [15]. The authors' system converts all of the hand symbols and associated movements of the ASL sign box. A speech-to-sign movement translation based on Spanish Sign Language (SSL) has been proposed in [12]. The authors used two types of translation techniques (rule-based and statistical) of the Natural Language Processing (NLP) toolbox to generate SSL. A linguistic model-based 3D avatar (for British Sign Language) has been proposed for implementing the visual realization of sign language [16]. A web-based interpreter from text to sign language was developed in [17]. The interpreter tool has been created from a large dictionary of ISL such that it can be shared among multilingual communities. An android app-based translation system has been designed to convert sign movements from hand gestures of ISL [18]. In [19], the authors designed a Malayalam text to ISL translation system using a synthetic animation approach. Their model has been used to promote sign language education among the common people of Kerala, India. A Hindi text to the ISL conversion system has been implemented in [20]. Their model used the dependency parser and Part-of-Speech (PoS) tagger, which correctly categorize the input words into their syntactic forms. An interactive 3D avatar-based math learning system of American Sign Language (ASL) has been proposed in [21]. The math-learner model can increase the effectiveness of parents of hearing-impaired children in teaching mathematics to their children. A brief description of existing sign language learning systems is presented in Table 1. It can be observed that some sign language translation models work on speech-to-sign conversion, whereas some models translate the text to signs and represent the signs using a 3D avatar. Our proposed model successfully converts the input speech to corresponding text and then renders the signs movements using the 3D avatar.

**Table 1.** Brief description of previous studies of sign language learning systems. Note: Arabic Sign Language (ArSL), Chinese Sign Language (CSL), Spanish Sign Language (SSL), American Sign Language (ASL), and Indian Sign Language (ISL). "Sentence-wise sign" represents the continuous signs corresponding to a sentence in its correspondent spoken language.

| Study | Sign Language | Input: Speech | Input: Text | 3D Avatar | Sentence-Wise Sign |
|---|---|---|---|---|---|
| Al-Barahamtoshy, O.H. et al. [11] | ArSL | ✗ | ✓ | ✓ | ✗ |
| Li et al. [22] | CSL | ✓ | ✗ | ✗ | ✓ |
| Halawani et al. [14] | ArSL | ✓ | ✗ | ✗ | ✗ |
| Lopez-Ludena et al. [23] | SSL | ✗ | ✓ | ✓ | ✗ |
| Bouzid, Y. et al. [24] | ASL | ✗ | ✗ | ✓ | ✗ |
| Dasgupta et al. [8] | ISL | ✗ | ✓ | ✗ | ✗ |
| Nair et al. [19] | ISL | ✗ | ✓ | ✗ | ✓ |
| Vij et al. [20] | ISL | ✗ | ✓ | ✗ | ✗ |
| Krishnaraj et al. [6] | ISL | ✓ | ✓ | ✗ | ✗ |
| Duarte et al. [25] | ASL | ✓ | ✗ | ✗ | ✓ |
| Patel et al. [26] | ISL | ✓ | ✗ | ✓ | ✓ |
| Proposed | ISL | ✓ | ✓ | ✓ | ✓ |

### 2.2. Performance Analysis of the Sign Language Translation System

This section discussed the effectiveness of the different sign language translation systems based on different evaluation metrics such as Sign Error Rate (SER), Bilingual Evaluation Understudy (BLEU), and the National Institute of Standards and Technology (NIST). In [27], the authors have designed an avatar-based model that can generate sign movements from spoken language sentences. They achieved a 15.26 BLEU score with a Recurrent Neural Network (RNN)-based model. In [28,29], the authors have proposed a "HamNoSys" system that converts the input words to corresponding gestures of ISL. "HamNoSys" represents the syntactic representation of each sign using some symbols, which can be converted into the respective gestures (hand movement, palm orientation). Apart from "HamNoSys", Signing Gesture Markup Language (SiGML) [30] also has been used for transforming sign visual representations into a symbolic design. In [31], the authors have used BLEU and the NIST score, which are relevant for performance analysis of language translation. Speech to SSL translation has been implemented with two types of natural language-based translations (rule-based and statistical) [12]. The authors have identified that rule-based translation outperforms statistical translation with a 31.6 SER score and a 0.5780 BLEU score.

## 3. Materials and Methods

This section illustrates the framework of the proposed sign language learning system for ISL. The proposed model is subdivided into three modules, as depicted in Figure 2. The first module corresponds to the conversion of speech to an English sentence, which is then processed using NLP to obtain the corresponding ISL sentence. Lastly, we feed the extracted ISL sentence to the avatar model to produce the respective sign language. We discuss the detailed description of each module in Sections 3.1–3.3.

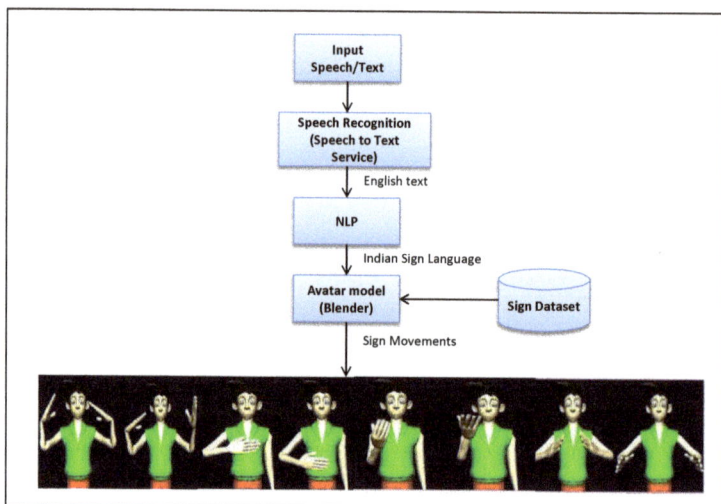

**Figure 2.** Framework of the proposed sign language learning system using text/speech. NLP: Natural Language Processing.

### 3.1. Speech to English Sentence Conversion

We used the IBM-Watson service (available online: https://cloud.ibm.com/apidocs/speech-to-text (accessed on 20 December 2020)). to convert the input speech into text. The service is classified into three phases, i.e., input features, interfaces, and output features. The first phase illustrates the input audio format (.wav, .flac, .mp3, etc.) and settings (sampling rate, number of communication channels) of the speech signal. Next, an HTTP request is generated for each speech signal. The input speech signal interacts with the speech-to-text service using various interfaces (web socket interface, HTTP interface, and asynchronous HTTP interface) using the communication channel. Finally, in the third phase, the output text is constructed based on the keyword spotting and word confidence metrics. The confidence metrics indicate how much of the transcribed text is correctly converted from input speech based on the acoustic evidence [32,33].

### 3.2. Translation of ISL Sentence from English Sentence

This section provides the details of the conversion process from English text to its corresponding ISL text. The words in the ISL sentence have been identified to generate corresponding sign movements. For the conversion from English to ISL, we use the Natural Language Toolkit (NLTK) [34]. The model of converting the ISL sentence from an English sentence is plotted in Figure 3. A detailed discussion of the translation process is presented below.

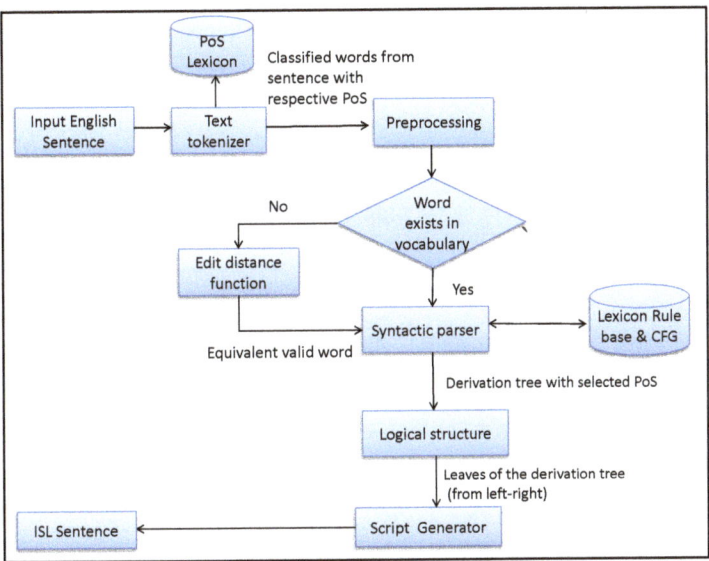

**Figure 3.** Model of ISL sentence generation from English sentence. PoS: Part-of-Speech; CFG: Context-Free Grammar.

3.2.1. Preprocessing of Input Text Using Regular Expression

If a user mistakenly enters an invalid/misspelled word, the "edit distance" function is used to obtain an equivalent valid word. A few examples of the misspelled words, along with the corresponding valid words, are presented in Table 2.

**Table 2.** Mapping of misspelled/invalid word into equivalent valid word.

| Misspelled/Invalid Word | Equivalent Valid Word |
| --- | --- |
| "Hellllo" | "Hello" |
| "Halo" | "Hello" |
| "Hapyyyy" | "Happy" |
| "Happppyyy" | "Happy" |
| "Noooooooo" | "No" |

The edit distance function takes two strings (source and target) and modifies the source string such that both source and target strings become equal. NLTK divides the English sentence into separate word–PoS pairs using the text tokenizer. The regular expression identifies the meaningful English sentence using the lexicon rule. During the preprocessing of input text, we define the regular expression (1) using the PoS tokens of the NLTK module. The regular expression starts with at least one verb phrase (VP) and is terminated with one noun phrase (NP). In the middle part, the regular expression can take zero or more number of any words that match PoS tokens (preposition (PP) or a pronoun (PRP) or adjective (JJ)). In a regular expression, + refers to one or more symbols, whereas ∗ refers to zero or more symbols. Therefore $(VP)^+$ represents one or more verb phrases. For example, our first sentence, "Come to my home", starts with the verb phrase ("come"), followed by a preposition ("to"), pronoun ("my"), and ends with a noun phrase ("home").

$$(VP)^+(PP|PRP|JJ)^*(NP) \qquad (1)$$

where VP ∈ (VB,VBN), NP ∈ (NN), VB ∈ ("hello","Thank","Please"), VBN ∈ ("come"), PP ∈ ("to","with"), PRP ∈ ("my","you","me"), JJ ∈ ("Good"), NN ∈ ("home", "morning").

### 3.2.2. Syntactic Parsing and Logical Structure

After the preprocessing step, the NLTK module returns the parse tree based on the grammatical tokens (VP, PP, NP, etc.). Then, we construct the derivation tree of the Context-Free Grammar (CFG), which is similar to the parse tree of the NLTK module. CFG consists of variable/nonterminal symbols, terminal symbols, and a set of production rules. The nonterminal symbols generally appear on the left-hand side of the production rules, though they can also be introduced on the right-hand side. The terminal symbols appear on the right-hand side of the rules. The production rule generates the terminal string from the nonterminal symbol. The derivation tree can represent the derivation of the terminal string from the nonterminal symbol. In the derivation tree, terminal and non-terminal symbols refer to the leaves and intermediate nodes of the tree. Each meaningful ISL sentence has its own derivation tree. After the creation of the derivation tree, the leaves of the tree are combined to make a logical structure for the sign language. We have plotted the different derivation trees for a few sentences in Figure 4A–D.

*Context-free grammar*

S→VP PP NP|VP NP|VP VP PP NP
VP→VB|VBN
PP→"to"|"with"
NP→PRP NN|JJ NN|PRP
VB→"hello"|"Thank"|"please"
VBN→"Come"
PRP→"my"|"you"|"me"
JJ→"Good"
NN→"home"|"morning"
where ("hello", "Thank", "please", "Come", "my", "you", "me", "Good", "morning", "to", "with", "home") ∈ terminals and (VP, PP, NP, VB, VBN, PRP, NN, JJ) ∈ nonterminals of the context-free grammar.

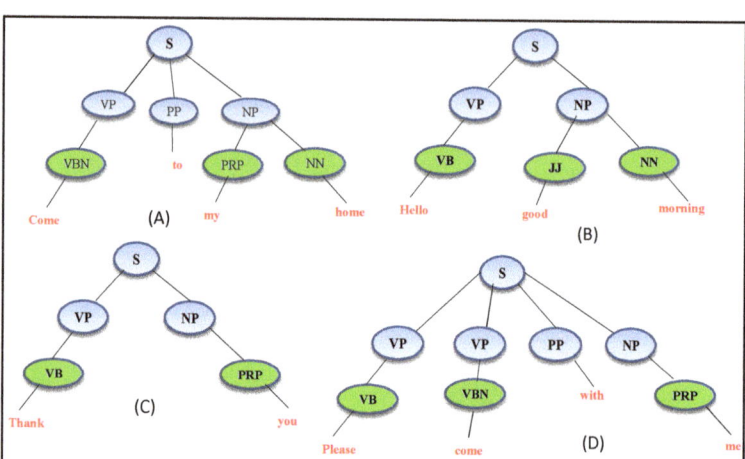

**Figure 4.** Derivation tree for the sentences: (**A**) Come to my home, (**B**) Hello good morning, (**C**) Thank you, (**D**) Please come with me. Note: terminal symbols are represented in red, whereas green and blue refer to the nonterminal symbols of the derivation tree (generated from the above CFG). The start symbol (often represented as S) is a special nonterminal symbol of the grammar.

### 3.2.3. Script Generator and ISL Sentence

The script generator creates a script for generating an ISL sentence from the English sentence. The script takes a valid English sentence (after semantic parsing) as input and generates the sequence tree, where each node of the tree is related to different gestures that are associated with the avatar movement. The sequence tree maintains the order of the motion performed by the avatar model.

The structures of the English and ISL sentences are quite different. The representation of ISL from the English sentences is done using Lexical Functional Grammar (LFG). The f-structure of LFG encodes the grammatical relation, like a subject, object, and tense of an input sentence. ISL follows the word order "Subject–Object–Verb", whereas the English language follows the word order "Subject–Verb–Object" [35]. Moreover, the ISL sentence does not consider any conjunction and preposition in the sentence [36]. Some examples of mapping from English to ISL sentences are represented in Table 3.

**Table 3.** English sentence–ISL sentence mapping.

| English Sentence | ISL Sentence |
| --- | --- |
| I have a pen. | I pen have. |
| The child is playing. | Child playing. |
| The woman is blind. | Woman blind. |
| It is cloudy outside. | Outside cloudy. |
| I see a dog. | I dog see. |

*3.3. Generation of Sign Movement*

The generation of sign movements based on the input text is accomplished with the help of an animation tool called Blender [37]. The tool is popularly used for designing games, 3D animation, etc. The game logic and game object are the key components of the Blender game engine. We developed the 3D avatar by determining its geometric shape. The whole process for creating the avatar is divided into three steps. First, the skeleton and face of the avatar are created. In the second step, we define the viewpoint or orientation of the model. In the third step, we define the movement joints and facial expressions of the avatar. Next, we provide the sequence of frames that determine the movement of the avatar over the given sequence of words over time. Finally, motion (movement like walking, showing figures, moving hands, etc.) is defined by giving solid animation. The game engine was written from scratch in C++ as a mostly independent part and includes support for features such as Python scripting and OpenAI 3D sound. In this third module, we generate sign movements for the ISL sentence (generated in the second module). The entire framework of the movement generation of the avatar from the ISL sentence is described in Figure 5. For the generation of sign movement from ISL, initially, the animation parameters are extracted from the ISL sentence. Once the animation parameters are identified, the motion-list of each sign is performed using a 3D avatar model. In the proposed 3D avatar model, each movement is associated with several motions, and all such motions are listed in the "motionlist" file. A counter variable is initialized for tracking current motion. Each motion has its timestamps mentioning how long the action of different gestures will be performed. Each motion is generated by a specific motion actuator when some sensor event has occurred. The controller acts as a linker between the sensor and actuator. The conditional loop checks the maximum bounds of the number of motions, and it performs the motion one by one using a specific actuator (e.g., the actuator[counter]). The next valid motion is performed by incrementing the counter variable. If the value of the counter variable exceeds the number of motions in the "motionlist" file, then the counter variable along with the "motionlist" is reset to the default value, and the next movement will be performed.

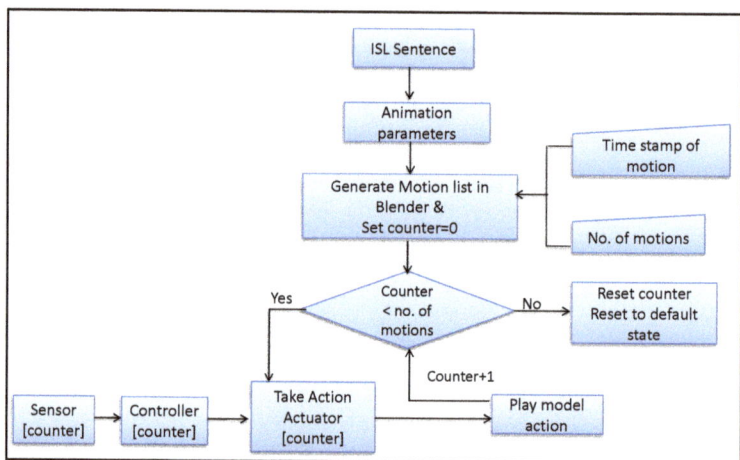

**Figure 5.** Movement generation of avatar from ISL sentence.

## 4. Results

Here, we present a detailed analysis of the proposed sign language learning system using the 3D Avatar model. This section consists of four sections, namely: sign database (Section 4.1), speech recognition results (Section 4.2), results of translation process (Section 4.3), and generation of sign movement from ISL sentence (Section 4.4).

### 4.1. Sign Database

We created a sign language database that contains sentences based on 50 daily used ISL words (e.g., I, my, come, home, welcome, sorry, rain, you, baby, wind, man, woman, etc.) and other dialogues between different users. We create 150 sentences that contain 763 different words, including the most used words in ISL. For each word, the sign movements were defined in the blender toolkit. The description of the sign database is depicted in Table 4. The vocabulary items were created based on the unique words in ISL. For better understanding, we represented four animation sequences of each word. For the sake of simplicity, we present some example sequences of sign movement (Table 5) using two animated series. The table shows the sign movements along with their actual words in English. All such sign movements were defined with the help of a sign language expert from a hearing-impaired school ('Anushruti') in the Indian Institute of Technology Roorkee.

**Table 4.** Dataset description.

| Total Sentences | Total Words | Vocabulary | Running Words |
|---|---|---|---|
| 150 | 763 | 365 | 50 |

**Table 5.** English word to sign movements (each sign movement consists of 2 sequential motions).

| English Word | Sign Movement | English Word | Sign Movement |
| --- | --- | --- | --- |
| Home | | Night | |
| Morning | | Work | |
| Welcome | | Bye | |
| Rain | | Baby | |
| Please | | Sorry | |

## 4.2. Speech Recognition Results

The speech recognition is performed using the "IBM-Watson speech to text" service that converts English audio recordings or audio files into the respective text. The service takes a speech or audio file (.wav, .flac, or .mp3 format) with a different sampling frequency and converts the resulting text as output. The sampling frequency of our audio files is 16 KHz. The results of the speech recognition module for both isolated words (discrete speech) and complete sentences (continuous speech) are presented in Table 6. The X-axis denotes the time in seconds, and the Y-axis represents the amplitude of the signal. For the sake of simplicity, we take two discrete and two continuous speech signals for conversion.

Table 6. Mapping of the speech signal to text for different types of speech.

| Speech Type | Speech Signal | Output Text (Using IBM-Watson Service) |
|---|---|---|
| Discrete | | Hello |
| Discrete | | Come |
| Continuous | | Do you like it? |
| Continuous | | Thank you |

### 4.3. Results of Translation Process

This section illustrates the translation process of the proposed model. The translation process includes English to ISL sentence conversion and ISL sentence-to-sign representation. The evaluation of the proposed system was performed by dividing the generated sentences into a 80:20 ratio between the training and testing sets, respectively, and the Word Error Rate (WER) of the input word was recorded. The result of the text processing system is presented in Table 7, where the WER metric is derived from Levenshtein distance (edit distance function). Here we compare the word from the reference sentence and the output sentence. The distance calculates the number of edits/changes (insertion/deletion and modifications) required to convert the input text to the correct reference text. In Table 7, Ins, Del, and Sub refer to the number of insertion, deletion, and modification/substitution operations for converting source text to the proper target text, respectively.

Table 7. Text processing results based on Word Error Rate (WER).

| WER (%) | Ins (%) | Del (%) | Sub (%) |
|---|---|---|---|
| 25.2 | 3.3 | 7.1 | 14.8 |

For evaluating the performance of the translation system, some metrics have been considered: SER, BLEU, and NIST. SER computes the sign error rate during the generation of each sign from the ISL sentence. In this work, we recorded SERs of 10.50 on the test data. This error occurred due to WER happening during text entry input, which resulted into wrong sign generation by the avatar. BLEU and NIST are used for evaluating the quality of text during the translation from English (source language) to ISL (target language). The translation is done based on the multiple reference text (used from the vocabulary), and it calculates the precision score based on the unigram, bigram, ..., $n$-gram model where $n$ is the number of words in the reference text.

BLEU assigns equal weights to all $n$-grams, whereas NIST gives more importance to the rare words and small weights to the frequently used words in the sentence, so the overall score of NIST is better than BLEU. The result of the rule-based translation process is presented in Table 8. From Table 8, it can be concluded that the NIST score outperforms the BLEU score.

**Table 8.** Performance analysis of proposed translation system. SER: Sign Error Rate; BLEU: Bilingual Evaluation Understudy; MIST: National Institute of Standards and Technology.

| SER | BLEU | NIST |
|---|---|---|
| 10.50 | 82.30 | 86.80 |

*4.4. Generation of Sign Movement from ISL Sentence*

After converting the ISL sentence from the English sentence in module 2 (Section 3.2), we proceeded to generate the sign movement for the ISL sentence. The avatar generates animated sign movements for each meaningful word. Here, we have used the avatar using blender software. We have plotted all the movements of ISL corresponding to English sentences. Figure 6A describes the actions (action 1, action 2, and action 3) of sign language representation for the English sentence "Come to my home". Figure 6B depicts the sign language representation of the English sentence "Hello, Good morning" (action 1, action 2, and action 3), and Figure 6C describes the sign language representation of the English sentence "Bye baby" (action 1, action 2). Figure 6D represents the sign language representation (action 1, action 2) of the English sentence "Please come".

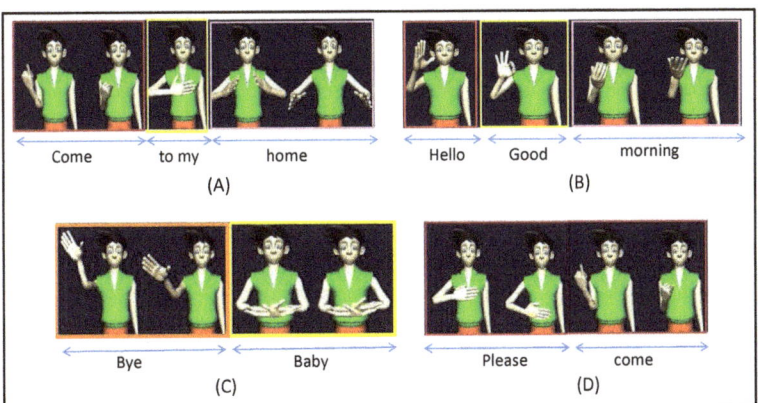

**Figure 6.** Sign language representations of English sentences: (**A**) Come to my home, (**B**) Hello, good morning, (**C**) Bye baby, (**D**) Please come.

The quality of the proposed system was evaluated by adopting the Absolute Category Rating (ACR) [38] scheme. The ratings presented to the users are sorted by quality in decreasing order: Excellent, Good, Fair, Poor, and Bad. The performance is measured based on the output sign movements produced using various input speech or text. A majority rating of "Good" was recorded among the rater population of 25. A prototype video representation of the system was also made (available online: https://www.youtube.com/watch?v=jTtRi8PG0cs&ab_channel=PradeepKumar (accessed on 22 December 2020)) on Youtube.

## 5. Conclusions

In this work, we developed a 3D avatar-based sign language learning system that converts the English speech or text into corresponding ISL movements. Initially, the input speech is converted into an equivalent English sentence using the IBM-Watson service. The converted English text is further translated into the corresponding ISL sentence using regular expression and script generator. Finally, each word of the ISL sentence is transformed into an equivalent sign movement, represented by a 3D avatar. The translation method (English–ISL and ISL–sign movement) has been evaluated by the SER, BLEU, and NIST metrics. The proposed model achieved an 82.30% BLEU score that represents the translation accuracy from English to ISL sentences. The text-to-sign translation model (from ISL sentence to sign movement) achieved a 10.50 SER score, which signifies that 89.50% of signs were correctly generated by the 3D avatar model for the respective ISL sentence. It may be noted that the proposed system has been developed for a limited corpus, and no facial expressions were included, which can be considered as an important part of any sign language system. The transition between the signs while performing a sign sentence can be improved further by learning specific transitions based on hand positions while signing. Moreover, the sign language recognition system that converts a sign to text/speech is significantly more difficult to develop. Such a system can be added within the proposed framework to build a complete sign language interpretation system.

**Author Contributions:** Conceptualization, D.D.C. and P.K.; methodology, D.D.C., P.K. and S.M.; software, S.M.; validation, D.D.C. and P.K.; formal analysis, D.D.C. and P.K.; investigation, P.P.R. and M.I.; data curation, D.D.C. and P.K.; writing—D.D.C., P.K.; writing—review and editing, P.P.R. and M.I.; visualization, M.I.; supervision, P.P.R. and B.-G.K.; project administration, B-G.K. All authors have read and agreed to the published version of the manuscript.

**Funding:** No research funding has been received for this work.

**Institutional Review Board Statement:** Not applicable.

**Informed Consent Statement:** Informed consent was obtained from all subjects involved in the study.

**Data Availability Statement:** Not applicable.

**Conflicts of Interest:** The authors declare no conflict of interest.

## References

1. Kumar, P.; Roy, P.P.; Dogra, D.P. Independent bayesian classifier combination based sign language recognition using facial expression. *Inf. Sci.* **2018**, *428*, 30–48. [CrossRef]
2. Mittal, A.; Kumar, P.; Roy, P.P.; Balasubramanian, R.; Chaudhuri, B.B. A modified LSTM model for continuous sign language recognition using leap motion. *IEEE Sens. J.* **2019**, *19*, 7056–7063. [CrossRef]
3. Kumar, P.; Gauba, H.; Roy, P.P.; Dogra, D.P. A multimodal framework for sensor based sign language recognition. *Neurocomputing* **2017**, *259*, 21–38. [CrossRef]
4. Kumar, P. Sign Language Recognition Using Depth Sensors. Ph.D. Thesis, Indian Institute of Technology Roorkee, Roorkee, India, 2018.
5. Kumar, P.; Saini, R.; Behera, S.K.; Dogra, D.P.; Roy, P.P. Real-time recognition of sign language gestures and air-writing using leap motion. In Proceedings of the IEEE 2017 Fifteenth IAPR International Conference on Machine Vision Spplications (MVA), Nagoya, Japan, 8–12 May 2017; pp. 157–160.
6. Krishnaraj, N.; Kavitha, M.; Jayasankar, T.; Kumar, K.V. A Glove based approach to recognize Indian Sign Languages. *Int. J. Recent Technol. Eng. IJRTE* **2019**, *7*, 1419–1425.
7. Zeshan, U.; Vasishta, M.N.; Sethna, M. Implementation of Indian Sign Language in educational settings. *Asia Pac. Disab. Rehabil. J.* **2005**, *16*, 16–40.
8. Dasgupta, T.; Basu, A. Prototype machine translation system from text-to-Indian sign language. In ACM Proceedings of the 13th International Conference on Intelligent User Interfaces, Gran Canaria, Spain, 13–16 January 2008; pp. 313–316.
9. Kishore, P.; Kumar, P.R. A video based Indian sign language recognition system (INSLR) using wavelet transform and fuzzy logic. *Int. J. Eng. Technol.* **2012**, *4*, 537. [CrossRef]
10. Goyal, L.; Goyal, V. Automatic translation of English text to Indian sign language synthetic animations. In Proceedings of the 13th International Conference on Natural Language Processing, Varanasi, India, 17–20 December 2016; pp. 144–153.
11. Al-Barahamtoshy, O.H.; Al-Barhamtoshy, H.M. Arabic Text-to-Sign Model from Automatic SR System. *Proc. Comput. Sci.* **2017**, *117*, 304–311. [CrossRef]
12. San-Segundo, R.; Barra, R.; Córdoba, R.; D'Haro, L.; Fernández, F.; Ferreiros, J.; Lucas, J.M.; Macías-Guarasa, J.; Montero, J.M.; Pardo, J.M. Speech to sign language translation system for Spanish. *Speech Commun.* **2008**, *50*, 1009–1020. [CrossRef]
13. López-Ludeña, V.; San-Segundo, R.; Morcillo, C.G.; López, J.C.; Muñoz, J.M.P. Increasing adaptability of a speech into sign language translation system. *Exp. Syst. Appl.* **2013**, *40*, 1312–1322. [CrossRef]
14. Halawani, S.M.; Zaitun, A. An avatar based translation system from arabic speech to arabic sign language for deaf people. *Int. J. Inf. Sci. Educ.* **2012**, *2*, 13–20.
15. Papadogiorgaki, M.; Grammalidis, N.; Tzovaras, D.; Strintzis, M.G. Text-to-sign language synthesis tool. In Proceedings of the IEEE 13th European Signal Processing Conference, Antalya, Turkey, 4–8 September 2005; pp. 1–4.
16. Elliott, R.; Glauert, J.R.; Kennaway, J.; Marshall, I.; Safar, E. Linguistic modelling and language-processing technologies for Avatar-based sign language presentation. *Univers. Access Inf. Soc.* **2008**, *6*, 375–391. [CrossRef]

17. ELGHOUL, M.J.O. An avatar based approach for automatic interpretation of text to Sign language. *Chall. Assist. Technol. AAATE* **2007**, *20*, 266.
18. Loke, P.; Paranjpe, J.; Bhabal, S.; Kanere, K. Indian sign language converter system using an android app. In Proceedings of the IEEE 2017 International conference of Electronics, Communication and Aerospace Technology (ICECA), Coimbatore, India, 20–22 April 2017; Volume 2, pp. 436–439.
19. Nair, M.S.; Nimitha, A.; Idicula, S.M. Conversion of Malayalam text to Indian sign language using synthetic animation. In Proceedings of the IEEE 2016 International Conference on Next Generation Intelligent Systems (ICNGIS), Kottayam, India, 1–3 September 2016; pp. 1–4.
20. Vij, P.; Kumar, P. Mapping Hindi Text To Indian sign language with Extension Using Wordnet. In Proceedings of the International Conference on Advances in Information Communication Technology & Computing, Bikaner, India, 12–13 August 2016; p. 38.
21. Adamo-Villani, N.; Doublestein, J.; Martin, Z. The MathSigner: An interactive learning tool for American sign language. In Proceedings of the IEEE Eighth International Conference on Information Visualisation, London, UK, 16 July 2004; pp. 713–716.
22. Li, K.; Zhou, Z.; Lee, C.H. Sign transition modeling and a scalable solution to continuous sign language recognition for real-world applications. *ACM Trans. Access. Comput.* **2016**, *8*, 7. [CrossRef]
23. López-Ludeña, V.; González-Morcillo, C.; López, J.C.; Barra-Chicote, R.; Córdoba, R.; San-Segundo, R. Translating bus information into sign language for deaf people. *Eng. Appl. Artif. Intell.* **2014**, *32*, 258–269. [CrossRef]
24. Bouzid, Y.; Jemni, M. An Avatar based approach for automatically interpreting a sign language notation. In Proceedings of the IEEE 13th International Conference on Advanced Learning Technologies, Beijing, China, 15–18 July 2013; pp. 92–94.
25. Duarte, A.C. Cross-modal neural sign language translation. In Proceedings of the 27th ACM International Conference on Multimedia, Nice, France, 21–25 October 2019; pp. 1650–1654.
26. Patel, B.D.; Patel, H.B.; Khanvilkar, M.A.; Patel, N.R.; Akilan, T. ES2ISL: An advancement in speech to sign language translation using 3D avatar animator. In Proceedings of the 2020 IEEE Canadian Conference on Electrical and Computer Engineering (CCECE), London, ON, Canada, 30 August–2 September 2020; pp. 1–5.
27. Stoll, S.; Camgöz, N.C.; Hadfield, S.; Bowden, R. Text2Sign: Towards Sign Language Production Using Neural Machine Translation and Generative Adversarial Networks. *Int. J. Comput. Vis.* **2020**, *128*, 891–908. [CrossRef]
28. Kaur, K.; Kumar, P. HamNoSys to SiGML conversion system for sign language automation. *Proc. Comput. Sci.* **2016**, *89*, 794–803. [CrossRef]
29. Hanke, T. HamNoSys-representing sign language data in language resources and language processing contexts. *LREC* **2004**, *4*, 1–6.
30. Kennaway, R. Avatar-independent scripting for real-time gesture animation. *arXiv* **2015**, arXiv:1502.02961.
31. Zhang, Y.; Vogel, S.; Waibel, A. Interpreting bleu/nist scores: How much improvement do we need to have a better system? In Proceedings of the Fourth International Conference on Language Resources and Evaluation, Lisbon, Portugal, 26–28 May 2019; pp. 1650–1654.
32. Larson, M.; Jones, G.J. Spoken content retrieval: A survey of techniques and technologies. *Found. Trends Inf. Retriev.* **2012**, *5*, 235–422. [CrossRef]
33. Lee, L.S.; Glass, J.; Lee, H.Y.; Chan, C.A. Spoken content retrieval—Beyond cascading speech recognition with text retrieval. *IEEE/ACM Trans. Audio Speech Lang. Process.* **2015**, *23*, 1389–1420. [CrossRef]
34. Bird, S.; Klein, E.; Loper, E. *Natural Language Processing with Python: Analyzing Text with the Natural Language Toolkit*; O'Reilly Media Inc.: Newton, MA, USA, 2009.
35. Mehta, N.; Pai, S.; Singh, S. Automated 3D sign language caption generation for video. *Univers. Access Inf. Soc.* **2020**, *19*, 725–738. [CrossRef]
36. Zeshan, U. Indo-Pakistani Sign Language grammar: A typological outline. *Sign Lang. Stud.* **2003**, *3*, 157–212. [CrossRef]
37. Roosendaal, T.; Wartmann, C. *The Official Blender Game Kit: Interactive 3d for Artists*; No Starch Press: San Francisco, CA, USA, 2003.
38. Tominaga, T.; Hayashi, T.; Okamoto, J.; Takahashi, A. Performance comparisons of subjective quality assessment methods for mobile video. In Proceedings of the IEEE 2010 Second international workshop on quality of multimedia experience (QoMEX), Trondheim, Norway, 21–23 June 2010; pp. 82–87.

Article

# Context-Based Structure Mining Methodology for Static Object Re-Identification in Broadcast Content

Krishna Kumar Thirukokaranam Chandrasekar * and Steven Verstockt

IDLab, Ghent University—imec, 9052 Ghent, Belgium; steven.verstockt@ugent.be
* Correspondence: krishnakumar.tc@ugent.be; Tel.: +32-9-33-14920

**Abstract:** Technological advancement, in addition to the pandemic, has given rise to an explosive increase in the consumption and creation of multimedia content worldwide. This has motivated people to enrich and publish their content in a way that enhances the experience of the user. In this paper, we propose a context-based structure mining pipeline that not only attempts to enrich the content, but also simultaneously splits it into shots and logical story units (LSU). Subsequently, this paper extends the structure mining pipeline to re-ID objects in broadcast videos such as SOAPs. We hypothesise the object re-ID problem of SOAP-type content to be equivalent to the identification of reoccurring contexts, since these contexts normally have a unique spatio-temporal similarity within the content structure. By implementing pre-trained models for object and place detection, the pipeline was evaluated using metrics for shot and scene detection on benchmark datasets, such as RAI. The object re-ID methodology was also evaluated on 20 randomly selected episodes from broadcast SOAP shows *New Girl* and *Friends*. We demonstrate, quantitatively, that the pipeline outperforms existing state-of-the-art methods for shot boundary detection, scene detection, and re-identification tasks.

**Keywords:** object detection; logical story unit detection (LSU); object re-ID

Citation: Thirukokaranam Chandrasekar, K.K.; Verstockt, S. Context-Based Structure Mining Methodology for Static Object Re-Identification in Broadcast Content. *Appl. Sci.* **2021**, *11*, 7266. https://doi.org/10.3390/app11167266

Academic Editors: Byung-Gyu Kim and Dongsan Jun

Received: 27 June 2021
Accepted: 4 August 2021
Published: 6 August 2021

**Publisher's Note:** MDPI stays neutral with regard to jurisdictional claims in published maps and institutional affiliations.

**Copyright:** © 2021 by the authors. Licensee MDPI, Basel, Switzerland. This article is an open access article distributed under the terms and conditions of the Creative Commons Attribution (CC BY) license (https://creativecommons.org/licenses/by/4.0/).

## 1. Introduction

Due to advances in storage and digital media technology, videos have become the main source of visual information. The recording and accumulation of a large number of videos has also become very easy, and many popular websites, including YouTube, Yahoo Video, Facebook, Flickr, and Instagram, allow users to share and upload video content globally. Today, we have arrived at the point where the volume of video that arrives on the internet increases exponentially on a daily basis. Apart from this, there are very many broadcast channels with enormous amounts of video content—shot and stored every second. With such large collections of videos, it is very difficult to locate the appropriate video files and extract information from them effectively. Moreover, with such a vast quantity of data, even the suggestion list expands tremendously; thus, it is even more difficult to make an efficient and informed decision. Large file sizes, the temporal nature of the content, and the lack of proper indexing methods to leverage non-textual features, creates difficulty in cataloguing and retrieving videos efficiently [1]. To address these challenges, efforts are being made—in every direction—to bridge the gap between low-level binary video representations and high-level text-based video descriptions (e.g., video categories, types or genre) [2–7]. Due to the absence of structured intermediate representations, powerful video processing methodologies which can utilise scene, object, person, or event information do not yet exist. In this paper, we address this problem by proposing a framework involving an improved semantic content mining approach, which obtains frame-level location and object information across the video. The proposed architecture extracts semantic tags such as objects, actions and locations from the videos, using them not only to obtain scene/shot boundaries, but also to re-ID objects from the video.

Since this paper deals with several video features/aspects, it is important to clearly state the definitions for the various structures and components of a video as used in this paper. Any video can essentially be broken down into several units. First, a video is a collection of successive images; specific frames shown at a particular speed. Each frame is one of the many still images that make up the video. Next, a group of uninterrupted and coherent frames constitute a shot. Every frame belongs to a shot, which lasts for a minimum of 1 s and is based on the frame rate of the broadcast video (which can be anywhere between 20 to 60 frames per second). Enriching every frame of a video would be computationally expensive and practically inefficient. Thus, we find it logical to consider a shot as the fundamental unit of the video. Based upon these shots, the entire video can be iteratively enriched with data, such as scene types, actions and events.

Humans, on the other hand, tend to remember specific events or scenarios from a video that they view during a video-retrieval process. Such an event could be a dialogue, an action scene, or any series of shots unified by location or a dramatic incident [8]. Therefore, it is events themselves which should be treated as an elementary retrieval unit in future advanced video retrieval systems. Various terms denoting temporal video segments on a level above shots, but below sequences, appear in the literature [9]. These include scenes, logical units, logical story units, and topic units. The flow diagram on Figure 1 shows how this space could be well-defined [10]. A logical story unit (LSU) could thus be a scene or a topic unit, depending on the type of content. Our proposed pipeline can automatically segment videos into logical story units.

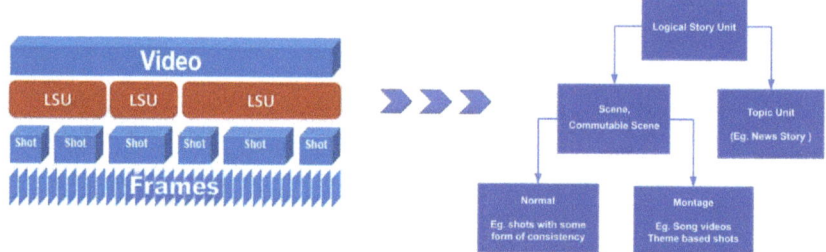

**Figure 1.** Pictorial representation of the structure of video, detailing the position and definition of a logical story unit (LSU). As shown in the flow diagram, an LSU can either be a scene or a topic unit. This paper predominately focuses on normal scene- and topic-unit-type videos.

Researchers often address semantic mining and structure mining problems separately, because they were historically applied to different domains. However, during the last decade, image recognition algorithms have improved exponentially, and deep learning models, together with GPU/TPU computational hardware, allow very accurate real-time detectors to be trained and served. This has paved the way to complex pipelines that can be defined and reused across multiple domains. We have made use of these technological advancements in defining a versatile semantic extraction pipeline that proves to address multiple video analytic problems simultaneously. In summary, the main contributions of this paper can be listed as follows:

1. We propose a flexible pipeline that can derive high-level features from detection algorithms and semantically enrich a video by performing automatic video structure mining. This pipeline consolidates the frame-level place and object tags using time-efficient deep neural networks in such a way that it could be used for further enrichment tasks, such as re-ID.
2. Within the pipeline, we have implemented a novel boundary-detection algorithm to cluster the temporally coherent, semantically closer segments into shots and LSUs.
3. We also propose a novel multi-object re-ID algorithm-based on context similarity in SOAP and broadcast content to generate object timelines.

The remainder of this paper is organised as follows. Section 2 reviews related work. Subsequently, Section 3 presents our methodology, which explains, in detail, the algorithms used for semantic extraction, boundary prediction and object re-ID. The experimental set up and model selection are presented in Section 4. Section 5 discusses the results, while Section 6 concludes this paper and discusses the future work.

## 2. Related Work

This work elaborates the role of semantics in video analysis tasks such as video structure mining and re-ID. Spatial semantics includes the objects and persons in, as well as the location of, a frame. Temporal semantics includes actions, events, and their interactions across the video. For a system to understand a video, therefore, the system requires the ability to automatically comprehend such spatio-temporal relationships. In the following subsections, we discuss various approaches for semantic extraction, LSU/shot boundary detection and re-ID methodologies.

*2.1. Semantic Extraction*

2.1.1. Image Classification and Localization

Image classification and object recognition tasks have been investigated for a long time. Yet, for much of this period, there were no suitable general solutions available. This was mainly attributed to the quality of training data and accessible computational hardware. Moreover, the classification accuracy when using a smaller, rather than a larger, number of classes was observed to be greater [11]. However, performance in image-classification tasks has been exponentially improved in open competitions, such as the Large Scale Visual Recognition Challenge (ILSVRC) and MIT-Places-365. These competitions encouraged the development of region proposal network (RPN)-based deep neural networks, including AlexNet, GoogleNet and Vision Geometry Group (VGG). These networks have revolutionised image classification and have opened doors, in all directions, for classification and annotation. We use the VGG-16 network trained on MIT-Places-365 for obtaining the place/location of a frame, because it is very generalised and the architecture could be reused for further tasks, including the Dense Captioning of a frame that also has VGG-16 as its base architecture.

In addition to classification tasks, the success of the above-mentioned challenges has also fuelled research on localisation and detection tasks. Speed and accuracy have been the major areas of focus and, based on these, there are two major types of object detection models: (1) region-based convolution models, such as R-CNN and Faster RCNN, that split the image into a number of sub-images, and (2) convolution models, such as Single Shot Detector (SSD) and You Only Look Once (YOLO), that detect objects in a single run [12]. Even though the Faster RCNN have slightly higher accuracy, the latest version of YOLO (YOLOv3 [12]) detects objects up to 20 times faster while retaining similar/acceptable accuracy. Thus, our pipeline has a pre-trained YOLOv3 model that has been used for detecting objects and persons in a frame.

2.1.2. Video Annotation

There has also been research pertaining to video annotation. [13] proposed an event-based approach to create text annotations, which infers high-level textual descriptions of events. This method does not take into account the temporal flow or correlations between different events in the same video. Thus, the approach does not have the ability to interact or fuse multiple events into scenes or activities. As explained in the previous section, it is important to search for and retrieve continuous blocks of video, often referred to as scenes or story units.

Stanislav Protasov et al. [14] proposed a pipeline with keyframe-based annotation of scene descriptions, while [15] proposed a sentence-generation pipeline which provides descriptions for keyframes based on the semantic information. Even though the techniques produced acceptable results, the annotations still lacked information and faced information losses. Torralba et al. [16], on the other hand, proposed a solution for semantic video annotation that consists of per-frame annotations of scene tags. The per-frame annotations are computationally expensive and often redundant. Therefore, we incorporated a pipeline that takes into account the drawbacks of these previous methodologies. The pipeline obtains all possible spatial information, ranging from the location to objects and persons, in the form of textual descriptions for every $n$th frame of the video. This $n$ depends on the frame rate of the video and is adjusted so that textual descriptions are obtained for a minimum of 4 frames per second.

*2.2. Boundary Detection*

Shot and scene detection is one of the long-studied problems in video structure mining. There have been a lot of different approaches based on the different features used and the different clustering methods available. In this subsection we discuss the latest approaches for shot and LSU detection.

In the existing works for shot boundary detection, there a prevailing and striking pattern of similarities. We have come to the conclusion that boundary detection is performed by calculating or learning the deviation of features over adjacent frames. Widely used features include RGB, HSV, or LUV colour histograms [17], background similarity [4], motion features [18], edge ratio change and SIFT [19], and spectral features. Ref. [17] uses a spectral clustering algorithm to cluster shots, while [18] proposes a new adaptive scene-segmentation algorithm that uses the adaptive weighting of colour and motion similarity to distinguish between two shots. They also propose an improved overlapping-links scheme to reduce shot grouping time. Recently, deep features, extracted using CNN, were employed to obtain significant state-of-the-art results [20]. This team used an end-to-end trainable CNN model that was trained using a cross entropy loss to detect shot transitions. In this work, we employ frame-level object-, person- and location-type semantic descriptions as features to estimate shot boundaries.

For scene detection, Stanislav Protasov et al. [14] proposed a pipeline that utilises scene descriptions for keyframes of shots, while [15] proposed a pipeline that generates sentences or captions based on objects in a keyframe. The former utilises a scene transition graph to cluster similar shots to scenes, while the latter proposes to use Jaccard-similarity for obtaining similarity between shots. As per survey [21], the LSU-detection task is understood as a three-stage problem. In the first step, frames are grouped into shots. In the second step, location, person and object descriptions are consolidated to obtain shot-level descriptions. In the third stage, shot-level descriptions are used to cluster the shots into story units, using a similarity metric and assumptions about the film structure. For shot boundary detection, we have proposed and utilised the shot-detection algorithm defined in our methodology.

## 3. Methodology

Based on the motivations explained in Section 1, we propose a pipeline that utilises semantic descriptions and their co-occurrences across a video to address the fundamental video processing challenges pertaining to structure mining and object re-ID tasks. The proposed pipeline is shown in Figure 2. We follow a step-wise approach to explain the implementation of the pipeline:

1. Semantic Extraction
2. Structure Mining
3. Similarity Estimation
4. Object Re-Identification

*3.1. Semantic Extraction: Recognizing Objects, Places and Their Relations*

In order to work with the high-level semantic features, it is important to have thorough information regarding the composition of each frame (e.g., objects, persons, and places in the frame). Since broadcast videos do not carry that much frame-level semantic information, it is necessary for our pipeline to have a good model that can predict, with high accuracy, the objects and places in a frame. As seen in Figure 2, frame-level semantic extraction is a common step for all the tasks dealt with in the paper—from shot/LSU boundary prediction to object timeline generation.

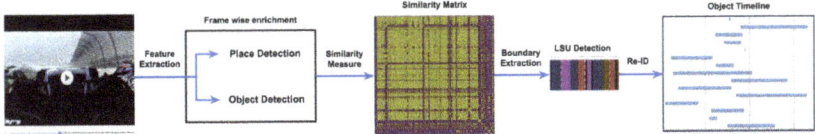

**Figure 2.** Overview of the proposed pipeline. Given the input video, the framework extracts visual features to obtain frame-level semantics. The enriched semantic information can then be used for search and retrieval of video segments, predict shot and scene boundaries, and also to create object timelines.

Feature Extraction

We make use of low-level and mid-level visual information for predicting the high-level features that are necessary to determine the semantic composition of a logical story unit. In our approach we use object, person and location tags as high-level features for detecting the LSU boundaries. To obtain the object and person annotations, the latest version of the YOLO object detector [12], pre-trained on the COCO dataset [22], is used. COCO stands for Common Objects in Context. The dataset comprises of 1.5 million object instances covering 80 object classes. Along with the object detector, the place or the location of the scenes are predicted using the ResNet-50 CNN architecture, pretrained on the places-365 dataset [11]. This dataset contains more than 10 million images in total, comprising 400+ unique scene categories [23].

3.2. Structure Mining: Shot Boundary Detection

Once we extract the visual features of the video frames, we utilise them to estimate the similarity between frames. This, in turn, is used to predict the overall structure of the video as shown in Figure 3. Broadcast videos generally have a frame rate of 24 fps. We process every sixth frame of our video for computational advantage (4 frames/s). Furthermore, we cluster temporally similar frames to form shot and story units.

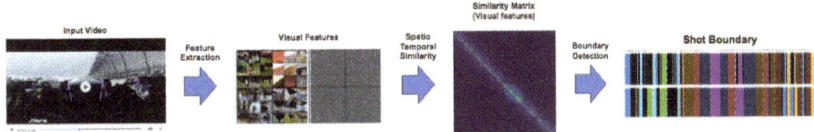

**Figure 3.** Overview of the framework for Shot Detection. Shot is defined as a group of continuous frames without a cut. To predict shot boundaries, the framework utilises only frame-level visual features from the given input video.

*Spatio-Temporal Visual Similarity Modelling*

In contrast to other approaches that use clustering for boundary detection, we construct a similarity matrix that jointly describes spatial similarity and temporal proximity. The generic element $S_{ij}$ defines the similarity between frames $i$ and $j$, as shown in Equation (1).

$$S_{ij} = exp\left(-\frac{d_1^2(\psi(x_i), \psi(x_j)) + \alpha \cdot d_2^2(x_i, x_j)}{2\sigma^2}\right) \quad (1)$$

where, $\psi(x_i)$ and $\psi(x_j)$ are the list of visual tags for the $i$th and $j$th frame, respectively. $d_1^2$ is the cosine distance between frame $x_i$ and $x_j$, while $d_2^2$ is the normalised temporal distance between frame $x_i$ and frame $x_j$. The parameter $\alpha$ tunes the relative importance of semantic similarity and temporal distance. The effect of alpha on the similarity matrix is shown in Figure 4.

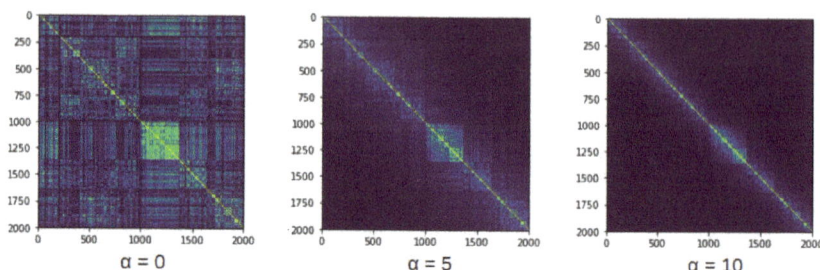

**Figure 4.** Effect of $\alpha$ (from left to right 0, 5, and 10) on similarity matrix $S_{ij}$. Higher values of $\alpha$ enforce temporal connections between nearby frames and increase the quality of the detected shots.

As shown in Figure 4, the effect of applying increasing values of $\alpha$ to the similarity matrix is to raise the similarities of adjacent frames, thereby boosting the temporal correlations of frames in the neighbourhood. At the same time, too high values of $\alpha$ would lead to the boosting of the temporal correlation of very close neighbouring frames, thereby failing to capture gradual shot changes. The final boundaries are created between frames that do not belong to the same cluster. An experiment was conducted with the videos of the RAI dataset, where values from 1 to 10 were provided for $\alpha$, and its effect was studied. We found that an $\alpha$ value of 5 performed well on average, for both gradual and sharp shot changes. Therefore, we use an $\alpha$ value of 5 for our shot boundary detection experiments, since it provides the right amount of local temporal similarity for the prediction of boundaries.

As seen in Equation (1), semantic composition-based frame-similarity estimation is composed of the following two sub parts:

- Semantic similarity scoring scheme
- Temporal model analysis

### 3.2.1. Semantic Similarity Scoring Scheme

We use the cosine similarity principle to measure inter-frame similarity; that is, we measure the cosine angle between the two frame vectors of interest. The cosine similarity between the $i$th and the $j$th frame is calculated by taking the normalised dot product as follows:

$$sim(x_i, x_j) = ||\psi(x_i)|| \cdot ||\psi(x_j)|| \qquad (2)$$

where, $\psi(x_i)$ is the normalised vector based on the list of visual tags for frame $x_i$. This results in a spatial similarity matrix. The similarity measure is converted into a distance measure based on the following Equation:

$$d_1^2(\psi(x_i), \psi(x_j)) = 1 - sim(x_i, x_j) \qquad (3)$$

An example of utilising the spatial similarity matrix to retrieve the top four similar frames from a video is shown in Figure 5.

**Figure 5.** An example of utilising the spatial similarity matrix to retrieve top four similar frames from a video. The video used is Season 5 Episode 21 of FRIENDS show.

### 3.2.2. Temporal Model Analysis

As per Equation (1) the temporal proximity is modelled using $d_2^2$, which is the normalised temporal distance between frames $x_i$ and $x_j$. The normalised temporal distance can be defined by Equation (4)

$$d_2^2(x_i, x_j) = \frac{|f_i - f_j|}{l} \tag{4}$$

where $f_i$ and $f_j$ are the index of frame $x_i$ and $x_j$, respectively, and $l$ is the total number of frames in the video.

### 3.2.3. Boundary Prediction

Based on Equation (1), the lower the value of $S_{ij}$, the more dissimilar frames $x_i$ and $x_j$ are. Thus, we calculate the shot boundary by thresholding $S_{ij}$. In our experiments, 0.4 was used as the threshold value. The entire shot boundary detection algorithm is shown in Algorithm 1.

---

**Algorithm 1:** Shot boundary detection

**Input:** List of frame-level objects and places tags
**Output:** Shot boundaries
1  shots = []
2  for $i = 1:n$ do
3     for $j = 1:n$ do
4        $place\_sim(x_i, x_j) = ||\psi(x_i)|| \cdot ||\psi(x_j)||$   // $\psi(x_i)$ = normalised vector of place tags for frame $x_i$
5        $obj\_sim(x_i, x_j) = ||\psi(x_i)|| \cdot ||\psi(x_j)||$   // $\psi(x_i)$ = normalised vector of object tags for frame $x_i$
6        $sim(x_i, x_j) = \frac{w_1(place\_sim) + w_2(obj\_sim)}{w_1 + w_2}$
7        $d_1^2(\psi(x_i), \psi(x_j)) = 1 - sim(x_i, x_j)$
8        $S_{ij} = exp\left(-\frac{d_1^2(\psi(x_i),\psi(x_j)) + \alpha \cdot d_2^2(x_i,x_j)}{2\sigma^2}\right)$

9  for $i = 1:n$ do
10    if $S_{i,i+1} <$ threshold then
11       shots.append($i$)

---

### 3.3. Similarity Estimation: Context Based Logical Story Unit Detection

Based on our experiments, we have deduced that normal broadcast content, such as a SOAP episode or the news, often make use of multiple angles pertaining to the same story unit. In more than 90% of the cases, these angles recur multiple times throughout the video. Therefore, as shown in Figure 6, the context-based similarity estimation begins with shot detection. Progressing from these estimated shot boundaries, frame-level semantic descriptions are merged as follows:

$$L_{ij} = \frac{w_1(place\_sim) + w_2(obj\_sim)}{w_1 + w_2} \tag{5}$$

where $w_1$ and $w_2$ are the weights for place and object descriptions. In our experiments, we have given more importance to place descriptions than to object descriptions, mainly because the state-of-the-art object detection models do not have the ability to predict all the objects in a frame. Moreover, the pre-trained place-detection model has the ability to capture the overall context of the shot location, and therefore has been deemed more important. Thus we have maintained $w_1$ and $w_2$ as 2 and 1, respectively, in all our experiments.

The shot-level similarity measure is calculated based on the joint similarity estimated using Equation (5). An example of the similarity matrix of a video from RAI is shown in Figure 7. The final similarity matrix is used along with the re-identification algorithm to generate object timelines.

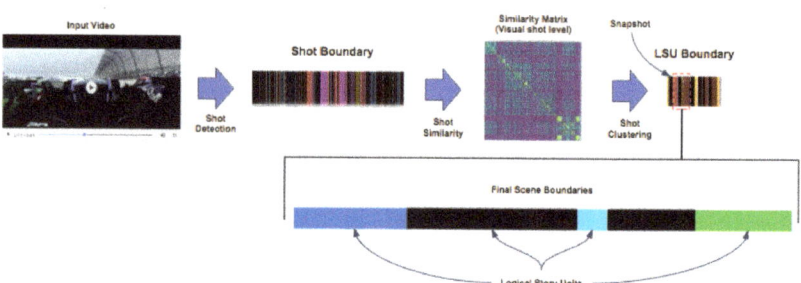

**Figure 6.** Overview of the LSU-detection module. Given the input video, the framework extracts audio–visual features to predict logical story unit boundaries based on semantic similarity between temporally coherent *shots*. The final decision boundary is based on thresholding the distance between consecutive *shots*.

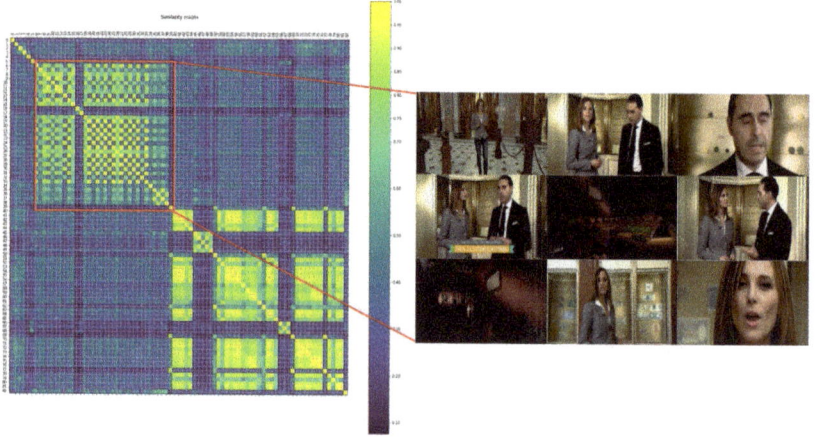

**Figure 7.** Estimated shot similarity for RAI video 23353. The figure also shows key frames of a selected LSU (red box).

*3.4. Object Re-Identification*

We propose an algorithm that formulates unique object IDs using LSUs and frame-level object detections, such that re-occurring objects are provided with the same ID. The algorithm we propose is based on the following hypothesis:

**Hypothesis 1.** *If two shots $S_a$ and $S_b$ are similar, then the objects present in $S_a$ and $S_b$ are also similar.*

3.4.1. Explanation

Multimedia broadcast content, such as SOAP, news, or talk shows, often reuse locations that conserve the objects that they contain. Then, based on the above hypothesis, the objects are the same if they are present in the same location. For example, in Figure 8, Image 1 is frame 26070 of the video and image 2 is frame 27604. Although they are approximately 1500 frames apart, they both pertain to the same location, and thus the objects in them are the same. An important point to note is that the hypothesis holds only for stationary/static objects; if there are dynamic objects present in the shots (e.g., persons) the hypothesis will fail. Our approach focuses only on static object re-identification—thus, the current paper will only address problems of this kind.

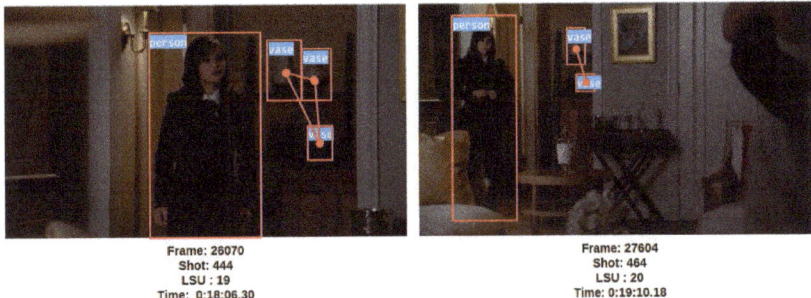

Figure 8. Example of multiple instance object class. This example is taken from the Season 4 Episode 16 of *New Girl* TV SOAP show. In the left side image (frame 26070) there are three different objects of the class (*vase*) detected, while in the right image (frame 27604), there are two objects of this class detected.

Based on the number of occurrences of a same class object in the same frame, the re-ID algorithm is composed of two sub-parts:

- Single instance
- Multiple instance

### 3.4.2. Single Instance

If there is just a single occurrence of the object in every frame it appears in throughout the video, then by Hypothesis 1, the *id* for object $O$ at frame $n$ is given by Equation (6) as follows:

$$O_{id}^n = \begin{cases} id = 1, & n = 0 \\ O_{id'}^a, & S_a^n > threshold, \\ id + 1, & S_a^n < threshold, \\ & where\ (a = 1{:}n-1) \end{cases} \quad (6)$$

where in $O_{id}^n$ is the object *id* at frame $n$, and $S_a^n$ is the context-similarity measure between the frame $a$ and $n$ as calculated in Equation (5).

### 3.4.3. Multiple Instance

If there are multiple occurrences of the object, we propose a graph-based approach to correctly localise the object in the frame. An example of this problem is shown in Figure 8. In such cases, where multiple objects of the same class exist, it is not only important to know whether shot/LSU of the frames are similar, but also to know the spatial position/location of the object in the frame, so that the object can be re-IDed correctly.

Therefore, based on the bounding box co-ordinates of the detected objects, a location graph is estimated using spatial distances between the objects, as shown in Figure 9. The idea here is to generate and compare the graphs such that the IDs of the objects can be matched.

*Spatial Distance Estimation*

Although, the 2-D Euclidean distance measure works well between frames with similar angles across similar LSUs, there are cases where the angle and zoom changes across similar LSUs. The topological information contained within the frame is also lost, making it impossible to obtain a realistic distance estimation. To compensate for the topological information, we propose to use depth maps, in combination with the location graph, to estimate a more realistic spatial distance between the objects in a frame. To obtain depth information, we use Dense Depth [24], pre-trained on NYU Depth V2 dataset [25]. The estimated depth is used as a third dimension, and thereby the Euclidean measure is recalculated as shown in Figure 10.

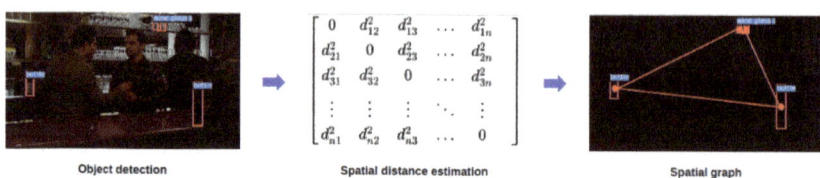

**Figure 9.** Spatial location graph generated for a frame using the centre of the bounding box coordinates and Euclidean distance between them.

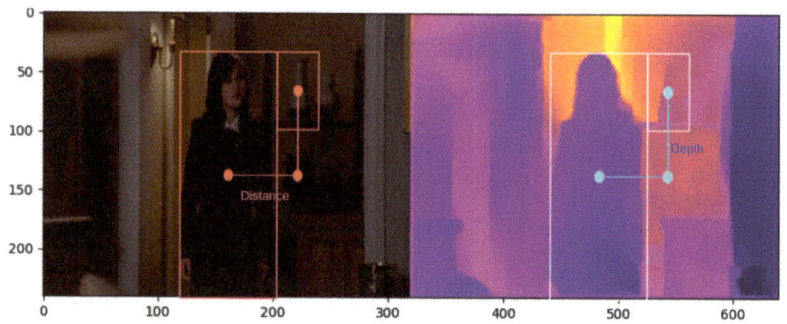

**Figure 10.** Comparison of frame 26070 with its estimated depth. Using the depth and distance measures, the actual distance between the two objects can be estimated.

Let $x$ and $y$ be the centre points of the objects $O_x$ and $O_y$, respectively, in a frame. Then the distance between them is given by:

$$distance = |x - y| \qquad (7)$$

The estimated depth has a range of values that are clipped between 10 and 1000, where 10 is the closest and 1000 is the farthest. If the depth values at points $x$ and $y$ can be represented as $\delta(x)$ and $\delta(y)$, the depth between the objects can be estimated by:

$$depth = |\delta(x) - \delta(y)| \qquad (8)$$

Finally, from Equations (7) and (8), the actual distance between the objects can be calculated as follows:

$$D_x^y = \sqrt{(distance)^2 + (depth)^2} \qquad (9)$$

*Spatial Location Graph*

For every frame with multiple instance objects, the spatial location graph is estimated based upon the pairwise distance between the objects in the frame, using Equation (9). Let $G_i(O, D)$ and $G_j(O, D)$ be the graphs with objects as nodes and their distances as edges for two similar frames $i$ and $j$. The objects in frame $j$ are matched with the objects in $i$, based on comparing the distances between the objects in $j$ and $i$ such that the difference between the distances is always minimal. For instance, if frame $i$ has 4 objects, $O_{i1}, O_{i2}, O_{i3}, O_{i4}$, of which $O_{i1}$ and $O_{i2}$ belong to the same class, and $D_i^{12}, D_i^{13}$ denotes the distance between objects, then to re-identify objects $O_1$ in frame $j$, the sub-graph distances of $G_i[O_1']$ and $G_i[O_2']$ are compared with $G_i[O_1']$. $O_{j1}$ is deduced to be the same as the object in $i$ for which the difference between distances is minimal. The overall object re-ID algorithm is shown in Algorithm 2 while the complete re-ID pipeline is shown in Figure 11.

**Algorithm 2:** Multi-object re-ID.

**Input:** Objects_list per frame, shot boundary and LSU similarity
**Output:** Object IDs per frame

1  shots = []
2  **for** *object = object_list[0]:object_list[len(object_list)]* **do**
3     **if** *count(objects) in all_frames <= 1* **then**
4        single_instance.append(object)
5     **else**
6        multi_instance.append(object)
7  **for** *object = single_instance[0]:single_instance[len(single_instance)]* **do**
8     id = 1
9     **for** *i = 1:class* **do**
10       **for** *i = 1:n* **do**
11          **if** *i==0* **then**
12             $object_{id}$ = id
13             id = id + 1
14          **else**
15             **if** $similarity(frame_n, 1 : frame_{n-1} > threshold,$ **then**
16                Let frame *a* be the frame most similar to frame *n*
17                $object_{id} = O^a_{id}$
18             **else**
19                $object_{id}$ = id
20                id = id + 1
21 **for** *object = multiple_instance[0]:multiple_instance[len(multiple_instance)]* **do**
22    id = 1
23    **for** *i = 1:class* **do**
24       **for** *i = 1:n* **do**
25          **if** *i==0* **then**
26             $object_{id}$ = id
27             id = id + 1
28          **else**
29             **if** $similarity(frame_n, 1 : frame_{n-1} > threshold,$ **then**
30                Let frame *a* be the frame most similar to frame *n*
31                $object\_list = graph\_compare(G_n[O'_{class}], G_a[O'_{class}])$
32                **for** *object_id in object_list* **do**
33                    $object_{id}$ = object_id
34             **else**
35                $object_{id}$ = id
36                id = id + 1

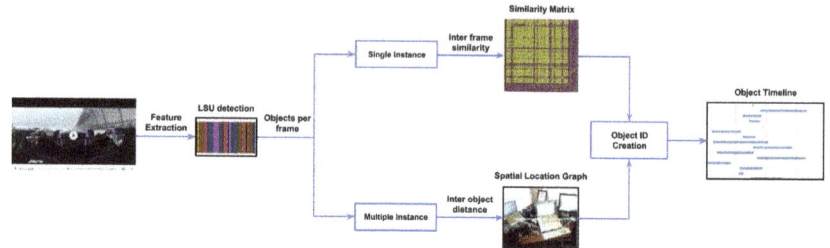

**Figure 11.** Proposed pipeline for multi-object re-ID. Given the input video, we estimate LSU and objects per frame for the video. Based on the number of occurrences of the object in a frame, the objects are categorised as single- and multi-instance objects. Subsequently using the inter-frame similarity and graph-based algorithms, object IDs are created and visualised.

## 4. Experiments

To provide a comprehensive overview of the strengths of the pipeline, it was separately evaluated on benchmark task-specific datasets. All the experiments were performed on a Linux Intel(R) Core(TM) i5-7440HQ CPU system with a RAM capacity of GB; the GPU was an NVidia GeForce 980 with 4 GB memory; and the operating system was Ubuntu version 16.04. The entire pipeline was implemented in Python 3.6 with the Pytorch deep learning library. The datasets and evaluation metrics used for evaluating our pipeline are explained in the following sections.

### 4.1. Dataset

In this work, a thorough, objective, and accurate performance evaluation has been carried out to evaluate the pipeline for shot boundary detection, LSU boundary detection and object re-ID.

To evaluate the proposed approach for shot and LSU boundary detection, we tested the pipeline on the benchmark RAI dataset. This dataset is a collection of ten challenging broadcasting videos from the Rai Scuola video archive, ranging from documentaries to talk shows constituted by both simple and complex transitions.

We evaluate our approach for object re-ID on randomly selected SOAP episodes. For fair evaluation, we chose to validate our approach on two different sets of SOAP broadcast content; namely, *New Girl* and *Friends*. We selected 10 episodes from Season 4 of *New Girl* and 10 episodes from Season 3 of *Friends* as our final dataset for object re-ID.

### 4.2. Evaluation Metrics

We evaluated the pipeline based on three tasks: (1) accuracy of the shot boundary detection; (2) accuracy of the LSU boundary detection; and (3) accuracy of the object re-ID algorithm.

For all the experiments, we use the precision, recall, and f1-score for the evaluation of our results. Precision, recall, and f1-score are computed based on the matched shots/LSU with the ground truth. Furthermore, the results were graphically visualised and analysed to promote insight.

The precision measure refers to the fraction of rightly predicted boundaries from total predictions, whereas recall measure denotes the fraction of boundaries rightly retrieved. If *groundtruth* refers to the list of ground-truth values and *prediction* refers to the list of automatically predicted values, then precision and recall can be expressed as in Equation (10).

$$precision = \frac{|groundtruth \cap prediction|}{|prediction|}$$
$$recall = \frac{|groundtruth \cap prediction|}{|groundtruth|} \quad (10)$$

F-score, on the other hand, combines precision and recall measures; it is the harmonic mean of the two. Traditional $F_{shot}$ can be defined as follows:

$$F_{shot} = 2 \cdot \frac{precision \cdot recall}{precision + recall} \tag{11}$$

As mentioned in earlier sections, the precision, recall, and f1 measure would not suffice to validate the accuracy of the LSU boundary detection algorithm. The reason for this is that humans and algorithms employ different ways of perceiving story units. Humans can relate changes in time and location to discontinuities in meaning, whereas an algorithm solely depends on visual dissimilarity to identify discontinuities. This semantic gap makes it impossible for algorithms to achieve fully correct detection results. Therefore, as suggested in [9], we use coverage and overflow metrics to measure how well our LSU boundary detection algorithm performs with respect to human labelled LSUs, using visual features. That is, in addition to the precision, recall, and f1 measures, we propose to use *coverage* and *overflow* measures to evaluate the number of frames that were correctly clustered together.

Coverage C measures the quantity of frames belonging to the same scene correctly grouped together, while Overflow O evaluates to what extent frames not belonging to the same scene are erroneously grouped together. Formally, given the set of automatically detected scenes $s = [s_1, s_2, ..., s_m]$, and the ground truth $g = [s_1, s_2, ..., s_n]$, where each element of s and g is a set of shot indexes, the coverage of scene s is proportional to the longest overlap between $s_i$ and $g_t$:

$$coverage = \frac{max_{i=1...n} \#(s_i \cap g_t)}{\#(g_t)} \tag{12}$$

$$overflow = \frac{\sum_{i=1}^{m} \#(s_i/g_t) \cdot min(1, (s_i \cap g_t))}{\#(g_{t-1}) + \#(g_{t+1})} \tag{13}$$

$F_{scene}$ combines the coverage and overflow measures and is the harmonic mean of the two. For coverage, values closer to 1 indicate better performance, and for overflow, values closer to 0 indicate better; thus we use $1 - overflow$ for calculating $F_{scene}$:

$$F_{scene} = 2 \cdot \frac{coverage \times (1 - overflow)}{coverage + (1 - overflow)} \tag{14}$$

For the experiments pertaining to object re-ID, we make use of *Accuracy* metrics. Accuracy is the most intuitive performance measure; it is simply the ratio of correctly predicted observations to total observations. In our scenario, the predicted observations are labelled as *True* if they are correctly predicted, and *False* otherwise. Therefore, if the total number *True* samples is denoted by *True*, and total number of *False* samples is denoted by *False*, then Accuracy can be calculated as follows:

$$Accuracy = \frac{True}{True + False} \tag{15}$$

## 5. Results and Discussion

### 5.1. Quantitative Results

#### 5.1.1. Shot Boundary Detection

In this study, to evaluate shot boundary detection, we have compared our framework with state-of-the-art CNN-based fast shot boundary detection[20]. We have used 10 random Internet Archive videos from the RAI dataset. Table 1 compares the precision, recall, and F-score of our pipeline with this state-of-the-art algorithm. These experimental results show that the state-of-the-art model performs extremely well on normal transitions, while performing comparatively poorly on complex transitions. Our approach, on the other hand, has obtained similar precision values for both complex and normal transitions. On average, our approach has outperformed the state-of-the-art with an f1 measure of 0.92.

5.1.2. LSU Boundary Detection

In this study, we also evaluated LSU boundary detection by comparing the results against two different algorithms for scene detection: [26], which uses a variety of visual and audio features that are integrated in a Shot Transition Graph (STG); and [27], which uses a spectral clustering algorithm and Deep Siamese network-based model to detect scenes. We used the same 10 videos from the RAI dataset for validation. Table 2 tabulates the coverage and overflow measures calculated based on the above methods. Our experimental results indicate that the model in [26] has the highest coverage value of 0.8—but it also has a very high overflow measure. Ref. [27] provides a comparatively better overflow result and overall performance than [26]. Although our approach achieved a lower coverage measure, it has obtained a very good overflow measure, which has resulted in a higher $F_{score}$. Our approach, with an average $F_{score}$ of 0.74, outperformed the other methods by more than 10%.

Table 1. Performance comparison for shot detection using boundary-level metrics.

|  | Gygli et al. [20] | | | Our Approach | | |
| --- | --- | --- | --- | --- | --- | --- |
| Video | Precision | Recall | $F_{shot}$ | Precision | Recall | $F_{shot}$ |
| 23353 | 0.95 | 0.99 | **0.96** | 0.877 | 0.99 | 0.945 |
| 23357 | 0.91 | 0.97 | 0.939 | 0.874 | 0.99 | **0.940** |
| 23358 | 0.92 | 0.99 | **0.954** | 0.775 | 0.99 | 0.873 |
| 25008 | 0.94 | 0.94 | **0.94** | 0.849 | 0.99 | 0.918 |
| 25009 | 0.97 | 0.96 | **0.965** | 0.726 | 0.98 | 0.841 |
| 25010 | 0.93 | 0.94 | 0.935 | 0.955 | 0.99 | **0.977** |
| 25011 | 0.62 | 0.9 | 0.734 | 0.863 | 0.99 | **0.927** |
| 25012 | 0.66 | 0.89 | 0.758 | 0.890 | 0.890 | **0.89** |
| Average | 0.853 | 0.948 | 0.899 | **0.861** | **0.986** | **0.912** |

Table 2. Performance comparison for LSU detection using frame-level metrics.

|  | Lorenzo et al. [27] | | | Sidiropoulos et al. [26] | | | Our Approach | | |
| --- | --- | --- | --- | --- | --- | --- | --- | --- | --- |
| Video | Coverage | Overflow | $F_{scene}$ | Coverage | Overflow | $F_{scene}$ | Coverage | Overflow | $F_{scene}$ |
| 23553 | 0.82 | 0.40 | 0.69 | 0.63 | 0.20 | 0.70 | 0.66 | 0.0083 | **0.79** |
| 23557 | 0.77 | 0.24 | **0.76** | 0.73 | 0.47 | 0.61 | 0.65 | 0.2016 | 0.72 |
| 23558 | 0.77 | 0.37 | 0.69 | 0.89 | 0.64 | 0.51 | 0.73 | 0.1346 | **0.80** |
| 25008 | 0.42 | 0.06 | 0.58 | 0.72 | 0.24 | **0.74** | 0.41 | 0.0100 | 0.58 |
| 25009 | 0.95 | 0.76 | 0.39 | 0.69 | 0.53 | 0.56 | 0.67 | 0.124 | **0.76** |
| 25010 | 0.66 | 0.40 | 0.63 | 0.89 | 0.92 | 0.15 | 0.66 | 0.012 | **0.79** |
| 25011 | 0.70 | 0.14 | **0.77** | 0.94 | 0.92 | 0.15 | 0.61 | 0.048 | 0.74 |
| 25012 | 0.53 | 0.15 | 0.65 | 0.93 | 0.94 | 0.11 | 0.63 | 0.0400 | **0.76** |
| Average | 0.70 | 0.30 | 0.66 | **0.8** | 0.63 | 0.43 | 0.63 | 0.074 | **0.74** |

5.1.3. Object re-ID

In this study, to evaluate object re-ID, we have applied the algorithm on 10 random episodes from Season 4 of *New Girl* and 10 random episodes from Season 3 of *Friends* TV shows. The dataset does not possess ground truth labels. Thus, the approach was manually validated—if the object was re-IDed correctly it was marked *True*; else it was marked *False*. The *True* and *False* values were consolidated per object class for all episodes of *New Girl* and *Friends* separately, and the object classes that had a minimum of 20 occurrences in all the episodes of SOAP put together were chosen to estimate the accuracy. The accuracy was then calculated for each SOAP separately. Table 3 shows the accuracy results for the object re-ID applied on the two SOAP series. These experimental results show that our object re-ID algorithm performs at an average accuracy of 0.87.

**Table 3.** Performance evaluation of object re-ID.

| Class | New Girl (10 episodes) | | Friends (10 episodes) | |
|---|---|---|---|---|
| | True | False | True | False |
| bed | 29 | 0 | 152 | 0 |
| bottle | 604 | 153 | 51 | 14 |
| refrigerator | 23 | 0 | 56 | 0 |
| sofa | 76 | 0 | 306 | 13 |
| dining table | 202 | 11 | 87 | 12 |
| vase | 43 | 8 | 143 | 45 |
| bowl | 59 | 0 | 78 | 39 |
| tv | - | - | 51 | 0 |
| cup | - | - | 69 | 20 |
| car | 74 | 13 | - | - |
| handbag | 61 | 0 | - | - |
| potted plant | 20 | 0 | - | - |
| **Count** | 1212 | 187 | 993 | 143 |
| **Accuracy** | | 0.866 | | 0.874 |

### 5.2. Ablation Study

In order to evaluate the importance of depth information in spatial distance estimation, tests were conducted by selecting random frames of different angles from similar LSUs, and distance was estimated with and without depth information. For example, as shown Figure 12, distance and depth were measured for two different frames. Depth-based distance using Equation (9) and normal Euclidean distance between the person object and the vase object were estimated. On comparing the depth-based distance and Euclidean distance between the two frames, it was seen that the error of the depth-based distance metric is much less than the error of the Euclidean distance metric. The experiment was repeated for 10 different scenarios from 10 different episodes; depth-based distance error was estimated to be at least six times smaller than the Euclidean distance error, on average.

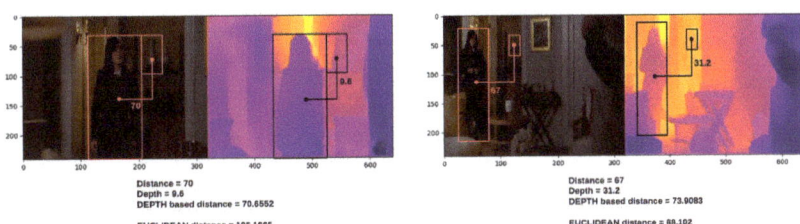

**Figure 12.** An example of ablation experiment to study the effect of depth in spatial distance estimation. Depth-based distance is found to be more comparable and less erroneous.

### 6. Conclusions and Future Work

We have proposed and presented a flexible pipeline for the annotation, structure mining, and re-ID of objects in broadcast videos by exploring the semantic composition of this pipeline. The high-level features extracted from low- and mid-level visual features provided useful information about various aspects of the analysed videos. A video-mining approach was used to infer high-level semantic concepts from the low-level features extracted from the videos. The results of this video data mining were further improved by exploiting temporal correlations within the video and constructing new features from them. Boundary prediction algorithms were proposed, which clustered and segmented each video based on its structure. Furthermore, object re-ID was explored and adapted to re-ID static objects in the videos. This helped us to create object timelines, which could be interesting for a variety of applications. Our experiments show that our approach is general enough for all broadcast videos, including different genres and languages. Upon inspecting the failure cases, it was found that the selection of similarity threshold played a vital role in the overall accuracy of the pipeline. Therefore, for future work, we would look into adapting the similarity threshold automatically, which would further improve the efficiency of the pipeline. Moreover, multi-modal features and effective methods to fuse multi-modal information will be investigated. In addition, we would also further optimise the spatial location graph to include dynamic/moving objects. Finally, the framework must be evaluated on a large scale and the models should be improved accordingly.

**Author Contributions:** Conceptualization, methodology, software, validation, formal analysis, investigation, resources, data curation and writing—original draft preparation: K.K.T.C.; writing—review and editing, visualization, supervision, project administration, and funding acquisition: S.V. Both authors have read and agreed to the published version of the manuscript.

**Funding:** The research activities as described in this paper were funded by Ghent University, IMEC, and the Flanders Innovation & Entrepreneurship (VLAIO) agency.

**Data Availability Statement:** Rai Dataset: https://aimagelab.ing.unimore.it/imagelab/researchActivity.asp?idActivity=019.

**Conflicts of Interest:** The authors declare no conflict of interest.

## References

1. Bagdanov, A.D.; Bertini, M.; Bimbo, A.D.; Serra, G.; Torniai, C. Semantic annotation and retrieval of video events using multimedia ontologies. In Proceedings of the International Conference on Semantic Computing (ICSC 2007), Irvine, CA, USA, 17–19 September 2007; pp. 713–720. [CrossRef]
2. Lu, Z.; Grauman, K. Story-driven summarization for egocentric video. In Proceedings of the IEEE Computer Society Conference on Computer Vision and Pattern Recognition, Portland, OR, USA, 23–28 June 2013; pp. 2714–2721. [CrossRef]
3. Mahasseni, B.; Lam, M.; Todorovic, S. Unsupervised Video Summarization with Adversarial LSTM Networks. In Proceedings of the 2017 IEEE Conference on Computer Vision and Pattern Recognition (CVPR), Honolulu, HI, USA, 21–26 July 2017; pp. 2982–2991. [CrossRef]
4. Goyal, P.; Hu, Z.; Liang, X.; Wang, C.; Xing, E.P. Nonparametric Variational Auto-encoders for Hierarchical Representation Learning. *arXiv* **2017**, arXiv:1703.07027.
5. Han, J.; Yang, L.; Zhang, D.; Chang, X.; Liang, X. Reinforcement Cutting-Agent Learning for Video Object Segmentation. In Proceedings of the IEEE Conference on Computer Vision and Pattern Recognition (CVPR), Salt Lake City, UT, USA, 18–23 June 2018.
6. Meng, J.; Wang, H.; Yuan, J.; Tan, Y. From Keyframes to Key Objects: Video Summarization by Representative Object Proposal Selection. In Proceedings of the 2016 IEEE Conference on Computer Vision and Pattern Recognition (CVPR), Las Vegas, NV, USA, 27–30 June 2016; pp. 1039–1048. [CrossRef]
7. Plummer, B.A.; Brown, M.; Lazebnik, S. Enhancing Video Summarization via Vision-Language Embedding. In Proceedings of the 2017 IEEE Conference on Computer Vision and Pattern Recognition (CVPR), Honolulu, HI, USA, 21–26 July 2017; pp. 1052–1060. [CrossRef]
8. Minerva, M.; Yeung, B.L.Y. Video content characterization and compaction for digital library applications. In *Storage and Retrieval for Image and Video Databases V*; International Society for Optics and Photonics: San Jose, CA, USA, 1997; Volume 3022, pp. 45–58. [CrossRef]
9. Vendrig, J.; Worring, M. Systematic evaluation of logical story unit segmentation. *IEEE Trans. Multimed.* **2002**, *4*, 492–499. [CrossRef]
10. Petersohn, C. *Temporal Video Segmentation*; Jörg Vogt Verlag: Dresden, Germany, 2010.
11. Zhou, B.; Khosla, A.; Lapedriza, A.; Oliva, A.; Torralba, A. Learning Deep Features for Discriminative Localization. In Proceedings of the IEEE Conference on Computer Vision and Pattern Recognition, Las Vegas, NV, USA, 27–30 June 2016.
12. Redmon, J.; Farhadi, A. YOLOv3: An Incremental Improvement. *arXiv* **2018**, arXiv:1804.02767.
13. Altadmri, A.; Ahmed, A. Automatic semantic video annotation in wide domain videos based on similarity and commonsense knowledgebases. In Proceedings of the 2009 IEEE International Conference on Signal and Image Processing Applications, Kuala Lumpur, Malaysia, 18–19 November 2009; pp. 74–79. [CrossRef]
14. Protasov, S.; Khan, A.M.; Sozykin, K.; Ahmad, M. Using deep features for video scene detection and annotation. *Signal Image Video Process.* **2018**, *12*, 991–999. [CrossRef]

15. Ji, H.; Hooshyar, D.; Kim, K.; Lim, H. A semantic-based video scene segmentation using a deep neural network. *J. Inf. Sci.* **2019**, *45*, 833–844. [CrossRef]
16. Torralba, A.; Murphy, K.P.; Freeman, W.T.; Rubin, M.A. Context-based Vision System for Place and Object Recognition. In Proceedings of the Ninth IEEE International Conference on Computer Vision, Nice, France, 13–16 October 2003; Volume 2, p. 273.
17. Odobez, J.M.; Gatica-Perez, D.; Guillemot, M. Spectral structuring of home videos. In *International Conference on Image and Video Retrieval*; Springer: Berlin/Heidelberg, Germany, 2003; pp. 310–320.
18. Kwon, Y.M.; Song, C.J.; Kim, I.J. A new approach for high level video structuring. In Proceedings of the 2000 IEEE International Conference on Multimedia and Expo. ICME2000 Proceedings, Latest Advances in the Fast Changing World of Multimedia (Cat. No.00TH8532), New York, NY, USA, 30 July–2 August 2000; Volume 2, pp. 773–776.
19. Mitrović, D.; Hartlieb, S.; Zeppelzauer, M.; Zaharieva, M. Scene segmentation in artistic archive documentaries. In *Symposium of the Austrian HCI and Usability Engineering Group*; Springer: Berlin/Heidelberg, Germany, 2010; pp. 400–410.
20. Gygli, M. Ridiculously Fast Shot Boundary Detection with Fully Convolutional Neural Networks. *arXiv* **2017**, arXiv:1705.08214.
21. Del Fabro, M.; Böszörmenyi, L. State-of-the-art and future challenges in video scene detection: A survey. *Multimed. Syst.* **2013**, *19*, 427–454. [CrossRef]
22. Lin, T.; Maire, M.; Belongie, S.J.; Bourdev, L.D.; Girshick, R.B.; Hays, J.; Perona, P.; Ramanan, D.; Dollár, P.; Zitnick, C.L. Microsoft COCO: Common Objects in Context. *arXiv* **2014**, arXiv:1405.0312.
23. Zhou, B.; Lapedriza, A.; Khosla, A.; Oliva, A.; Torralba, A. Places: A 10 million Image Database for Scene Recognition. *IEEE Trans. Pattern Anal. Mach. Intell.* **2017**, *40*, 1452–1464. [CrossRef] [PubMed]
24. Alhashim, I.; Wonka, P. High Quality Monocular Depth Estimation via Transfer Learning. *arXiv* **2019**, arXiv:1812.11941.
25. Silberman, N.; Hoiem, D.; Kohli, P.; Fergus, R. Indoor Segmentation and Support Inference from RGBD Images. In *European Conference on Computer Vision*; Springer: Berlin/Heidelberg, Germany, 2012.
26. Sidiropoulos, P.; Mezaris, V.; Kompatsiaris, I.; Meinedo, H.; Bugalho, M.; Trancoso, I. Temporal Video Segmentation to Scenes Using High-Level Audiovisual Features. *IEEE Trans. Circuits Syst. Video Technol.* **2011**, *21*, 1163–1177. [CrossRef]
27. Baraldi, L.; Grana, C.; Cucchiara, R. A Deep Siamese Network for Scene Detection in Broadcast Videos. *arXiv* **2015**, arXiv:1510.08893.

Article

# Recommendations for Different Tasks Based on the Uniform Multimodal Joint Representation

Haiying Liu [1,2,†], Sinuo Deng [1,†], Lifang Wu [1,*], Meng Jian [1], Bowen Yang [1] and Dai Zhang [1]

1. Faculty of Information Technology, Beijing University of Technology, Beijing 100124, China; liuhaiying@emails.bjut.edu.cn (H.L.); dsn0w@emails.bjut.edu.cn (S.D.); mjian@bjut.edu.cn (M.J.); yangbowen108@emails.bjut.edu.cn (B.Y.); zhangd1@chinatelecom.cn (D.Z.)
2. College of Artificial Intelligence, North China University of Science and Technology, Tangshan 063009, China
* Correspondence: lfwu@bjut.edu.cn
† These authors contributed equally to this work.

Received: 27 July 2020; Accepted: 2 September 2020; Published: 4 September 2020

**Abstract:** Content curation social networks (CCSNs), such as Pinterest and Huaban, are interest driven and content centric. On CCSNs, user interests are represented by a set of boards, and a board is composed of various pins. A pin is an image with a description. All entities, such as users, boards, and categories, can be represented as a set of pins. Therefore, it is possible to implement entity representation and the corresponding recommendations on a uniform representation space from pins. Furthermore, lots of pins are re-pinned from others and the pin's re-pin sequences are recorded on CCSNs. In this paper, a framework which can learn the multimodal joint representation of pins, including text representation, image representation, and multimodal fusion, is proposed. Image representations are extracted from a multilabel convolutional neural network. The multiple labels of pins are automatically obtained by the category distributions in the re-pin sequences, which benefits from the network architecture. Text representations are obtained with the word2vec tool. Two modalities are fused with a multimodal deep Boltzmann machine. On the basis of the pin representation, different recommendation tasks are implemented, including recommending pins or boards to users, recommending thumbnails to boards, and recommending categories to boards. Experimental results on a dataset from Huaban demonstrate that the multimodal joint representation of pins contains the information of user interests. Furthermore, the proposed multimodal joint representation outperformed unimodal representation in different recommendation tasks. Experiments were also performed to validate the effectiveness of the proposed recommendation methods.

**Keywords:** multimodal joint representation; content curation social networks; different recommend tasks; content based recommend systems

## 1. Introduction

Content curation social networks (CCSNs) are booming social networks where users demonstrate, collect, and organize their multimedia contents. Pinterest is a typical CCSN. Since its inception in March 2010, Pinterest developed fast, broke through the 10 million user barrier [1] in January 2012, and CCSNs have gone on to become a very popular social network worldwide. In China, there are many Pinterest-like networks such as Huaban, Meilishuo, Mogu Street, and Duitang online from 2010–2012. The rapid growth of CCSNs has attracted much attention from multiple research fields, such as network characteristic analysis [1,2], user behavior research [3–5], social influence analysis [6], link analysis [7], word embedding [8], search engines [9], user modeling [10–12], and recommender systems [13–25].

As is well known, CCSNs are content-centric social networks [10]. Different from user-centric social networks, users on CCSNs pay more attention to the contents that users collect, which are not only communication carriers but also carriers of user interests. Taking one of the best known CCSNs, Pinterest, as an example, a "pin" is an image with a brief text description supplied by the users. A "board" is a collection of some similar style pins. In other words, the pins are curated into "boards" by categories [19]. On CCSNs, the collection of a user is composed of several boards, and a board is composed of pins. The relationships between users, boards, categories, and pins are shown in Figure 1. A "user" represents the users on CCSNs. A "board" represents a container of pins and is organized into different categories. A "category" is the category of the board, which is given by the user. A "pin", which is created by a user, is the basic unit, composed of an image and a corresponding brief text description. Users on CCSNs can "follow" the users they are interested in like on Twitter or Facebook. "Re-pin" is an action like a "repost" or a "retweet", whereby users can re-save pins and re-organize them with new descriptions and new categories to their own board. "Create" is similar to "post" on Twitter or Weibo, allowing users to post their original contents on CCSNs. From the figure, we can see that the pin is the basic unit in CCSNs.

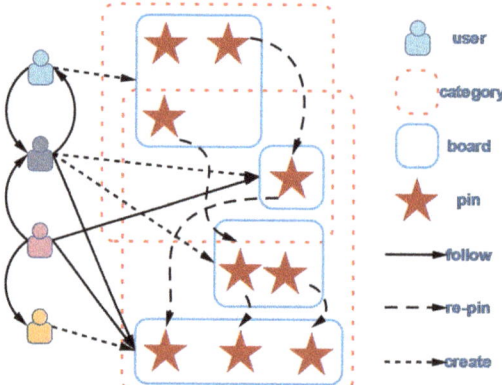

**Figure 1.** Items on content curation social networks (CCSNs) and their relationships.

Besides the content collections, there are also abundant social behaviors on CCSNs. Users can follow other users or other users' boards, users can also re-pin other users' pins and collect them into their own board. Furthermore, the re-pin path is recorded in CCSNs. All the users who have re-pinned a pin can be connected using a re-pin path. All users of the same re-pin path have collected the same image, but they have organized them into different boards and different categories. As shown in Figure 2.

On content-centric CCSNs, most user activities are related to the pins. Liu et al. [25] found that only 30% of pins are re-pinned from their followers by statistics on Huaban. Furthermore, users do not follow the users from whose boards they re-pin the pins [4]. A non-trivial number of pins are collected from non-followees [5], and those from native followees are more than those from cross-domain followees [3]. These observations suggest that social relationships are not the main motivation of content discovery on CCSNs. On the contrary, user interests represented by pins play an important role in user behaviors on CCSNs. It is possible that content-based recommender algorithms will be more effective than social behavior-based algorithms such as collaborative filtering. Inspired by this, we managed to implement recommendations for different tasks based on identical representations of pins. The problem can be broken down into two questions: how to represent a given pin effectively; and how to implement the different tasks with the obtained representation.

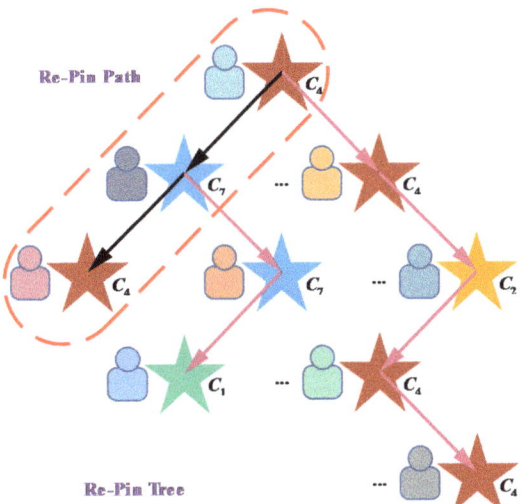

**Figure 2.** Illustration of a re-pin tree composed of some re-pin paths. Each star represents a pin and the $C_i$ next to it is the category given by the corresponding user. Note that all pins in the same re-pin tree have an identical image.

As shown in Figure 3, a pin is an image with its text description, hence it is obvious that both modalities should be utilized for complete representations. In order to fully utilize two modalities, we propose a framework that can learn the multimodal joint representation of pins. Image representations and text representations are obtained separately by deep models and are then fused to form multimodal joint representations. An intermediate layer of a convolutional neural network (CNN) is used to extract image representations. In order to establish the relation between image representations and user interests, some chosen images are annotated with their category distributions, which are the statistics of selections of users, to fine-tune the CNN. Text representations are means of word vectors in a word2vec model trained on public text corpora. Then, a multimodal deep Boltzmann machine (DBM) is trained with two modalities as inputs and the activation probabilities of the top layer are extracted as the final representation of pins.

**Figure 3.** Examples of pins on CCSNs: (**a**) Pinterest; (**b**) Huaban.

Recommendation tasks include recommending pins to users, recommending thumbnails to boards, recommending categories to boards, and recommending boards to users. On the basis of the representation of pins, pin recommendation becomes a problem of similarity measurement in the representation space, which can be solved by ranking the similarities between the candidates and the target pin. A board thumbnail consists of representative pins that can be selected by clustering pins in the board. The board category, which is the coarse interest selected by its owner, is considered to be the accumulation of the category distribution of its pins, and the category distribution can be obtained with a trained multidimensional logistic regression (LR). Boards and users are treated as pin collections, modeled as the Fisher vector (FV) of all their pin representations, and recommended to target users, similar to the pin recommendation method.

This paper makes the following contributions:

- On the basis of the characteristics of CCSNs, an easy-to-accomplish annotation method is proposed to automatically label the images by the category distributions on the re-pin tree of the corresponding pins. On the basis of the image and the corresponding labels, a multilabel CNN Network was fine-tuned, which significantly enhances the capability of image representation;
- We designed a framework which combines deep features of images and texts into a joint representation to maintain both consistent information and specific characteristic of different modalities. On this basis, a uniform recommendation scheme was designed for different tasks on CCSNs;
- The experimental results demonstrate that the proposed multimodal representation is more effective than representations learned from unimodal information. Furthermore, the proposed method performs better than existing multimodal representation learning methods on multiple recommendation tasks.

## 2. Related Work

With the rise of CCSNs, several studies have been performed, of which search engine, user modeling, and recommender systems are the most relevant. Most prior work only studied monomodal data. Yang et al. [16] recommended boards re-ranked with image representations based on boards with the text representations model. Liu et al. [22] recommended pins with two unimodal representations separately. Cinar et al. [11] predicted categories of pins with two kinds of unimodal representations and fused the two modality results using decision fusion. All the models are late fusion models that do not concern multimodal joint representations.

Multimodal joint representation includes unimodal representation models and multimodal fusion schemes. For image representation, CNNs have achieved remarkable performance in the field of computer vision. Creating a large labelled dataset is the key to train CNNs. Cinar et al. [11] and You et al. [12] directly used a pin's category as its label. However, this label may not be absolutely correct since the same image may have different categories selected by different users. Geng et al. [10] trained a multitask CNN with ontological concepts, but the ontology was constructed in the fashion domain and was difficult to extend to all other domains. Zhai et al. [21] extracted more detailed labels on Pinterest by taking top text search queries, but the quality and consumption of this annotation highly depends on the search engine. Inspired by the fact that the predefined categories on CCSNs are not independent objects but related notions, labels formed by statistics category distributions are used and a CNN is fine-tuned as a multilabel regressor. With regard to text representation, one-hot representations [13,16] and distributed representations, such as the word2vec tool [11], have been used. From the practical point of view, the word2vec tool [26], which can capture syntactic and semantic relationships between words in the corpus, is more scalable. In addition, mean vectors [27] of the word2vec tool can obtain usable text representation without further learning.

Several multimodal fusion studies are being performed on classification and retrieval. Except for directly concatenating modalities, most existing schemes are designed based on models such as CNNs [28] and recurrent neural networks [8]. These models mainly learn the consistency between multiple modalities and cannot deal with missing input modalities well. On the generative side, latent Dirichlet allocations (LDAs) [29], restricted Bolzmann machines (RBMs) [30], deep autoencoders (DAEs) [31], and deep Boltzmann machines (DBMs) [32] have been proven to be feasible methods for learning the consistency and complementarity between modalities and can easily deal with some missing modalities. However, limited studies have focused on fusing features obtained from these deep learning models. Zhang et al. [33] used a DAE for fusing the textual features extracted by training the Word2Vec tool [26] and visual features generated by the 6-th layer of AlexNet [34]. However, there are no existing studies that have used information from all modalities from CCSNs for recommendation tasks. In this paper, we trained a multimodal DBM to handle a situation in which the data from CCSNs are unlabeled and some modality inputs are missing, and we used features obtained by deep learning as the input to make our multimodal representation more accurate and compact.

Compared to pin and board category recommendation, few studies have been performed on board and user recommendations. Kamath et al. and Wu et al. [13,23] model boards and users, respectively, with text data and some collaborative filtering methods [15,20,25] to recommend users with user behaviors, but they do not take images, which are the essential content on CCSNs, into account. Yang et al. [17] represent boards by sparse coding the descriptors of images, but similarly to Yang et al. [16] as mentioned above, their methods require cross-domain information. Moreover, the information loss of the sparse code based on a cluster dictionary is more than the FV based on a Gaussian mixed model (GMM). Furthermore, no studies on board thumbnails have been published.

Existing research only focuses on one recommendation task, while the method in this paper uses identical pin representation to accomplish different recommendations such that the problems are simplified and resource saving.

## 3. Multimodal Joint Representation of Pins

A pin, that is, an image with text descriptions, is the basic item and the carrier of user interests on CCSNs. The purpose of this section is to learn the representation of pins from both modalities. As the foundation of further applications, the representation should contain the information of user interests.

The proposed framework of learning multimodal joint representation of pins is shown in Figure 4. The proposed process can be divided into three parts: image representation learning, text representation learning, and the multimodal fusion. For an input pin, the image representation of the pin is extracted by our modified CNN and the text representation is generated by the pre-trained Word2Vec model. Finally, we can obtain the joint representation by fusing both the image and text representations with a modified multimodal DBM model. The whole process comprises three parts: the image representation, the text representation, and the multimodal fusion. For given pins, their images are loaded by a CNN that is fine-tuned on an image dataset that is annotated automatically, and one of the intermediate layers of the CNN is extracted as image representations. Meanwhile, text representations are computed by applying mean pooling on word vectors derived from the word2vec tool, which are trained on some text corpora. Then, a multimodal DBM is trained on both image and text representations. Finally, the activation probabilities of the last hidden layer of the multimodal DBM are inferred as the expected multimodal joint representation of pins.

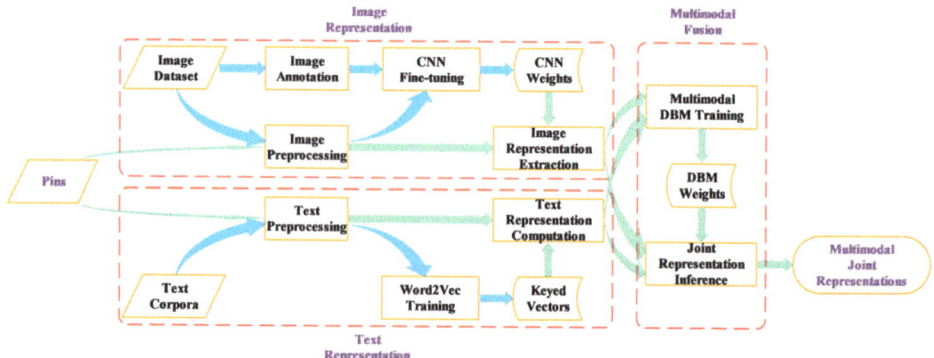

**Figure 4.** Framework of learning multimodal joint representation of pins (CNN: convolutional neural network).

### 3.1. Image Representation

Image representation aims to learn image features which not only maintain intrinsic characteristics, but also reflect user interests on CCSNs. CNNs have become the dominant approach in computer vision. Top layers of CNNs can extract high-level image features interpreted as color, material, scene, texture, object, and so on by various means. Intermediate layers of CNNs, especially fully connected (FC) layers, are often used for image representation and for further applications. As supervised learning models, CNNs can capture the relationships between user interests and images if user interests are trained as labels during the training process.

As a typical deep learning framework, a CNN requires a training set with large number of images with corresponding labels. Social networks are good sources for collecting the images, but noisy labels are always a primary problem. On CCSNs, all of the pins are collected by users and the categories are assigned by users, therefore the categories can be seen as labels with a high level of confidence. Users can create boards, and then create or collect pins into the boards to exhibit their interests. When a user creates a board, he is asked to choose one of the predefined categories on CCSNs, and the chosen category is the category of all the pins on the board. This is to say, every pin has a user-selected category. Since the category can reflect the theme of the board and the pin, it can describe coarse-grained user interests and can be trained as the label of an image.

On CCSNs, different users may select different categories for the same image. For example, the pin in Figure 3a is re-pinned by 50 users. Because the image is a poster of the video game NBA 2K12, 26 users categorized the pin into category 'sports', 16 users categorized it into category 'entertainments', and the other eight users categorized it into 'design'. On the basis of the statistical distribution of predefined categories given by users, the category distribution of pins can be computed as

$$Interest_I = \left( p_{C_i} = \frac{f_{C_i}}{\sum_{i=1}^{N_C} f_{C_i}} \right) \in [0,1]^{N_C}, \quad (1)$$

where $f_{C_i}$ denotes the $i$-th category ($C_i$) frequency, and $N_C$ is the total number of predefined categories. As a result of the fact that the minority opinion is sometimes hard to understand and spammers exist, in practice, before the computation of category distribution, we set

$$f_{C_i} = 0 \quad \text{if} \quad f_{C_i} < \frac{\sum_{i=1}^{M_C} f_{C_i}}{M_C}, \quad (2)$$

where $M_C$ is the total number of chosen categories that appear in the re-pin tree of $I$, to remove spam and make the sequence represent the majority opinion. Using the proposed annotation method, we were able to acquire labels of collected images without any additional human labor. Furthermore, compared to expensive human-labeled data, we believe the category distribution contributed by the collective user intelligence from re-pin trees is more suitable as the label of a pin. In contrast to existing image representation learning methods, which rely on high-quality label supervision, our category distribution of pins is acquired by mining the rich re-pin relationships from inexhaustible CCSN contents.

We then fine-tuned a pretrained CNN model to accelerate the training process. A deeper and wider architecture commonly performs better, while it is usually more time and space consuming. Thus, AlexNet [34] was chosen as a basis. The core visual deep model could be replaced by any of the other state-of-the-art models, such as GoogLeNet and ResNet. AlexNet, with weights pretrained by ImageNet [35], is commonly used to classify independent objects, though we needed a multilabel regressor model. Accordingly, the loss layer from softmax was changed from a logarithmic loss layer to a sigmoid with a cross entropy loss layer. We define loss function as

$$E = -\sum_{i=1}^{N_C} [p_{C_i} \ln \hat{p}_{C_i} + (1 - p_{C_i}) \ln (1 - \hat{p}_{C_i})], \qquad (3)$$

where $p_{C_i}$ denotes the percentage in Equation (1), and $\hat{p}_{C_i}$ is the corresponding sigmoid output. After fine-tuning the CNN, its weights are stored for feature extraction. Then, the image representations are the activation values of the FC layer.

*3.2. Text Representation*

The text description is an important personalized supplement to the image representation. Similar to the image representation, we generate the text representation for the purpose of discovering the relationships between the descriptions and categories of the pins. Contrary to the case of images, descriptions of the same pin may be different. Therefore, it is not easy to build a large high-quality labelled dataset on CCSNs.

Since words used on CCSNs have no obvious difference with those in common situations, we trained a word2vec [26] model on some public corpora for encoding words. The efficient shallow model was designed for studying word representations. The learned word vectors capture a large amount of syntactic word relationships and meaningful semantic relationships. The training dictionary should include words from the category words and the text description to represent the relationships between the text representation and the categories. In addition, word vectors, which encode words into compact vector spaces, are more scalable than one-hot representations, because the vocabulary of natural language is extremely wide. Both the training speed and the quality of the vectors could be improved by several extensions including the hierarchical softmax, negative sampling, noise contrastive estimation, and subsampling of frequent words [36]. For details of the Word2Vec model, please refer to the original paper.

Because of the diverse lengths of the texts, it is necessary to generate vectors with a constant dimension from a set of word vectors to represent a complete text. Some pooling methods, such as mean pooling [27], have been proven feasible in solving this problem. For a text $T = \{Word_1, Word_2 \cdots, Word_{M_T}\}$, we compute the mean vector in Equation 4 as its text representation,

$$V_T = \frac{1}{M_T} \sum_{i=1}^{M_T} KeyedVector_{Word_i}, \qquad (4)$$

where $KeyedVector_{Word_i}$ denotes the $i$-th word ($Word_i$) vector, and $M_T$ is the text length.

## 3.3. Multimodal Fusion

Different modalities can provide both consistent and complementary information, while their distinct statistical properties make it difficult to combine them into a joint representation that maintains their specific characteristics using a shallow architecture. A multimodal DBM [32] can effectively model a joint distribution over modalities, which adds a shared hidden layer on top of DBMs to combine them.

As illustrated in Figure 5, a multimodal DBM is an undirected graphical model with fully bipartite connections between adjacent layers. Each pathway of it is a DBM, which is structured by stacking two restricted Bolzmann machines (RBMs) in a hierarchical manner. All layers, except the two bottom layers, use standard binary units. An RBM with hidden units $H = (h_j) \in \{0,1\}^F$ and visible units $V = (v_i) \in \{0,1\}^D$ defines the energy function as follows:

$$E(V, H; \theta) = -\sum_{i=1}^{D}\sum_{j=1}^{F} v_i w_{ij} h_j - \sum_{i=1}^{D} a_i v_i - \sum_{j=1}^{F} b_j h_j, \quad (5)$$

where $\theta = \{(w_{ij}) \in \mathbb{R}^{D \times F}, (a_i) \in \mathbb{R}^D, (b_j) \in \mathbb{R}^F\}$ are model parameters comprising the symmetric interaction term $w_{ij}$ between the hidden unit and the visible unit, the visible unit bias term $a_i$, and the hidden unit bias term $b_j$. RBMs can be considered autoencoders, and one of their applications is dimensionality reduction by reducing F. Both bottom layers of our model change to Gaussian–Bernoulli RBMs which use Gaussian distribution to model real-valued inputs. The energy function of a Gaussian–Bernoulli RBM with visible variables $V = (v_i) \in \mathbb{R}^D$ and hidden variables $H = (h_j) \in \{0,1\}^F$ is defined as

$$E(V, H; \theta) = \sum_{i=1}^{D} \frac{(v_i - a_i)^2}{2\sigma_i^2} - \sum_{i=1}^{D}\sum_{j=1}^{F} \frac{v_i}{\sigma_i} w_{ij} h_j - \sum_{j=1}^{F} b_j h_j, \quad (6)$$

where $\sigma_i$ denotes the standard deviation of the $i$-th visible unit and $\theta = \{(w_{ij}) \in \mathbb{R}^{D \times F}, (a_i) \in \mathbb{R}^D, (b_j) \in \mathbb{R}^F, (\sigma_i) \in \mathbb{R}^D\}$. During the unsupervised pretraining process of the multimodal DBM, modalities can be thought of as labels for each other. Each of the multimodal DBM layers has a small contribution to eliminating modality-specific correlations. Therefore, in contrast to the modality-full input layers, the top layer can learn representations that are relatively modality free. The joint representation of the image and text inputs can be represented as follows:

$$P(V_I, V_T; \theta) = \sum_{H_{I2}, H_{T2}, H_3} P(H_{I2}, H_{T2}, H_3) \left( \sum_{H_{I1}} P(V_I, H_{I1}, H_{I2}) \right) \left( \sum_{H_{T1}} P(V_T, H_{T1}, H_{T2}) \right), \quad (7)$$

where $\theta$ denotes all model parameters. The reader may refer to the original paper for more details of multimodal DBMs.

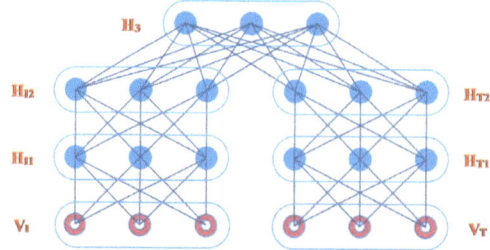

**Figure 5.** Architecture of a multimodal deep Boltzmann machine (DBM) for fusing image and text representations.

An advantage of multimodal DBMs is that they can deal with the absence of some modalities. After training our multimodal DBM, even though some pins may have no descriptions, the activation probabilities of $H_3$, which are used as our final multimodal joint representation of pins, could be inferred from different conditional distributions with the standard Gibbs sampler. In addition, the multimodal DBMs are used to generate the missing text representation in a similar manner. Moreover, multimodal DBMs can be trained supervised by connecting additional label layers on top of them.

## 4. Implementation of Recommendations for Different Tasks

Once the representations of pins were obtained, we then aimed to apply them to the recommender system. According to the practical applications on CCSNs, there are four recommendation tasks: recommending pins or boards to users, recommending thumbnails to boards, and recommending board categories to boards. All the recommendation methods are content-based.

### 4.1. Pin Recommendation

Pin recommendation is a crucial function for content discovery on CCSNs. It can be inferred that pins with similar interests are close in the representation vector space. Considering that the different boards a user collects have different characteristics, accordingly, given a target user, the similarity between pins in a board and the candidate pins is computed in the vector space, and the pins are ranked by similarities in descending order. For different boards, different pins are recommended. Most similarity metrics can be used; cosine similarities were computed in our work. Pins are ranked according to the similarity score and the most similar pins are selected as candidates.

### 4.2. Board Thumbnail Recommendation

Boards are displayed as thumbnails on all public and personal home pages. A thumbnail includes a cover and two/three small images or just six small images. A well-designed thumbnail can attract other users to access the board. Both Pinterest and Huaban allow users to select a cover from pins of the board, but they do not recommend candidates to users. As illustrated in Figure 6a, if a cover is selected, the small images will automatically be selected from the two latest pins. If the user has not selected an image for the cover, the thumbnail will be composed of the six latest pins. It is difficult for a user to select a suitable image to represent the board without any recommendation. Furthermore, the thumbnail consists of the latest pins possible that could not represent the boards. Boards like the bottom two have such wide interests that images in the thumbnail cannot fully express them. Similarly, thumbnails on Huaban, one of which consists of the cover and the three latest pins, have same drawbacks, as respectively shown in Figure 6b,c.

In view of the above, we defined a new task for recommending board thumbnails. The mean vector of pins in the board are computed, which is the center of the boards. The pins nearest to the center of the board are selected as the cover candidates. Then, we implement clustering, and the closest images with respect to the cluster centers are selected as substitutions for the latest pins.

**Figure 6.** Examples of board thumbnail examples on CCSNs: (**a**) includes examples from Pinterest; (**b**,**c**) are from Huaban.

*4.3. Board Category Recommendation*

On CCSNs, every board should be assigned a category, though some boards with no category were created before the constraint of the forced-choice approach. However, it is illogical because even if it is difficult to choose a board category from different user interests, users can select the category "other". Board category recommendation is convenient for category choice, not only in terms of first selection but also for further editing.

As mentioned in Section 3.1, interests associated with an image can be spread over the categories which occur in the re-pin tree. The only way to estimate the user preference on one image is to analyze its description and category. Because individual understanding of certain notions differs, even if manual analyzing cannot determine which single category the user intends to describe, it is common sense in this condition. We consider that personalization on CCSNs is mainly formed by the way the user organizes his or her boards. Hence, similarly to how user interests are reflected by pins, user interests reflected by boards should be more than one category. With the increasing number of pins, the category preference of the board in the majority opinion is reinforced. A board interest distribution $B$ can be calculated by the average of all its pin interest distributions as

$$Interest_B = \frac{1}{N_B} \sum_i^{N_B} Interest_{I_i} = \left( \frac{1}{N_B} \sum_i^{N_B} p_{iC_j} \right) \in [0,1]^{N_C}, \tag{8}$$

where $N_B$ denotes the pin count of $B$, $Interest_{I_i} = \left( p_{iC_j} \right) \in [0,1]^N$ is the interest distribution of the $i$-th pin $I_i$. In order to infer $Interest_{I_i}$, we trained a multidimensional LR between the representation of pins and the labels obtained in Equation (1). The generated $Interest_B$ should be normalized immediately. The recommended category is the category which is the highest number in terms of the board interest distribution.

This method can also be used for computing the interest distribution of a user. As an important part of the user profile, the interest distribution of a user can be intuitively represented by normalizing a frequency distribution of categories of boards or pins. However, this distribution has many limitations. First, it cannot deal with the absence of some categories, however, this does not mean that the user is not interested in those categories. Secondly, the ratio between categories may not be accurate, not only because categories are related, but also because images related to certain fine-grained interests are rarer than others and the user cannot collect enough pins related to these interests. Thirdly, it cannot be used to represent the interests of a board, as the categories of pins in it are the same. Our interest distribution of a target user $U$ is computed as

$$Interest_U = \frac{1}{N_U} \sum_i^{N_U} Interest_{X_i} = \left( \frac{1}{N_U} \sum_i^{N_U} p_{iC_j} \right) \in [0,1]^{N_C}, \quad (9)$$

where $N_U$ denotes the pin count of $U$. Because $Interest_{I_i}$ actually spreads over all the categories, $Interest_U$ does not suffer from the absence of some categories. In addition, to some extent, the ratio error, which is caused by the imbalance between pin counts of boards, is reduced, since the strong categories have faster accumulation processes than the weak categories.

### 4.4. Board and User Recommendation

As the pins are assembled, the theme of a board emerges. Users can easily collect pins with well-organized boards. For this reason, board recommendation is another important function for content discovery on CCSNs. In this section, we discuss how to model boards and users using the acquired multimodal joint representations of pins.

There is an analogy between user contents on CCSNs and articles, as user contents consist of boards which consist of pins, while articles are composed of paragraphs or sentences that are composed of words. One clear difference between them is that the order of pins or boards may not be that important. Therefore, the loss of order information is not an issue when modeling. Inspired by this, we consider that applying pooling methods to transform a different number of pins into a constant dimension vector, as we mentioned in Section 3.2, is reasonable for board and user modeling. Among pooling methods, the Fisher vector (FV) was chosen as our solution for board and user modeling.

The FV [37] was designed for encoding patch descriptors of an image into a high-dimensional vector. Since boards and users are image collections, a pin can be treated as a descriptor of them. A common method to encode a set of descriptors is to assign them into a visual dictionary, which is composed of prototypical elements such as cluster centers, while the FV approximates the distribution of descriptors with a GMM, whose Gaussian distributions can be treated as a universal probabilistic visual dictionary. As for representation of pins $V_{X_i} = (v_{ij}) \in \mathbb{R}^J$, the GMM is defined as

$$GMM(V_X) = \sum_{k=1}^{K} \omega_k norm_k(V_X), \quad (10)$$

where $norm_k$ denotes the $k$-th multivariate normal distribution, $\omega_k$ is the weight of the $k$-th mixture component and is subject to the following constraints: $\forall_k : \omega_k \geq 0$ and $\sum_{k=1}^{K} \omega_k = 1$; and $K$ is the number of mixture components. The parameters of the GMM also include $(\mu_{kj}) \in \mathbb{R}^J$ and $\Sigma_k$, which are the mean vector and covariance matrix of the $k$-th mixture component, respectively. The FV first computes the partial derivatives with respect to the parameters of the logarithm of the GMM, and then it normalizes them with the Fisher information matrix. The simplified normalized partial derivatives of a board $B$ are given by

$$g_{\omega_k} = \frac{1}{N_B\sqrt{\omega_k}} \sum_{i=1}^{N_B} (\gamma_{ik} - \omega_k), \tag{11}$$

$$g_{\mu_k} = \left( \frac{1}{N_B\sqrt{\omega_k}} \sum_{i=1}^{N_B} \gamma_{ik} \left( \frac{v_{ij} - \mu_{kj}}{\sigma_{kj}} \right) \right) \in \mathbb{R}^J, \tag{12}$$

$$g_{\sigma_k} = \left( \frac{1}{N_B\sqrt{2\omega_k}} \sum_{i=1}^{N_B} \gamma_{ik} \left[ \frac{(v_{ij} - \mu_{kj})^2}{\sigma_{kj}^2} - 1 \right] \right) \in \mathbb{R}^J, \tag{13}$$

where $\sigma_{kj}$ denotes the standard deviation of the $j$-th dimension of the $k$-th mixture component, and $\gamma_{ik}$ is the soft assignment of $V_{P_i}$ to the $k$-th mixture component, which is written as

$$\gamma_{ik} = \frac{\omega_k \text{norm}_k(V_{X_i})}{\sum_{k=1}^{K} \omega_k \text{norm}_k(V_{X_i})} = \frac{\omega_k \text{norm}_k(V_{X_i})}{\text{GMM}(V_{X_i})}, \tag{14}$$

and is also known as the posterior probability or responsibility. All partial derivatives are concatenated to compose the FV. Since one of $\omega_k$ is redundant because of the constraints, the dimension of the FV is $(2J+1)K - 1$. Power normalization and L2-normalization [38] are applied to improve the quality of the FV as follows:

$$g_i \leftarrow \text{sgn}(g_i)|g_i|^\rho, \tag{15}$$

$$g_i \leftarrow \frac{g_i}{\sqrt{\sum_i^{(2J+1)K-1} g_i^2}}, \tag{16}$$

where $\rho \in [0,1]$ is the normalization parameter. The FV of a user can be computed in the same manner. Please refer to the original paper for more details regarding the FV.

In essence, the FV is the gradient of the log-likelihood of a board. Notice that the computations of Equations (11)–(13) can be simplified with

$$S_k^0 = \sum_i^{N_B} \gamma_{ik}, \tag{17}$$

$$S_k^1 = \left( \sum_i^{N_B} \gamma_{ik} v_{ij} \right) \in \mathbb{R}^J, \tag{18}$$

$$S_k^2 = \left( \sum_i^{N_B} \gamma_{ik} v_{ij}^2 \right) \in \mathbb{R}^J, \tag{19}$$

where $S_k^0$, $S_k^1$, and $S_k^2$ are the zeroth order, first order, and second order statistics of the board, respectively. Accordingly, the FV preserves more information than other pooling methods, such as the vector of aggregate locally descriptor and sparse coding, with the same dictionary capacity. It actually measures not only which words in the visual dictionary the pins belong to, but also the differences between the mean vectors of the GMM and the board or user. On the other hand, the FV uses a relatively small dictionary to generate the same dimension vector as the others, such that the computational complexity is lower. In addition, the FV is interpretable. If we consider the mean vectors as the center of interests, improving $K$ will make the FV more fine-grained, while the curse of dimensionality is a significant limitation of the FV. For the sake of large-scale applications, the FV could be lossless compressed by sparsity encoding with product quantization [39].

After modeling, boards can be recommended according to the similarity metrics between them and the target board. Because users can be considered image collections with wider interests than boards, user recommendation done in this same manner is also helpful for content discovery, although users on CCSNs are not very interested in following.

## 5. Experiments and Results

In this section, the datasets and implementation details are firstly introduced. Then, the performance of our representation of pins are evaluated in an interest analysis. Thereafter, the results of experiments on real-world datasets are presented to verify the feasibility and effectiveness of our recommendation methods.

*5.1. Datasets and Implementation Details*

We crawled data used in experiments from Huaban, a typical Chinese CCSN. Huaban provides certain applications similar to those in Pinterest, while the main differences between the two networks are as follows: There are "like" pins or board operations on Huaban but not on Pinterest; Huaban records both users and the paths in a re-pin tree, while Pinterest only records all the users and the initially created user.

We first crawled the pins of 5957 users without images, and then sampled 88 users according to board categories and pin counts. Some extremely active and cold-start users had been confirmed among them to make our dataset diverse and to take the influence of pin counts into account. We then crawled all images of the sampled users and all their "like" pins. In addition, we crawled the top 1000 pins recommended by the system of each category to fine-tune AlexNet and their re-pin paths for automatic annotation. The dataset for recommendation included 151,631 pins, which were categorized into 33 categories from 1694 boards, and the number of unique images for both fine-tuning and recommendation was 167,747. All pins were used as supplement elements for obtaining distributions of the recommended pin categories. The average re-pin path length was 47.57.

After a little manual label balancing, labelled images were split into 80% for training and validating and the remaining 20% for testing. Because the input dimension of AlexNet should be constant, every image was firstly rescaled so that the shorter side was 256 pixels, and then the central 256 × 256 patch of the processed image was cropped out. The loss layer of our AlexNet was replaced. As a comparison, the most frequent category was used as the label to fine-tune a multiclass AlexNet. The dimensions of the FC8 layers of both Alexnets were changed to 33. Image representations were generated from the FC7 layer of the multilabel Alexnet.

We trained our Word2Vec model on Wikipedia dumps (https://dumps.wikimedia.org/) and Sougou Lab dataset (http://www.sogou.com/labs/resource/list_news.php) with the CBOW (Continuous Bag of Words) model and negative sampling. In addition, the vector dimension was 300. The words with a frequency lower than five were ignored. Word preprocessing, such as removing punctuation, traditional and simplified Chinese conversion, word tokenization, machine translation, and removing stop words, was applied on pin descriptions.

All image and text representations were exploited for the multimodal DBM training. The dimensions of $H_{T1}$, $H_{T2}$ and $H_{V1}$ were the same as their corresponding visible inputs, and dimensions of $H_{V2}$ and $H_3$ were set to 2048 to compress the vectors, as the FV would increase the dimension. Each layer was pretrained using a contrastive divergence strategy to accelerate the training of the DBM. Then, missing text representations were extracted using Gibbs sampler and the multimodal joint representation of pins was inferred.

K in Equation (10) was set to 1 such that the dimension of the FV of a board was twice that of the pin vector. $\alpha$ in Equation (15) was set to 0.5.

To evaluate the effectiveness of the proposed model, we compared it with the following multimodal deep architectures: the Multimodal Autoencoder (MAE), which was proposed in [31] and connects two deep autoencoders of multimodalities by a shared hidden layer; and ICMAE, which imposes Independent Component Analysis (ICA) constraints in the MAE architecture to de-correlate the relationships among the variables. All the baseline methods had the same number of layers, and we used the same features as inputs to ensure that the comparisons were fair.

*5.2. Analysis of Interests Represented by Pins*

Analysis of interests based on pins is the prerequisite of analysis of interests based on boards and users. As mentioned above, it is hard to measure the interest distribution of one pin. Hence, we treated the interest distribution of its image as an approximation, even though some categories would be improved by its text description.

Multidimensional LRs were trained on the dataset to fine-tune for all unimodal representations and multimodal representations. Table 1 illustrates the results, together with those of the multiclass classification with softmax. The mean nonzero error was the average error between all nonzero categories and corresponding predictions. The accuracy of the dominant category checks the consistency of the most frequent category between labels and predictions. The comparison of multiclass and multilabel CNNs shows that our method with multilabel annotation improves the accuracy significantly. This is not only because the interference of related categories could be eliminated by category distributions, but also because more information from the users' collective intelligence was provided for learning. Although the performance of text representations and image representations was not comparable, the performance of the multimodal joint model was better than that of image representations that are complementary between two modalities.

**Table 1.** Comparison of Pin Category Prediction(MAE: Multimodal Autoencoder).

| Model | Dimension | Dominant Category Accuracy | Mean Nonzero Error | Mean Error |
| --- | --- | --- | --- | --- |
| Multiclass | 4096 | 45.85% | — | — |
| Multilabel | 4096 | 82.71% | 0.1320 | 0.0141 |
| Word2Vec | 300 | 42.88% | 0.3249 | 0.0415 |
| MAE | 1778 | 83.02% | 0.1307 | 0.0136 |
| ICMAE | 1898 | 83.96% | 0.1223 | 0.0128 |
| *Ours* | 2048 | 84.13% | 0.1181 | 0.0119 |

From the results, we can see that our method had the best performance because all unimodal and multimodal representations contained information about user interests and our joint representation contains richer information than other methods. Our method could also analyze interests of images on other networks. The comparison of MAE/ICMAE shows that the joint representation of pins learned by our method has a higher correlation with their categories.

*5.3. Pin Recommendation*

We invited 10 users to engage in the evaluation of pin recommendation. Each user was given 200 randomly selected target images and corresponding recommendation results of different methods. They were required to decide whether to pin some images of three candidates, but not if they were the owner of the target pin. Table 2 shows the precision of recommendations. A simple content-based filtering, which randomly selects an image with the same category as the target image, was implemented as a reference. All other methods achieved higher accuracies than the category-based method, simply because they utilized more information to reduce the affect of related categories. Object-based and interest-based methods used the probability layer from the original AlexNet and multilabel AlexNet, respectively. The results of those two methods were comparable, while interest distributions were more compact than object distributions. This indicates that even coarse-grained interests of an image were a little more important than what this image was on CCSNs. The other methods computed cosine similarity between representations. We note that using only dominant categories as the label to fine-tune AlexNet led to a decline, which may have been caused by confusion of similar images with different categories. Notice that the performance of multimodal features was worse than that of image features. We believe that the descriptions could not completely describe all the interests and characteristics that images have. Our text representations were clearly not as effective as our image representations, therefore, image representations were more suitable for image recommendation.

Table 2. Comparison of image recommendation.

| Model | Dimension | Top-1 | Top-2 | Top-3 |
|---|---|---|---|---|
| Category Based | 1 | 8.75% | 9.30% | 9.67% |
| Object Based | 1000 | 46.55% | 40.80% | 37.00% |
| Interest Based | 33 | 47.20% | 39.95% | 35.67% |
| AlexNet | 4096 | 81.30% | 75.33% | 72.57% |
| Fine-tuned AlexNet | 4096 | 78.55% | 72.65% | 69.60% |
| *Multilabel* | **4096** | **88.50%** | **86.40%** | **85.67%** |
| Text | 300 | 52.00% | 49.45% | 47.63% |
| MAE | 1778 | 80.21% | 77.36% | 73.58% |
| ICMAE | 1898 | 83.17% | 82.68% | 79.95% |
| Ours | 2048 | 85.90% | 85.08% | 82.27% |

Figure 7 illustrates 10 images and their recommendation results. Obviously, intrinsic characteristics such as background, scene, pattern, texture, color, object, material, and so forth are maintained in the image representation and usually had an effect on the recommendation, especially for images in the left panel. Images in the right panel show that some abstract notions, for example, style and user interest, influenced the results. All these high-level image features learned from CNNs could significantly improve the accuracy and diversity of recommendations.

**Figure 7.** Examples of recommendation results based on image representation. Images with red borders are the target images.

From the recommendation results, we can clearly see that our model recommended similar styles and types of images. This means that our model could achieve a good recommendation effect in terms of content-based recommendations. This further illustrates that the features extracted by our multimodal joint representation model were effective. The recommendation data were different from the training data, which proves that our model had a good generalization ability.

*5.4. Board Thumbnail Recommendation*

In this experiment, we recommended the board thumbnail according to the interest distributions of pins and the representation of pins. Because Huaban does not yet offer the function of editing thumbnails, we manually re-pinned all pins from the original board and changed the orders of the pins to display our results.

Figure 8a,b are the recommendation results in Figure 6b regarding narrow interests. As shown in Figure 6b, pins from the board are album covers of a music group. Four pins in the original thumbnail were all from the same album. Strong categories for this board were "file music book" (20.87%), "design" (16.24%), and "architecture" (11.50%), while those for the cover in Figure 8a are "film music books" (20.44%), "design" (14.15%), and "architecture" (10.81%). Three clusters, whose centers mainly belonged to "photography" (15.72%), "film music books" (80.93%), and "architecture" (47.19%), contained 30, 7, and 4 pins, respectively. This indicates that the recommendation results are consistent with the target board thumbnail. On the other hand, those four components of the thumbnail were from different albums. Similar to the result generated with interest distributions, the result generated with image featured comprise pins from different albums, partly owing to the fact that image representations were also related to interests. Our results also indicated that even narrow interests could be divided. It is obvious that recommending thumbnails for a board about wide interests was easier, the recommendations for Figure 6c are shown in Figure 8c,d. We believe that our recommended thumbnails, which depicted more interests, were more attractive.

**Figure 8.** Results of the board thumbnail recommendation: (**a,c**) are generated with interest distributions of pins; (**b,d**) are generated with representation of pins.

### 5.5. Board Category Recommendation

The ground truth of board category recommendation is the crawled board category. The performance metric of the experiment was mean reciprocal rank (MRR). We only give the top MRR because there was only one accurate selection of the board category recommendation. The results are shown in Table 3.

From the table, we can see that our model had the highest MRR. Because the board category recommendation results were based on different features but the same classifier, the best result meant the best features. Our best recommendation results illustrate that multimodal representations with the benefit of personalized text representations had a better performance than other baselines.

**Table 3.** Comparison of board category recommendation (MRR: mean reciprocal rank).

| Model | Top-1 MRR | MRR |
|---|---|---|
| Random | 3.03% | 12.39% |
| Text + Cosine Similarity | 25.65% | 38.78% |
| Image + Multidimensional LR | 60.10% | 73.43% |
| Text + Multidimensional LR | 38.00% | 54.30% |
| MAE + Multidimensional LR | 60.89% | 73.86% |
| ICMAE + Multidimensional LR | 61.76% | 74.62% |
| *Ours + MultidimensionalLR* | **63.41%** | **76.13%** |

## 5.6. Board Recommendation

Every board was divided into two parts based on the order of pins. One part must be similar to the other part. The user of each half part should be interested in another and naturally further like or follows or re-pin from it. Depending on this fact, half of the board was treated as the only accurate recommendation result, and we retrieved the index in the similarity sequence. Because there were five pins in the top row exhibited on Huaban with common resolution screens, the top five MRR was also demonstrated. Table 4 shows the experimental results.

Table 4. Comparison of board recommendation.

| Model | Dimension | Top-5 MRR | MRR |
|---|---|---|---|
| Category Based | 1 | 2.12% | 3.85% |
| Image + LR | 33 | 16.08% | 18.61% |
| Text + LR | 33 | 15.93% | 17.95% |
| MAE + LR | 33 | 16.85% | 19.39% |
| ICMAE + LR | 33 | 17.62% | 20.27% |
| Ours + LR | 33 | 17.58% | 20.13% |
| Image + Mean Vector | 4096 | 33.66% | 35.97% |
| Text + Mean Vector | 300 | 25.93% | 27.49% |
| MAE + Mean Vector | 1778 | 34.31% | 36.48% |
| ICMAE + Mean Vector | 1898 | 35.02% | 37.23% |
| Ours + Mean Vector | 2048 | 35.76% | 37.88% |
| Image + FV | 8192 | 35.19% | 37.50% |
| Text + FV | 600 | 24.39% | 25.64% |
| MAE + FV | 3556 | 35.81% | 37.97% |
| ICMAE + FV | 3796 | 37.76% | 39.84% |
| Ours + FV | 4096 | 38.97% | 41.22% |

From the table we can see two things. Firstly, the same feature encoded with FV performed the best. For example, the method with pin vectors, except for text vectors encoded with the FV, performed better than that with the corresponding pin vectors combined with the mean vector. The better performance is due to the utilization of higher order statistics. Secondly, our representation demonstrated the best performance when different features were encoded with the same method. The results also illustrate that multimodal joint representations have a better board modeling performance than the unimodal representations with lower dimensions.

## 6. Conclusions

We propose a framework for multimodal joint representation learning of pins on CCSNs. The obtained representation contains the information of user interests, which is useful for recommender systems and user modeling. We modeled boards and users with the FV and propose a series of recommendation methods for different recommendation tasks, including a novel board thumbnail recommendation defined by us and based on our pin recommendation. The experimental results show that the obtained representations perform better in terms of interpreting pin-level interests than unimodal representations with lower dimensions, and our recommendation methods based on our multimodal representation are effective in terms of recommending pins, board thumbnails, board categories, and boards.

**Author Contributions:** Conceptualization, H.L. and S.D.; methodology, H.L. and S.D.; software, B.Y. and D.Z.; validation, B.Y. and D.Z.; formal analysis, H.L. and S.D.; investigation, B.Y. and D.Z.; resources, B.Y.; data curation, H.L. and D.Z.; writing—original draft preparation, H.L., S.D. and B.Y.; writing—review and editing, L.W. and M.J.; visualization, H.L.; supervision, L.W.; project administration, M.J. All authors have read and agreed to the published version of the manuscript.

**Funding:** This research received no external funding.

**Conflicts of Interest:** The authors declare no conflict of interest.

**References**

1. Gilbert, E.; Bakhshi, S.; Chang, S.; Terveen, L. "I Need to Try This!": A Statistical Overview of Pinterest. In Proceedings of the SIGCHI Conference on Human Factors in Computing Systems, Paris, France, 27 April–2 May 2013; pp. 2427–2436. [CrossRef]
2. Bernardini, C.; Silverston, T.; Festor, O. A Pin is Worth a Thousand Words: Characterization of Publications in Pinterest. In Proceedings of the 2014 International Wireless Communications and Mobile Computing Conference (IWCMC), Nicosia, Cyprus, 4–8 August 2014; pp. 322–327.
3. Zhong, C.; Sastry, N. Copy Content, Copy Friends: Studies of Content Curation and Social Bootstrapping on Pinterest. *SIGWEB Newsl.* **2014**, *4*, 1–6. [CrossRef]
4. Gelley, B.; John, A. Do I Need To Follow You?: Examining the Utility of The Pinterest Follow Mechanism. In Proceedings of the 18th Acm Conference on Computer Supported Cooperative Work & Social Computing, Vancouver, BC, Canada, 14–18 March 2015; pp. 1751–1762.
5. Zhong, C.; Kourtellis, N.; Sastry, N. Pinning Alone? A Study of the Role of Social Ties on Pinterest. In Proceedings of the International AAAI Conference on Web and Social Media, Cologne, Germany, 17–20 May 2016.
6. Liakos, P.; Papakonstantinopoulou, K.; Sioutis, M.; Tsakalozos, K.; Delis, A. Pinpointing Influence in Pinterest. In Proceedings of the 2nd International Workshop on Social Influence Analysis Co-Located with 25th International Joint Conference on Artificial Intelligence (IJCAI 2016), New York, NY, USA, 9–10 July 2016; pp. 26–37.
7. Venkatadri, G.; Goga, O.; Zhong, C.; Viswanath, B.; Gummadi, K.P.; Sastry, N. Strengthening Weak Identities Through Inter-Domain Trust Transfer. In Proceedings of the 25th International Conference on World Wide Web, Montreal, QB, Canada, 12 April 2016; pp. 1249–1259.
8. Mao, J.; Jiajing, X.; Jing, Y.; Yuille, A. Training and Evaluating Multimodal Word Embeddings with Large-scale Web Annotated Images. In *Advances in Neural Information Processing Systems 29*; NIPS Proceedings: Lake Tahoe, Nevada, 2016; pp. 442–450.
9. Jing, Y.; Liu, D.; Kislyuk, D.; Zhai, A.; Xu, J.; Donahue, J.; Tavel, S. Visual Search at Pinterest. In Proceedings of the 21th ACM SIGKDD International Conference on Knowledge Discovery and Data Mining, Sydney, Australia, 8–9 August 2015; pp. 1889–1898.
10. Geng, X.; Zhang, H.; Song, Z.; Yang, Y.; Luan, H.; Chua, T.S. One of a Kind: User Profiling by Social Curation. In Proceedings of the 22nd ACM International Conference on Multimedia, Orlando, FL, USA, 3–7 November 2014; pp. 567–576.
11. Cinar, Y.G.; Zoghbi, S.; Moens, M.F. Inferring User Interests on Social Media From Text and Images. In Proceedings of the 2015 IEEE International Conference on Data Mining Workshop (ICDMW), Atlantic City, NJ, USA, 14–17 November 2015; pp. 1342–1347.
12. You, Q.; Bhatia, S.; Luo, J. A picture tells a thousand words—About you! User interest profiling from user generated visual content. *Signal Proces.* **2016**, *124*, 45–53. [CrossRef]
13. Kamath, K.Y.; Popescu, A.M.; Caverlee, J. Board Recommendation in Pinterest. In Proceedings of the Late-Breaking Results, Project Papers and Workshop Proceedings of the 21st Conference on User Modeling, Adaptation, and Personalization, Rome, Italy, 10–14 June 2013.
14. Zhong, C.; Karamshuk, D.; Sastry, N. Predicting Pinterest: Automating a Distributed Human Computation. In Proceedings of the 24th International Conference on World Wide Web, Geneva, Switzerland, 18–22 May 2015; pp. 1417–1426.
15. Jing, Y.; Wu, L.; Zhang, X.; Wang, D. Recommending Users to Follow based on User Taste Homophily for Content Curation Social Networks. In Proceedings of the 6th International Workshop on Social Recommender Systems (SRS 2015), Sydney, Australia, 12 June–15 July 2015; pp. 1–6.
16. Yang, X.; Li, Y.; Luo, J. Pinterest Board Recommendation for Twitter Users. In Proceedings of the 23rd ACM International Conference on Multimedia, Brisbane, Australia, 26–30 October 2015; pp. 963–966.
17. Li, Y.; Cong, Y.; Mei, T.; Luo, J. User-Curated Image Collections: Modeling and Recommendation. In Proceedings of the 2015 IEEE International Conference on Big Data (Big Data), Santa Clara, CA, USA, 29 October–1 November 2015; pp. 591–600.

18. Geng, X.; Zhang, H.; Bian, J.; Chua, T.S. Learning Image and User Features for Recommendation in Social Networks. In Proceedings of the 2015 IEEE International Conference on Computer Vision (ICCV), Santiago, Chile, 7–13 December 2015; pp. 4274–4282.
19. Wu, L.; Wang, D.; Guo, C.; Zhang, J.; Chen, C.W. User Profiling by Combining Topic Modeling and Pointwise Mutual Information (TM-PMI). In Proceedings of the International Conference on Multimedia Modeling, Miami, FL, USA, 4–6 January 2016; pp. 152–161.
20. Wu, L.; Zhang, D.; Zhang, X.; Jing, Y.; Liu, H.; Chen, C.W. Recommending Followees Based on Content Weighted User Interest Homophily. In Proceedings of the International Conference on Internet Multimedia Computing and Service, Xi'an, China, 19–21 August 2016; pp. 146–151.
21. Zhai, A.; Kislyuk, D.; Jing, Y.; Feng, M.; Tzeng, E.; Donahue, J.; Du, Y.L.; Darrell, T. Visual Discovery at Pinterest. In Proceedings of the 26th International Conference on World Wide Web Companion, Geneva, Switzerland, 3 April 2017; pp. 515–524.
22. Liu, D.C.; Rogers, S.; Shiau, R.; Kislyuk, D.; Ma, K.C.; Zhong, Z.; Liu, J.; Jing, Y. Related Pins at Pinterest: The Evolution of a Real-World Recommender System. In Proceedings of the 26th International Conference on World Wide Web Companion, Geneva, Switzerland, 3 April 2017; pp. 583–592.
23. Wu, L.; Wang, D.; Zhang, X.; Liu, S.; Zhang, L.; Chen, C.W. MLLDA: Multi-level LDA for modelling users on content curation social networks. *Neurocomputing* **2017**, *236*, 73–81. [CrossRef]
24. Wu, L.; Zhang, L.; Jian, M.; Zhang, D.; Liu, H. Image Recommendation on Content-based Bipartite Graph. In Proceedings of the Internet Multimedia Computing and Servic, Qingdao, China, 23–25 August 2017; pp. 339–348.
25. Liu, H.; Wu, L.; Zhang, D.; Jian, M.; Zhang, X. Multi-perspective User2Vec: Exploiting re-pin activity for user representation learning in content curation social network. *Signal Proces.* **2018**, *236*, 450–456. [CrossRef]
26. Mikolov, T.; Chen, K.; Corrado, G.; Dean, J. Efficient Estimation of Word Representations in Vector Space. In Proceedings of the 2013 International Conference on Learning Representations Workshop, Scottsdale, AZ, USA, 2–4 May 2013.
27. Lev, G.; Klein, B.; Wolf, L. In Defense of Word Embedding for Generic Text Representation. In Proceedings of the Natural Language Processing and Information Systems—20th International Conference on Applications of Natural Language to Information Systems, Passau, Germany, 17–19 June 2015; pp. 35–50.
28. Ma, L.; Lu, Z.; Shang, L.; Li, H. Multimodal Convolutional Neural Networks for Matching Image and Sentence. In Proceedings of the 2015 IEEE International Conference on Computer Vision (ICCV) (2015), Santiago, Chile, 7–13 December 2015; pp. 2623–2631.
29. Qian, S.; Zhang, T.; Xu, C. Multi-view Topic-opinion Mining for Social Event Analysis. In Proceedings of the 2016 ACM on Multimedia Conference, Amsterdam, The Netherlands, 15–19 October 2016; pp. 2–11.
30. Jia, X.; Wang, A.; Li, X.; Xun, G.; Zhang, A. Multi-modal learning for video recommendation based on mobile application usage. In Proceedings of the IEEE International Conference on Big Data, Santa Clara, CA, USA, 29 October–1 November 2015.
31. Ngiam, J.; Khosla, A.; Kim, M. Multimodal Deep Learning. In Proceedings of the 28th International Conference on Machine Learning, Washington, DC, USA, 28 June–2 July 2011; pp. 689–696.
32. Srivastava, N.; Salakhutdinov, R. Multimodal Learning with Deep Boltzmann Machines. *J. Mach. Learn. Res.* **2014**, *15*, 2949–2980.
33. Zhang, H.; Yang, Y.; Luan, H.; Yan, S.; Chua, T.S. Start from Scratch: Towards Automatically Identifying, Modeling, and Naming Visual Attributes. In Proceedings of the 22Nd ACM International Conference on Multimedia, Orlando, FL, USA, 3–7 November 2014; pp. 187–196.
34. Krizhevsky, A.; Sutskever, I.; Hinton, G.E. ImageNet Classification with Deep Convolutional Neural Networks. In *Advances in Neural Information Processing Systems 25*; NIPS Proceedings: Lake Tahoe, Nevada, 2012; pp. 1106–1114.
35. Russakovsky, O.; Deng, J.; Su, H.; Krause, J.; Satheesh, S.; Ma, S.; Huang, Z.; Karpathy, A.; Khosla, A.; Bernstein, M.; et al. ImageNet Large Scale Visual Recognition Challenge. *Int. J. Comp. Vis.* **2015**, *115*, 211–252. [CrossRef]
36. Mikolov, T.; Sutskever, I.; Chen, K.; Corrado, G.; Dean, J. Distributed Representations of Words and Phrases and their Compositionality. In *Advances in Neural Information Processing Systems 26*; NIPS Proceedings: Lake Tahoe, Nevada, 2013; pp. 3111–3119.

37. Perronnin, F.; Dance, C. Fisher Kernels on Visual Vocabularies for Image Categorization. In Proceedings of the 2007 IEEE Conference on Computer Vision and Pattern Recognition, Minneapolis, MN, USA, 17–22 June 2007; pp. 1–8.
38. Perronnin, F.; Sánchez, J.; Mensink, T. Improving the Fisher Kernel for Large-Scale Image Classification. In Proceedings of the 11th European Conference on Computer Vision: Part IV, Heraklion, Greece, 5–11 September 2010; pp. 143–156.
39. Sánchez, J.; Perronnin, F.; Mensink, T.; Verbeek, J. Image Classification with the Fisher Vector: Theory and Practice. *Int. J. Comp. Vis.* **2013**, *105*, 222–245. [CrossRef]

© 2020 by the authors. Licensee MDPI, Basel, Switzerland. This article is an open access article distributed under the terms and conditions of the Creative Commons Attribution (CC BY) license (http://creativecommons.org/licenses/by/4.0/).

*Article*

# A Novel 1-D CCANet for ECG Classification

Ian-Christopher Tanoh [1,†] and Paolo Napoletano [2,*,†]

1 Department of Applied Mathematics, Ecole Polytechnique, 91120 Palaiseau, France; ian-christopher.tanoh.2017@polytechnique.org
2 Department of Informatics, System and Communication, University of Milan-Bicocca, 20126 Milan, Italy
* Correspondence: paolo.napoletano@unimib.it
† These authors contributed equally to this work.

**Abstract:** This paper puts forward a 1-D convolutional neural network (CNN) that exploits a novel analysis of the correlation between the two leads of the noisy electrocardiogram (ECG) to classify heartbeats. The proposed method is one-dimensional, enabling complex structures while maintaining a reasonable computational complexity. It is based on the combination of elementary handcrafted time domain features, frequency domain features through spectrograms and the use of autoregressive modeling. On the MIT-BIH database, a 95.52% overall accuracy is obtained by classifying 15 types, whereas a 95.70% overall accuracy is reached when classifying 7 types from the INCART database.

**Keywords:** heartbeat classification; convolutional neural network (CNN); canonical correlation analysis (CCA)

## 1. Introduction and Related Work

Cardiovascular diseases are the first cause of death in the world, with an estimated 17.9 million deaths each year. Among them, heart arrhythmia qualifies as an abnormal heart rhythm that can result in serious complications such as stroke or cardiac deaths. Early detection of arrhythmia is a major challenge for our society.

With electrocardiograms (ECGs), heartbeats can be visually labelled according to several classes such as Normal beat, Supraventricular escape beat, etc. An ECG is a graph of voltage versus time of the electrical activity of the heart using electrodes placed on the skin. To assess the condition of the heart from different angles, an ECG has several leads, each of them being the signal generated by a pair of electrodes.

In the last decades, researchers employed machine learning methods for the automatic classification of heartbeats contained in long-duration recordings of human ECGs [1,2]. A traditional heartbeat classification pipeline includes data preprocessing, data segmentation, feature extraction, feature selection, and classification [3].

Data preprocessing is used to remove noise from the ECG raw signal. The most used techniques are median filters [4], discrete wavelet transform (DWT) [5,6], adaptive filters [4,7], and frequency selective filters [8–10].

Data segmentation is used to isolate heartbeats from the whole ECG recording. Once a time segment including the heartbeat is available, time domain [11–16] or frequency domain [13,16–18] or morphological [11–13,15,16] or statistical [13,19] or neural features [20] are extracted.

Feature selection is used to reduce the number of features used by the classifier thus reducing the complexity and time required for computation. Several approaches have been adopted: principal and independent component analysis [5,6,21,22], linear discriminant analysis [6], and genetic algorithm [23].

Random forest [24,25], support vector machines (SVMs) [13–16,18,19], neural networks (NNs) [5,6] or deep neural networks (DNNs) [2,26–32] are employed to classify extracted features in one of the heartbeat classes.

As discussed above, ECGs can be recorded in different locations of the body thus obtaining the so-called multilead ECGs. Up to 12 leads can be recorded and each lead represents a specific characteristic of the heart. Multilead ECGs better reflect the state of the heart compared with single lead ECGs. Taking into account multi leads may bring performance improvement. Existing literature is mainly focused on the processing of single lead ECGs [20].

In this paper, we focus on two-lead ECGs: we use lead V1, that is a chest lead, and lead II, that is a limb lead. We propose the combination of hand-crafted features with a canonical correlation analysis network (CCANet) and SVMs for two-lead heartbeats classification. The analysis of the correlation between two leads of the ECG is exploited to increase heartbeat classification performance [20]. Proposed CCANet is a 1-D variant of the original 2-D CCANet proposed by Yang et al. [20] that allows to explore a deeper CCANet while maintaining a reasonable computational complexity and providing better results. CCANet has been originally proposed by Yang et al. [33] for the processing of two-view images in 2017. Compared to one-view image-based PCANet and RandNet, CCANet demonstrated to perform better [33]. CCANet has also been employed in other computer vision tasks such as remote sensing scene classification [34] as well as ECG interpretation [20].

There are two types of CNNs that are commonly used for ECG classification: the 1-D CNN and 2-D CNN [35]. 2-D CNNs usually operate on transformed ECG data, such as spectrograms, gray-level co-occurrence matrices, combined features and others. 1-D CNNs operate directly on the raw ECG signal. Our one-dimensional variant takes as input a combination of elementary hand crafted time domain features, frequency domain features through spectrograms, and the use of autoregressive modeling.

For the sake of comparison, we evaluate a suitable implemented 1-D convolutional neural network (CNN) solution based on residual networks (ResNet) [36]. ResNet demonstrated to be one of the most performing CNN for visual recognition [37]. The proposed method outperforms the state of the art on both the MIT-BIH and INCART arrhythmia databases.

*Our Contribution*

The main novel contributions of this paper are summarized as follows:

- We have designed a novel one-dimensional canonical correlation analysis network (1-D CCANet) to exploit two-lead ECGs for automatic classification of heartbeats that outperforms the state of the art;
- We have explored the use of handcrafted features in combination with a 1-D CCANet for ECG classification;
- Our proposal outperforms a solution based on a suitable one-dimensional ResNet that we have implemented for the sake of comparison.

## 2. Materials

### 2.1. MIT-BIH Database

The MIT-BIH database contains 48 sets of two-lead ECG signals (lead II and mostly V1). Each signal is approximately 30 min long, has been collected at a 360 Hz sampling frequency, and has been independently annotated by at least two cardiologists. Annotations include the 15 types listed in Table 1. In our study, we use the signals for which both II and V1 leads are available (see PhysioBank for further details).

### 2.2. INCART Database

The St. Petersburg Institute of Cardiological Techniques 12-lead arrhythmia database (INCART) contains 75 sets of 12-lead ECG signals (leads I, II, III, aVR, aVL, aVF, V1, V2, V3, V4, V5, V6). Each signal is approximately 30 min long and has been collected at a 257 Hz sampling frequency. We only consider leads II and V1 of each record. Annotations include the 7 types listed in Table 2.

**Table 1.** Details of the categories for MIT-BIH database.

| Type | Name | Quantity |
| --- | --- | --- |
| r | Rhythm change | 200 |
| N | Normal beat | 1000 |
| A | Atrial premature beat | 200 |
| V | Premature ventricular | 200 |
| P | Paced beat | 200 |
| x | Non-conducted P-wave | 100 |
| F | Fusion of ventricular contraction | 200 |
| j | Nodal (junction) escape beat | 200 |
| L | Left bundle branch block beat | 200 |
| a | Aberrated atrial premature beat | 100 |
| J | Nodal (junction) premature beat | 50 |
| R | Left bundle branch block beat | 200 |
| ! | Ventricular flutter | 200 |
| E | Ventricular escape beat | 100 |
| f | Fusion of paced and normal beat | 200 |
| Tot | | 3350 |

**Table 2.** Details of the categories for INCART database.

| Type | Name | Quantity |
| --- | --- | --- |
| N | Normal beat | 500 |
| A | Atrial premature beat | 200 |
| V | Premature ventricular | 500 |
| n | Supraventricular escape beat | 30 |
| F | Fusion of ventricular contraction | 200 |
| j | Nodal (junction) escape beat | 90 |
| R | Left bundle branch block beat | 200 |
| Tot | | 1720 |

## 3. Proposed Method

The input of the proposed method is a two-channel ECG segment obtained after a preliminary segmentation that consists in the isolation of heartbeats in each record. Given the R-peak positions, any heartbeat is isolated by retaining $T_1$ and $T_2$ samples to the left and to the right of the R-peak, respectively. For each of the two leads, a vector denoted as $x_h$ ($h = 1, 2$) is built with the values of the ECG (in Volts), of size ($T_2 + T_1$). The values of $T_1$ and $T_2$ are 160–200 and 120–136 for MIT-BIH and INCART, respectively. These values are given in Table 3, and are similar to the one used in [20].

**Table 3.** Sampling rates, values of $T_1$ and $T_2$ for both MIT-BIH and INCART databases.

| Database | Sampling Rate (Hz) | $T_1$ | $T_2$ |
| --- | --- | --- | --- |
| MIT-BIH | 360 | 160 | 200 |
| INCART | 257 | 120 | 136 |

The architecture of the whole process is shown Figure 1a,b. The first stage is feature extraction. The input of the process is, for each lead, a vector $x_h$ (with $h = 1, 2$) containing raw values of the segmented heartbeat. Each lead $x_h$ ($h = 1, 2$) is normalized (see the "Normalization" module in Figure 1a by using a rescaling procedure so that the resulting vector $x_{h,\text{norm}}$ has an intensity that ranges from 0 to 1, as per equation:

$$x_{h,\text{norm}} = \frac{x_h - \min(x_h)}{\max(x_h) - \min(x_h)}. \tag{1}$$

At the same time, hand-crafted features are extracted from each lead $x_h$ ($h = 1, 2$) frequency-domain features $x_{h,spec}$, and autoregression features $x_{h,ar}$. A single time-domain features vector $x_{time}$ is also computed for both leads. Frequency-domain features, autoregression features and the normalized segmented heartbeat $x_{h,norm}$ are concatenated to obtain the vector $x_{h,cat} = [x_{h,ar}\ x_{h,spec}\ x_{h,norm}]$ (see Figure 1a). The $x_{h,cat}$ ($h = 1, 2$) vector is processed by the neural module to produce a single output vector $f_{neur}$ for the two leads (see Figure 1b). The vector $f_{neur}$ is then reduced in dimensions by using Principal Component Analysis (PCA) thus obtaining the vector $f_{pca}$. The concatenation of the time-domain features $x_{time}$ and $f_{pca}$ is the input of a Support Vector Machine classifier. The output of the classifier is the predicted heartbeat class. In the following subsections the feature extraction and neural module are discussed more in detail.

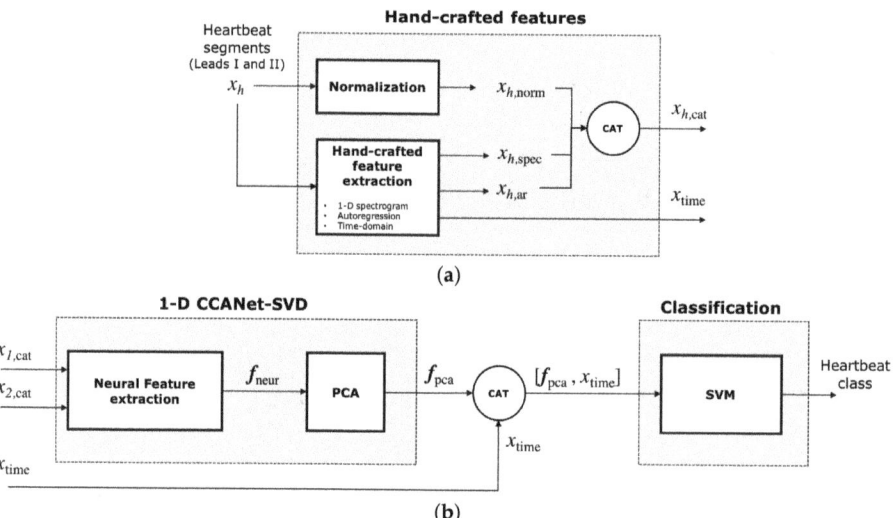

**Figure 1.** Classification processing pipeline of our method. (**a**) Feature extraction: for each of the two leads $x_h$ ($h = 1, 2$), hand-crafted features are extracted to build $x_{h,cat}$ ($h = 1, 2$) while $x_{time}$, a vector of time-domain features, is built for both leads. (**b**) Part of these features, $x_{h,cat}$ ($h = 1, 2$), feeds the 1-D CCANet-SVD module, which first outputs the neural features $f_{neur}$ and then a reduced version of the neural features $f_{pca}$. The concatenation of $x_{time}$ and $f_{pca}$ feeds the classification module. The output is the predicted heartbeat class.

### 3.1. Hand-Crafted Feature Extraction

Given an isolated heartbeat $x_h$ ($h = 1, 2$), hand-crafted features are extracted with three different methods: frequency-domain, time-domain, autoregressive modeling.

#### 3.1.1. One-Dimensional Spectrogram

For the frequency domain, we use a one-dimensional spectrogram, which is a representation of the spectrum of frequencies of a signal as it varies with time. It is built through a short-time Fourier transform (STFT) of each of the two non-normalized leads $x_h$ ($h = 1, 2$). A window slides through the signal (with potential overlapping) and computes at each step the squared magnitudes of the STFT of the portion of the signal belonging to the window. The Hamming windowing is used for this process. The spectrogram is then obtained by concatenating, along the time axis, the squared magnitudes acquired for each window. The squared magnitudes obtained for each frequency (up to half of the sampling rate) at each time step can be reported in a matrix where axis 0 and 1 are the frequency and time axis, respectively. Since the range of the squared magnitudes varies significantly, the resulting matrix is rescaled to $[0, 1]$ to yield $X_{h,spec}$ ($h = 1, 2$).

A weighted average along the frequency axis is performed thus turning the $X_{h,\text{spec}}$ matrix into a one-dimensional vector, $x_{h,\text{spec}}$. The equation used is the following (2):

$$x_{h,\text{spec}} = \frac{1}{S(n)} \sum_k \frac{1}{k^n} X_{h,\text{spec}}[k], \quad (2)$$

where $S(n) = \sum_k \frac{1}{k^n}$.

This feature extraction method requires three parameters: the number of samples in the window of the STFT ($N_{\text{wind}}$ = 64 and 46 for MIT-BIH and INCART respectively), the number of samples in the overlap between two consecutive steps ($N_{\text{overlap}}$ = 32 and 23 for MIT-BIH and INCART respectively) and $n$ (0.25 for both MIT-BIH and INCART), the weight parameter in Equation (2). Suitable parameters are found with a greedy search. The feature vector $x_{h,\text{spec}}$ is of size 10.

### 3.1.2. Autoregressive Modeling

Autoregressive (AR) modeling specifies that a time series value depends linearly on its own previous values and a stochastic term, as per Equation (3):

$$X_t = \sum_{i=2}^{p+1} \varphi_i X_{t-i+1} + \epsilon(t), \quad (3)$$

where $X_t$ is the time series, $\varphi_i$ are the AR coefficients computed with Yule-Walker's method and $p$ is the order of the AR model. Since the choice of the order $p$ depends highly on the sampling rate, non-normalized ECGs from both databases are resampled to 360 Hz [38]. The order was then chosen by performing best parameter search on the training data for both the MIT-BIH and INCART databases. We chose the order that maximized the average of our performance metrics (accuracy, specificity, sensitivity, ppv) on a validation set. Figure 2 shows that for the INCART data, the best order is 2 while the best order is 3 for the MIT-BIH data. Since the performance for MIT-BIH is quite comparable for orders 2 and 3, we chose an order equal to 2 for both datasets. We preferred a lower order to reduce the computational cost. The vector of AR coefficients obtained for each lead of one heartbeat, $x_h$ ($h = 1, 2$), is denoted as $x_{h,\text{ar}}$ and is of size 2.

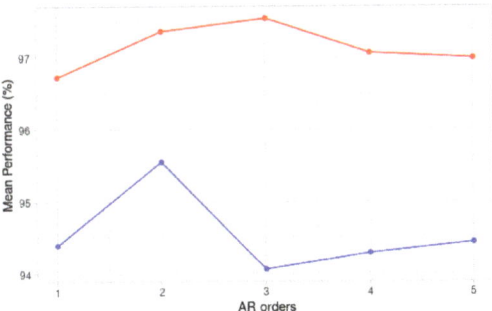

**Figure 2.** Mean performance for different AR orders on validation sets for MIT-BIH signals (red) and INCART signals (blue).

### 3.1.3. Time-Domain Features

For each of the two leads $x_h$ ($h = 1, 2$) of one segmented heartbeat, we compute the following time-domain features: the median value of $x_h$, its fourth order and fifth order central moments and the kurtosis of $x_h$. Finally, for both leads, we build a single vector of time-domain features including the previous features for each lead and the heartbeat rate of the patient to whom the heartbeat belongs. The resulting vector is denoted as $x_{time}$ and is of size 9.

### 3.2. Neural Feature Extraction

To exploit the correlation between two ECG leads, we use a one-dimensional variant of the canonical correlation analysis network (CCANet). First introduced in the field of image recognition by Yang et al. [33], CCANet has been employed in two-view image recognition tasks. Recently, CCANets, which are intrinsically two-dimensional, have been successfully employed in the signal processing field for the classification of two and three lead heartbeats [20]. A CCANet is usually composed of two cascaded convolutional layers and an output layer: (1) in the convolutional layers, the CCA technique is used to extract dual-lead filter banks; (2) in the output layer, the features extracted from the second convolutional layer are mapped into the final feature vector [20].

In this paper, with the aim of increasing performance, we design a new 1-D canonical correlation analysis network that is composed of four 1-D convolutional layers and an output layer. Contrary to CCANet, the filters are found by combining a CCA with a singular value decomposition (SVD), and features are extracted after each layer. The use of 1-D convolutions instead of 2-D permits to limit computational cost, thus allowing to increase the number of layers from two to four and, consequently, to increase performance.

The processing pipeline is shown in Figure 3. The input of the proposed 1-D CCANet-SVD is the concatenation of autoregressive features, spectrogram features, and the original normalized heartbeat, resulting in the following vector $x_{h,cat} = [x_{h,ar}\ x_{h,spec}\ x_{h,norm}] \in \mathbb{R}^m$, $h = 1, 2$. The 1-D CCANet-SVD is trained with $N$ two-lead heartbeats and then used as neural feature extractor in combination with a linear SVM for heartbeat classification. The network is trained separately for the MIT-BIH and INCART databases.

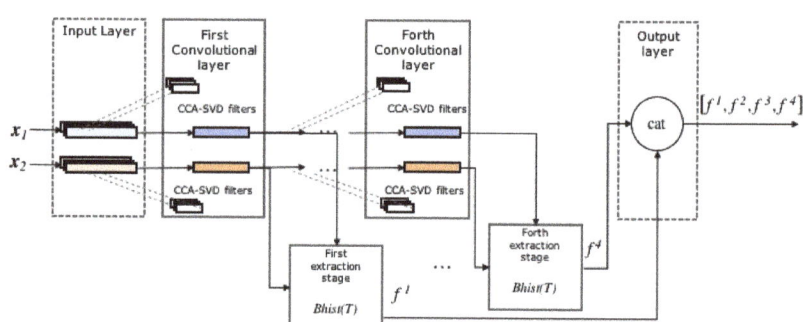

**Figure 3.** Proposed 1-D CCANet-SVD.

#### 3.2.1. First Convolutional Layer

We denote $x_{h,cat}^{(i)}$ the $i$-th element ($i \in \{1, \ldots, m\}$) of an input vector $x_{h,cat}$. We selected a series of segments of size $k$ centered on each value $x_{h,cat}^{(i)}$, to obtain the $m$ following segments, $b_{h,1}, \ldots, b_{h,m} \in \mathbb{R}^k$. The latter are then zero-centered and concatenated to build a matrix of the segments $[\bar{b}_{h,1}, \ldots, \bar{b}_{h,m}] \in \mathbb{R}^{k \times m}$. This procedure is performed on each of the $N$ training heartbeats and the resulting matrices of segments are finally concatenated to obtain $X_h \in \mathbb{R}^{k \times Nm}$, $h = 1, 2$. Note that our network is simultaneously fed with all the training heartbeats in order to build the two matrices $X_1$ and $X_2$.

Let us address the filter extraction stage. In [20], the filters are found with a CCA, thus by maximizing the correlation between pairs of projected variables. The first projection direction can be obtained by optimizing Equation (4):

$$\max \rho(a_1, b_1) = a_1^T S_{12} b_1 \qquad (4)$$

with the constraints $a_1^T S_{11} a_1 = 1$, $b_1^T S_{22} b_1 = 1$, where $S_{hh'} = (\mathbf{X}_h)(\mathbf{X}_{h'})^T$, and $a_1$ and $b_1$ are the first canonical vectors for each of the two leads. The Lagrange multiplier technique shows that $a_1$ and $b_1$ are eigenvectors of $M_1 = S_{11}^{-1} S_{12} S_{22}^{-1} S_{21}$ and $M_2 = S_{22}^{-1} S_{21} S_{11}^{-1} S_{12}$, respectively. Given the first $l-1$ directions, the $l$-th projection direction can be calculated by solving problem (4) with the additional constraints $a_i^T S_{11} a_l = b_i^T S_{22} b_l = 0$, $(i < l)$. In the end, the $L_1$ filters for the first lead are built by taking the $L_1$ primary eigenvectors of $M_1$ (i.e., associated with the $L_1$ biggest eigenvalues), whereas the $L_1$ filters for the second lead are built by considering the $L_1$ primary eigenvectors of $M_2$.

In this paper, we use a slightly different approach, referred to as the CCA-SVD filter extraction technique. We perform an SVD of both $M_1$, and $M_2$, as per $M_1 = U_1 D_1 V_1^T$ and $M_2 = U_2 D_2 V_2^T$, where the $U$ and $V$ matrices are unitary, and the $D$ matrices are diagonal with singular values on the diagonals. Using an SVD allows to retrieve the directions, which explain the most the variance of $M_1$ and $M_2$. Since these two matrices derive from the CCA, they capture the correlations between the two leads. Therefore, we use the directions found by performing an SVD on them to have the best explanation of the correlation between the two leads. Consequently, the $L_1$ filters for the first lead are built by taking the columns of $U_1$ that are associated with the $L_1$ biggest singular values of $D_1$, whereas the $L_1$ filters for the second lead are built by considering the columns of $U_2$ that are associated with the $L_1$ biggest singular values of $D_2$. Such an approach yields better results than the traditional CCA filter extraction technique (see *Experiments*). We denote as $W_{1,l}$ and $W_{2,l}$, $l = 1, \ldots, L_1$, the $L_1$ filters of size $k$ corresponding to the first and second lead, respectively.

As for the convolutions, for each lead $h$, each input signal $x_{h,cat}$ yields $L_1$ outputs $x_{h,cat,l} = x_{h,cat} * W_{h,l}$, $l = 1, \ldots, L_1$. The length of the input and output signal were kept identical, thanks to a zero-padding of the input.

### 3.2.2. First Extraction Stage

The extraction stage follows the same steps as in [20]. First, for each heartbeat, the output of the first convolution is converted to a decimal one-dimensional signal as per $T = \sum_{l=1}^{L_1} 2^{l-1} H([x_{1,cat,l}, x_{2,cat,l}]) \in \mathbb{R}^{2m}$, where $H$ is the Heaviside step function. Therefore, the range of each component of $T$ is $[0, 2^{L_1} - 1]$. $T$ is then divided in $B$ blocks of size $u_1$. Each block can overlap with its neighbor, according to $R_1 \in [0, 1]$, an overlapping proportion parameter. For each of these blocks, a histogram with $2^{L_1}$ bins is built. The values of the resulting histogram for each block is embedded in a $2^{L_1}$-long vector and the vectors provided by each block are then concatenated to obtain $Bhist(T) \in \mathbb{R}^{2^{L_1} B}$. The first feature vector, for the heartbeat, is $f^1 = Bhist(T)$.

### 3.2.3. Second Convolution Layer and Extraction Stage

The second layer is identical to the previous one, except for the fact that the input is different. Indeed, before the first convolution, each lead of a heartbeat was represented by a single vector of length $m$. After the first convolution, each lead is now represented by $L_1$ vectors of length $m$. Let's walk through the second layer with the notations used so far.

The $x_{h,cat,l} = x_{h,cat} * W_{h,l}$, $l = 1, \ldots, L_1$ produced after the first convolutional layer are the input of the second layer. Since we initially considered $N$ training heartbeats, it means that this layer has a total number of $N \times L_1$ input vectors corresponding to lead 1 and $N \times L_1$ input vectors corresponding to lead 2. The same segmentation and zero-centering process as in the first layer gives $Y_h \in \mathbb{R}^{k \times mNL_1}$ ($h = 1, 2$), the matrices of the concatenated segments for all the input vectors, for each lead.

Applying the CCA-SVD filter extraction technique with $\tilde{S}_{hh'} = (Y_h)(Y_{h'})^T$ leads us to perform the SVD of $\tilde{M}_1 = \tilde{S}_{11}^{-1} \tilde{S}_{12} \tilde{S}_{22}^{-1} \tilde{S}_{21}$ and $\tilde{M}_2 = \tilde{S}_{22}^{-1} \tilde{S}_{21} \tilde{S}_{11}^{-1} \tilde{S}_{12}$, for the first and second lead, respectively. The filters are then found exactly as in the first convolutional layer and we denote as $\tilde{W}_{1,\ell}$ and $\tilde{W}_{2,\ell}$, $\ell = 1, \ldots, L_2$, the $L_2$ filters of size $k$ extracted for the first and second lead, respectively.

As for the convolutions, for each initial lead $h = 1, 2$ and channel $l \in \{1, \ldots, L_1\}$, the signal $x_{h,cat,l}$ yields $L_2$ outputs $x_{h,cat,l,\ell} = x_{h,cat,l} * \tilde{W}_{h,\ell}$, $\ell = 1, \ldots, L_2$. At this stage, each initial lead of a heartbeat is now represented by $L_1 \times L_2$ vectors of size $m$.

The second extraction step is the same as after the first convolutional layer except for a few points. First, for each heartbeat, the output of the second convolutional layer is converted to a decimal signal as per $\tilde{T}_l = \sum_{\ell=1}^{L_2} 2^{\ell-1} H([x_{1,cat,l,\ell}, x_{2,cat,l,\ell}]) \in \mathbb{R}^{2m}$, $l \in \{1, \ldots, L_1\}$. The second feature vector for the heartbeat is obtained as per $f^2 = [Bhist(\tilde{T}_1), Bhist(\tilde{T}_2), \ldots, Bhist(\tilde{T}_{L_1})]$. The $Bhist$ are built with a block size and an overlapping parameter equal to $u_2$ and $R_2$, respectively.

The third and fourth convolutional layers are built similarly. $f^3$ and $f^4$ refer to the third and fourth feature vectors extracted for a heartbeat after each layer. We denote as $L_3$ and $L_4$, the number of filters for the third and fourth layers, respectively. $u_3$ and $u_4$ are the block sizes for the construction of $Bhist$ after the third and fourth convolutional layers, respectively. Finally, we denote as $R_3$ and $R_4$, the overlapping parameters for the last two layers.

### 3.2.4. Final Output and PCA

For a given heartbeat, the final output of the network is obtained by concatenating the four feature vectors, as per $f_{neur} = [f^1, f^2, f^3, f^4]$. Given the significant size of the final feature vector, a PCA is carried out to reduce dimensionality. The number of components is chosen such that the explained variance is over 99.99% thus obtaining $f_{pca}$. The final feature vector $\mathcal{F}$ is obtained by concatenating this vector to the vector of time-domain features corresponding to the heartbeat. $\mathcal{F} = [f_{pca}, x_{time}]$ is a vector of size 1382 or 3020, for INCART or MIT-BIH heartbeats, respectively.

The classification step is performed by a linear SVM, with a regularization parameter $C = 1$.

## 4. Experiments

### 4.1. Experimental Setup

To assess the performance of our method, we classified 15 and 7 different types of heartbeats from the MIT-BIH and INCART databases, respectively. One major obstacle of our databases is that they are not well balanced. For instance, the normal types are over-represented while the supraventricular escape beats from INCART have few samples in comparison. To address this issue, we randomly sampled (without repetition), as in [20], 3350 heartbeats from the MIT-BIH database and 1720 heartbeats from INCART, in the proportions given by Tables 1 and 2 respectively.

We used k-fold cross validation on the resampled heartbeats to fit the parameters of 1-D CCANet-SVD. The parameters are shown in Table 4.

**Table 4.** Parameters for 1-D CCANet-SVD.

| Layer 1 | Layer 2 | Layer 3 | Layer 4 |
|---|---|---|---|
| | $k = 7$ | | |
| $L_1 = 2$ | $L_2 = 3$ | $L_3 = 3$ | $L_4 = 5$ |
| $u_1 = 35$ | $u_2 = 35$ | $u_3 = 35$ | $u_4 = 50$ |
| $R_1 = 0.5$ | $R_2 = 0.4$ | $R_3 = 0.4$ | $R_4 = 0.3$ |

The results provided in the *Results and discussion* subsection derive from an overall confusion matrix obtained after summing the k confusion matrices given after each fold. As in [20], we performed 10 and 5-cross validation for the data from the MIT-BIH and INCART databases, respectively.

The code is written in Python 3.7 and we ran all the experiments on a personal computer equipped with Ubuntu 18.04. The hardware specifications of the computer are the following: 16 GB RAM, and i7-7700 CPU with a clock speed of 3.60 GHz.

1-D ResNet

To further validate our approach, we added a one-dimensional residual network (1-D ResNet) to our experiments. The input is the same as for 1-D CCANet-SVD. The 1-D ResNet has been implemented as follows:

- *Initial layer*: the input of the network undergoes an initial convolution with 2 input channels (one for each lead) and 16 output channels. This convolution is followed by a max-pooling step. This initial layer is followed by 4 identical residual blocks.
- *Residual blocks*: each of them contains two convolutional layers and, for each block, the output of the second convolutional layer is finally added to the block's input. For each block, the first convolution doubles the number of channels, while the second convolution has the same number of input and output channels. Consequently, the last convolution has 256 output channels. The output of the last block then undergoes average-pooling to obtain the feature vector.
- *Classification layer*: the feature vector, of size 256 serves as the input of a fully connected neural network. The classification is then performed thanks to the Softmax function. The loss used is the cross-entropy.

During the feature extraction process, Batch-Normalization is performed after each convolution and the Rectified Linear Unit (ReLU) is used as the activation function. Table 5 shows the architecture of the network and the various parameters fitted for each layer. All the parameters, including the number of layers, residual blocks and the number of channels for the convolutions were found through cross-validation.

**Table 5.** Architecture of 1-D ResNet with the parameters for the initial layer and the convolutions of each residual block.

| Output Size | | Layers |
|---|---|---|
| MIT-BIH | INCART | Initial Layer |
| | | Conv (kernel size = 7, stride = 1, padding = 3) |
| | | BatchNorm |
| | | ReLU |
| 184 × 16 | 132 × 16 | MaxPool (kernel size = 5, stride = 2, padding = 0) |
| 184 × 32 | 132 × 32 | ResBlock (kernel size = 5, stride = 1, padding = 2) |
| 184 × 64 | 132 × 64 | ResBlock (kernel size = 5, stride = 1, padding = 2) |
| 184 × 128 | 132 × 128 | ResBlock (kernel size = 5, stride = 1, padding = 2) |
| 184 × 256 | 132 × 256 | ResBlock (kernel size = 5, stride = 1, padding = 2) |
| 1 × 256 | 1 × 256 | AvgPool |
| 15 | 7 | Fully Connected Network |

### 4.2. Evaluation Metrics

Several measures have been employed for the evaluation of the goodness of the proposed approach: (1) Overall Accuracy (OACC) defined as (TP + TN)/(TP + TN + FP + FN); (2) Mean Accuracy (MACC) defined as the average of the class accuracies; (3) Specificity (SPE) defined as TN/(TN + FP); (4) Sensitivity (SENS) defined as TP/(TP + FN); (5) Positive Predictive Values (PPV) defined as TP/(TP + FP). TP, TN, FP, FN are the number of True Positive, True Negative, False Positive and False Negative, respectively. Note that in Table 6, the values of SPE, SENS, and PPV are averaged across the classes.

### 4.3. Results and Discussion

Table 6 shows the results obtained for various classification methods. It includes the previously described model, variations of it (e.g., without adding the time-domain features), the 1-D ResNet and the best dual-leads method in the state of the art [20].

**Table 6.** Results for various methods (%). For each method, first line reports results for the MIT-BIH database while the second line reports for the INCART database. OACC stands for Overall Accuracy; MACC for Mean Accuracy; SPE for Specificity; SENS for Sensitivity; PPV for Positive Predictive Values; AVG is the Average of the first 5 columns of the table. Best results are highlighted in boldface.

| Method | OACC | MACC | SPE | SENS | PPV | AVG |
|---|---|---|---|---|---|---|
| DL-CCANet [20] | 95.25 | 99.40 | 99.60 | 94.60 | 96.30 | 97.03 |
| | 94.01 | 98.31 | 98.85 | 90.89 | 94.11 | 95.23 |
| 1-D ResNet | 91.88 | 98.92 | 99.36 | 90.11 | 90.14 | 94.08 |
| | 86.25 | 96.07 | 97.55 | 85.05 | 80.66 | 89.12 |
| 1-D CCANet-SVD | 94.75 | 99.30 | 99.57 | 93.77 | 95.81 | 96.64 |
| (w/o SVD) | 93.60 | 98.17 | 98.80 | 90.63 | 93.33 | 94.91 |
| 1-D CCANet-SVD | 95.40 | 99.39 | 99.62 | 94.43 | 96.54 | 97.08 |
| (w/o time-domain feat.) | 95.35 | 98.67 | 99.11 | 93.26 | 96.22 | 96.52 |
| 1-D CCANet-SVD | 95.22 | 99.36 | 99.6 | 94.03 | 96.61 | 96.96 |
| (w/o 1D-spec) | 94.77 | 98.50 | 99.02 | 92.66 | 94.59 | 95.91 |
| 1-D CCANet-SVD | 95.43 | 99.39 | 99.62 | 94.53 | 96.73 | 97.14 |
| (w/o ar) | 95.12 | 98.60 | 99.09 | 93.68 | 95.13 | 96.32 |
| 1-D CCANet-SVD | 94.99 | 99.33 | 99.59 | 93.85 | 96.21 | 96.79 |
| (w/o stack) | 94.83 | 98.52 | 99.03 | 92.43 | 94.95 | 95.95 |
| **1-D CCANet-SVD** | **95.52** | **99.40** | **99.63** | **94.60** | **96.65** | **97.16** |
| **(proposed)** | **95.70** | **98.77** | **99.19** | **93.78** | **95.89** | **96.67** |

Our method and [20] demonstrated comparable performances on the MIT-BIH database, though our overall accuracy and mean ppv were better by around 0.3%. As for the INCART database, our results proved to be better, especially the overall accuracy (+1.69%), mean sensitivity (+2.89%), and mean ppv (+1.78%). Contrary to [20], our approach is purely one-dimensional, allowing to explore a more complex version of CCANet while maintaining a reasonable computational complexity and providing better results: we opted for 4 layers and stacking features extracted after each convolution gave better results than without doing so (see seventh method of Table 6), especially increasing the sensitivity. Using frequency features with the one-dimensional spectrogram helped obtain a better classification by notably increasing the sensitivity (+1.12% for INCART) and the ppv (+1.3% for INCART). The addition of the AR coefficients and the time-domain features contributed to slightly increase the performance of our model. The performances were significantly better when using our CCA-SVD filter extraction technique instead of the CCA technique described in [20], with a sensitivity gaining more than 3% for INCART (see the third method of Table 6). Finally, our method provided significantly better results than the 1-D ResNet approach (+3.64% for overall ACC for MIT-BIH, +9.45% for INCART). Our analysis of the correlations of the two leads, using SVD, proved to be a good way of recognizing the various types of heartbeats.

Tables 7 and 8 show the comparison between the best of our proposals and similar works in the state of the art for MIT-BIH and INCART databases respectively. In the case of MIT-BIH, Table 7 confirms that the use of dual-leads-based approaches brings improvements in performance (more than 1%). Also in the case of INCART we see an improvement with respect to single-lead-based method (more than 3%). Here, the proposed approach is slightly better than a variant of the work by Yang et al. [20] that uses three leads. Although our approach explores more complex structures with respect to Yang et al. [20], it remains comparable, in terms of computational cost, with it. The inference time for each heart beat classification is about 0.05 s while in the case of Yang et al. [20], it is about 0.02 s.

**Table 7.** Results for various methods on the MIT-BIH database.

| Authors | Year | Method | Type | #Samples | Ov. ACC |
|---|---|---|---|---|---|
| Plawiak et al. [13] | 2018 | Evol.Neural (1 lead) | 15 | 1000 | 91.00% |
| Lee et al. [39] | 2018 | PCANet (1 lead) | 15 | 3350 | 94.59% |
| Yang et al. [20] | 2019 | RandNet (1 lead) | 15 | 3350 | 94.39% |
| Yang et al. [20] | 2019 | DL-CCANet (2 lead) | 15 | 3350 | 95.25% |
| Proposed method | 2021 | 1-D CCANet-SVD (2 lead) | 15 | 3350 | 95.52% |

**Table 8.** Results for various methods on the INCART database.

| Authors | Year | Method | Type | #Samples | Ov. ACC |
|---|---|---|---|---|---|
| Lee et al. [39] | 2018 | PCANet (1 lead) | 7 | 1720 | 93.72% |
| Yang et al. [20] | 2019 | RandNet (1 lead) | 7 | 1720 | 92.91% |
| Yang et al. [20] | 2019 | DL-CCANet (2 lead) | 7 | 1720 | 94.01% |
| Yang et al. [20] | 2019 | TL-CCANet (3 lead) | 7 | 1720 | 95.52% |
| Proposed method | 2021 | 1-D CCANet-SVD (2 lead) | 7 | 1720 | 95.70% |

Our method presents a few limitations. First, the CCANet technique requires the network to be fed simultaneously with all the training data, in order to determine the filters and this may cause a growth in the computational cost as the size of the training data increases. This limitation is common to all the CCANet-based architectures. In our study, we only considered 2-lead signals as input, it could be interesting to include more leads with the hope of increasing the performance, especially for classes with fewer samples. Following the work by Yang et al. [20], our approach can be quite naturally extended to 3-lead signals. The number of layers might need to be reduced to compensate for the additional cost added by the addition of a third lead. Another interesting perspective would be to include some of the techniques we have used in our study in the original two-dimensional CCANet developed by Yang et al. [20]. Indeed, Table 6 shows that the use of the SVD significantly increases the performance, without adding additional computational cost compared to the original method. Therefore, we could also expect promising results when using SVD in the original 2-D CCANet. Likewise, it could be interesting to analyze how the spectrogram features influence the performance of the 2-D CCANet and allow to make significant improvement in the field of abnormal heartbeat recognition.

## 5. Conclusions

In this paper, we propose a novel heartbeat classification method based mainly on a new approach to the study of the correlation between the two ECG leads, to extract complex features. Our method also employ elementary hand-crafted time domain features, frequency domain features with a one-dimensional approach to spectrograms, and autoregressive coefficients. Our method is one-dimensional, allowing to explore a more complex neural architecture while maintaining a reasonable computational complexity, and providing better results. Our final model has an optimal structure and performs the classification of 15 and 7 heartbeat types for the MIT-BIH and INCART databases, respectively. Finally, our method outperforms [20] with a slightly better overall accuracy and mean ppv on the MIT-BIH database and a notably higher overall accuracy (+1.69%), mean sensitivity (+2.89%), and mean ppv (+1.78%) on the INCART database.

**Author Contributions:** All the authors contributed equally to the conceptualization; all the authors contributed equally to the methodology; all the authors contributed equally to the software; all the authors contributed equally to the validation; all the authors contributed equally to the draft preparation; all the authors contributed equally to the writing, review and editing. All authors have read and agreed to the published version of the manuscript.

**Funding:** This research received no external funding.

**Institutional Review Board Statement:** Not applicable.

**Informed Consent Statement:** Not applicable.

**Data Availability Statement:** Not applicable.

**Conflicts of Interest:** The authors declare no conflict of interest.

## References

1. Rim, B.; Sung, N.J.; Min, S.; Hong, M. Deep Learning in Physiological Signal Data: A Survey. *Sensors* **2020**, *20*, 969. [CrossRef] [PubMed]
2. Hannun, A.Y.; Rajpurkar, P.; Haghpanahi, M.; Tison, G.H.; Bourn, C.; Turakhia, M.P.; Ng, A.Y. Cardiologist-level arrhythmia detection and classification in ambulatory electrocardiograms using a deep neural network. *Nat. Med.* **2019**, *25*, 65. [CrossRef] [PubMed]
3. Bhirud, B.; Pachghare, V. Arrhythmia detection using ECG signal: A survey. In Proceeding of the International Conference on Computational Science and Applications, Pune, India, 7–9 August 2019; pp. 329–341.
4. Mathews, S.M.; Kambhamettu, C.; Barner, K.E. A novel application of deep learning for single-lead ECG classification. *Comput. Biol. Med.* **2018**, *99*, 53–62. [CrossRef]
5. Elhaj, F.A.; Salim, N.; Harris, A.R.; Swee, T.T.; Ahmed, T. Arrhythmia recognition and classification using combined linear and nonlinear features of ECG signals. *Comput. Methods Programs Biomed.* **2016**, *127*, 52–63. [CrossRef] [PubMed]
6. Martis, R.J.; Acharya, U.R.; Min, L.C. ECG beat classification using PCA, LDA, ICA and discrete wavelet transform. *Biomed. Signal Process. Control* **2013**, *8*, 437–448. [CrossRef]
7. Venkatesan, C.; Karthigaikumar, P.; Paul, A.; Satheeskumaran, S.; Kumar, R. ECG signal preprocessing and SVM classifier-based abnormality detection in remote healthcare applications. *IEEE Access* **2018**, *6*, 9767–9773. [CrossRef]
8. Alonso-Atienza, F.; Morgado, E.; Fernandez-Martinez, L.; García-Alberola, A.; Rojo-Alvarez, J.L. Detection of life-threatening arrhythmias using feature selection and support vector machines. *IEEE Trans. Biomed. Eng.* **2013**, *61*, 832–840. [CrossRef]
9. Xia, Y.; Zhang, H.; Xu, L.; Gao, Z.; Zhang, H.; Liu, H.; Li, S. An automatic cardiac arrhythmia classification system with wearable electrocardiogram. *IEEE Access* **2018**, *6*, 16529–16538. [CrossRef]
10. Hammad, M.; Maher, A.; Wang, K.; Jiang, F.; Amrani, M. Detection of abnormal heart conditions based on characteristics of ECG signals. *Measurement* **2018**, *125*, 634–644. [CrossRef]
11. Soria, M.L.; Martínez, J. Analysis of multidomain features for ECG classification. In Proceedings of the 2009 36th Annual Computers in Cardiology Conference (CinC), Park City, UT, USA, 13–16 September 2009; pp. 561–564.
12. De Lannoy, G.; François, D.; Delbeke, J.; Verleysen, M. Weighted conditional random fields for supervised interpatient heartbeat classification. *IEEE Trans. Biomed. Eng.* **2011**, *59*, 241–247. [CrossRef]
13. Pławiak, P. Novel methodology of cardiac health recognition based on ECG signals and evolutionary-neural system. *Expert Syst. Appl.* **2018**, *92*, 334–349. [CrossRef]
14. Huang, H.; Liu, J.; Zhu, Q.; Wang, R.; Hu, G. A new hierarchical method for inter-patient heartbeat classification using random projections and RR intervals. *Biomed. Eng. Online* **2014**, *13*, 90. [CrossRef] [PubMed]
15. Zhang, Z.; Dong, J.; Luo, X.; Choi, K.S.; Wu, X. Heartbeat classification using disease-specific feature selection. *Comput. Biol. Med.* **2014**, *46*, 79–89. [CrossRef]
16. Zhang, Z.; Luo, X. Heartbeat classification using decision level fusion. *Biomed. Eng. Lett.* **2014**, *4*, 388–395. [CrossRef]
17. Llamedo, M.; Martínez, J.P. Heartbeat classification using feature selection driven by database generalization criteria. *IEEE Trans. Biomed. Eng.* **2010**, *58*, 616–625. [CrossRef]
18. Pławiak, P.; Acharya, U.R. Novel deep genetic ensemble of classifiers for arrhythmia detection using ECG signals. *Neural Comput. Appl.* **2020**, *32*, 11137–11161. [CrossRef]
19. Park, K.; Cho, B.; Lee, D.; Song, S.; Lee, J.; Chee, Y.; Kim, I.Y.; Kim, S. Hierarchical support vector machine based heartbeat classification using higher order statistics and hermite basis function. In Proceedings of the 2008 Computers in Cardiology, Bologna, Italy, 14–17 September 2008; pp. 229–232.
20. Yang, W.; Si, Y.; Wang, D.; Zhang, G. A novel approach for multi-lead ECG classification using DL-CCANet and TL-CCANet. *Sensors* **2019**, *19*, 3214. [CrossRef]
21. Dohare, A.K.; Kumar, V.; Kumar, R. Detection of myocardial infarction in 12 lead ECG using support vector machine. *Appl. Soft Comput.* **2018**, *64*, 138–147. [CrossRef]

22. Rajagopal, R.; Ranganathan, V. Evaluation of effect of unsupervised dimensionality reduction techniques on automated arrhythmia classification. *Biomed. Signal Process. Control* **2017**, *34*, 1–8. [CrossRef]
23. Li, Q.; Rajagopalan, C.; Clifford, G.D. Ventricular fibrillation and tachycardia classification using a machine learning approach. *IEEE Trans. Biomed. Eng.* **2013**, *61*, 1607–1613.
24. Yang, W.; Si, Y.; Wang, D.; Guo, B. Automatic recognition of arrhythmia based on principal component analysis network and linear support vector machine. *Comput. Biol. Med.* **2018**, *101*, 22–32. [CrossRef]
25. Shimpi, P.; Shah, S.; Shroff, M.; Godbole, A. A machine learning approach for the classification of cardiac arrhythmia. In Proceedings of the 2017 International Conference on Computing Methodologies and Communication (ICCMC), Erode, India, 18–19 July 2017; pp. 603–607.
26. Park, J.; Kim, J.k.; Jung, S.; Gil, Y.; Choi, J.I.; Son, H.S. ECG-signal multi-classification model based on squeeze-and-excitation residual neural networks. *Appl. Sci.* **2020**, *10*, 6495. [CrossRef]
27. Nurmaini, S.; Umi Partan, R.; Caesarendra, W.; Dewi, T.; Naufal Rahmatullah, M.; Darmawahyuni, A.; Bhayyu, V.; Firdaus, F. An automated ECG beat classification system using deep neural networks with an unsupervised feature extraction technique. *Appl. Sci.* **2019**, *9*, 2921. [CrossRef]
28. Zubair, M.; Kim, J.; Yoon, C. An automated ECG beat classification system using convolutional neural networks. In Proceedings of the 2016 6th International Conference on IT Convergence and Security (ICITCS), Prague, Czech Republic, 26 September 2016; pp. 1–5.
29. Li, J.; Si, Y.; Lang, L.; Liu, L.; Xu, T. A spatial pyramid pooling-based deep convolutional neural network for the classification of electrocardiogram beats. *Appl. Sci.* **2018**, *8*, 1590. [CrossRef]
30. Acharya, U.R.; Oh, S.L.; Hagiwara, Y.; Tan, J.H.; Adam, M.; Gertych, A.; San Tan, R. A deep convolutional neural network model to classify heartbeats. *Comput. Biol. Med.* **2017**, *89*, 389–396. [CrossRef]
31. Yıldırım, Ö.; Pławiak, P.; Tan, R.S.; Acharya, U.R. Arrhythmia detection using deep convolutional neural network with long duration ECG signals. *Comput. Biol. Med.* **2018**, *102*, 411–420. [CrossRef]
32. Li, H.; Yuan, D.; Ma, X.; Cui, D.; Cao, L. Genetic algorithm for the optimization of features and neural networks in ECG signals classification. *Sci. Rep.* **2017**, *7*, 41011. [CrossRef]
33. Yang, X.; Liu, W.; Tao, D.; Cheng, J. Canonical correlation analysis networks for two-view image recognition. *Inf. Sci.* **2017**, *385*, 338–352. [CrossRef]
34. Yang, X.; Liu, W.; Liu, W. Tensor canonical correlation analysis networks for multi-view remote sensing scene recognition. *IEEE Trans. Knowl. Data Eng.* **2020**. [CrossRef]
35. Hong, S.; Zhou, Y.; Shang, J.; Xiao, C.; Sun, J. Opportunities and challenges of deep learning methods for electrocardiogram data: A systematic review. *Comput. Biol. Med.* **2020**, *122*, 103801. [CrossRef]
36. He, K.; Zhang, X.; Ren, S.; Sun, J. Deep residual learning for image recognition. In Proceedings of the IEEE Conference on Computer Vision and Pattern Recognition, Las Vegas, NV, USA, 27–30 June 2016; pp. 770–778.
37. Bianco, S.; Cadene, R.; Celona, L.; Napoletano, P. Benchmark analysis of representative deep neural network architectures. *IEEE Access* **2018**, *6*, 64270–64277. [CrossRef]
38. Ge, D.; Srinivasan, N.; Krishnan, S.M. Cardiac arrhythmia classification using autoregressive modeling. *Biomed. Eng. Online* **2002**, *1*, 5. [CrossRef] [PubMed]
39. Lee, J.N.; Byeon, Y.H.; Pan, S.B.; Kwak, K.C. An EigenECG network approach based on PCANet for personal identification from ECG signal. *Sensors* **2018**, *18*, 4024. [CrossRef] [PubMed]

MDPI  
St. Alban-Anlage 66  
4052 Basel  
Switzerland  
Tel. +41 61 683 77 34  
Fax +41 61 302 89 18  
www.mdpi.com

*Applied Sciences* Editorial Office  
E-mail: applsci@mdpi.com  
www.mdpi.com/journal/applsci

www.ingramcontent.com/pod-product-compliance
Lightning Source LLC
LaVergne TN
LVHW070740100526
838202LV00013B/1272